HAND-BOOK OF PRINTING TECHNOLOGY

(OFFSET, SCREEN, FLEXO, GRAVURE, INKJET & DIGITAL)

HAND-BOOK OF
PRINTING TECHNOLOGY

(OFFSET, SCREEN, FLEXO, GRAVURE, INKJET & DIGITAL)

Written By

EIRI Board of Consultants and Engineers

ENGINEERS INDIA RESEARCH INSTITUTE

4449, Nai Sarak, Main Road, Delhi - 110006 (India)
Ph: +91-11-43658117, Mob: +91-9289151047, +91-9811437895
* E-Mail: eiri@eiriindia.org
*Website: www.eiriindia.org

PRINTED & PUBLISHED BY
ENGINEERS INDIA RESEARCH INSTITUTE

4449, Nai Sarak, Main Road, Delhi - 110006 (India)
Ph: +91-11-43658117, Mob: +91-9289151047, +91-9811437895
* E-Mail: eiri@eiriindia.org
*Website: www.eiriindia.org

Distributors:

Industrial Technologies, India
Roop Nagar, Delhi-110007
and
EIRI PROJECT CONSULTANTS AND PUBLISHERS
Nai Sarak, Delhi-110006

ISBN: 9788186732755

Revised and Enlarged Edition

Printed in New Delhi (INDIA)

Rs. 1675/- or US$ 175

Printed & Published by "Engineers India Research Institute",
4449 Nai Sarak, Main Road, New Delhi -110006

Preface

Printing Industry generates a wide range of products which require in every step of our everyday life. Starting from newspapers, magazines, books, post cards to memo pads and business order forms each are the products of printing industry. Printing is a process for reproducing text and image, typically with ink on paper using a printing press. There are various types of printing process for example offset printing, modern printing, gravure printing, flexographic printing etc.

Offset printing is a widely used printing technique where the inked image is transferred from a plate to a rubber blanket, then to the printing surface. Printing technology is often carried out as a large scale industrial process, and is an essential part of publishing and transaction printing. This is the age of hi fi, jets and computers. Rapid advancements in science and technology have made their impact on the printing industry of the world too. The old techniques of printing have become obsolete and made way for the new technology. The printing industry is just one example of an entire industry movement that is changing while keeping up with the development of new technologies. The proliferation of emerging technologies has dictated a rebirth of the printing industry. The Indian Printing Industry is well established and presently growing at **12% per** annum.

This book majorly deals with typographic technology, photo scanning systems, sequence of steps in the printing processes, size and scope of the printing industry, high volume printing technologies for the production of polymer electronic structures, inking system, film high contrast printing, principle of planographic printing, modern printing process, ink zet etc. The book contains the latest printing processes like web, gravure, flexo, security and offset printing. This book is an invaluable resource for new entrants, technicians, craftsmen and executives working with printing industries.

The book covers *Printing Technology, Organization of Printing Services, Typographic Technology, Composing the Type, Desktop Publishing (Software), The Postscript Language, Proof Reading, Pre-Press Processes,*

Camera, Photographing Line & Continuous Tone Copy, Scanning, Offset Plate Making, Presses, Offset Press, Web Offset, Specifications for Offset Publications, Proof and Proofing Techniques, Modern Printing Process Ink Jet, Newspaper Printing, Letterpress, Offset, Flexo And Anilox, Newspaper Production Technology, Flexography & Packaging Printing Plates, Gravure & Packaging, Screen Printing, Screen Printing On Paper, Textiles and Other Printing Substrates, Screen Process Stencils, Screen Printing Automatic press, Choosing A Printing Process, Security Printing, Bar Coding, Ink-Jet Printing, Digital Printing, Plant Economics of Polythene Bags & Automatic Printing, Plant Economics of Textile Printing (Job Work), Plant Economics of Ceramic Tile Printing Medium for Screen Printing, Plant Economics of Cloth Printing (Job Work),

ENGINEERS INDIA RESEARCH INSTITUTE
4449, Nai Sarak, Main Road, Delhi - 110006 (India)
Ph: +91-11-43658117,
Mob: +91-9289151047, +91-9811437895
* E-Mail: eiri@eiriindia.org
*Website: www.eiriindia.org

EIRI PROCESS TECHNOLOGY BOOKS

CHEMICAL/ DYES/ PETROCHEMICALS/ LUBRICANT/ ELECTROPLATING

01 Small Medium & Large Chemical Industries
02 Industrial Chemicals Technology Hand Book
03 Modern Technology of Organic & Inorganic Chemicals
04 Hand Book of Electroplating Anodizing & Surface Finishing Technology
05 Hand Book of Agro Chemical Industries (Insecticides & Pesticides)
06 Technology of Synthetic Dyes, Pigments Intermediates
07 Petrochemicals, Lubricants, Greases & Petroleum Refining
08 Hand Book of Lubricants, Greases & Petrochemicals Technology

GUMS, ADHESIVES AND SEALANTS

01 Technology Of Gums, Adhesives & Sealants With Formulations
02 Hand Book Of Adhesives With Their Formulaes
03 Adhesives Technology & Formulations Hand Book (Hand Book of Adhesives)
04 Technology of Glue and Adhesives with Adhesives Bonding and Formulations

SMALL SCALE INDUSTRIES/ STATIONERY/PAPER/ INKS/ CANDLES & EXPORT BUSINESS

01 Start Your Own Export Business (How To Export)
02 Start Your Own Small Business And Industry
03 Candle Making Processes & Formulations Hand Book
04 Stationery, Paper Converting & Packaging Industries
05 Modern Inks Formulations & Manufacturing Industries
06 Profitable Businesses To Start For Entrepreneurs
07 Modern Small & Cottage Scale Industries
08 Profitable Small Cottage Tiny & Home Industries

BIO FUEL & BIOPROCESSING

01 Technology Of Bio-Fules (Ethanol & Biodiesel)
02 Modern Technology Of Bioprocessing

PRINTING, SCREEN PRINTING & PACKAGING

01 Hand Book of Printing Processes Technologies & Industries
02 Hand Book Of Printing Technology (Offset, Screen, Flexo, Gravure, Inkjet & Digital)
03 Hand Book Of Offset Printing Technology
04 Screen Printing With Processes & Tech.
05 Hand Book Of Prepress
06 Hand Book Of Packaging Technology
07 Modern Packaging Technology for Processed Food, Bakery, Snack Foods, Spices and Allied Food Products
08 Hand Book Of Food Packaging Technology
09 Modern Technology of Printing Inks

PAINT, VARNISH, SOLVENTS, COATING & LACQUERS, POWDER COATING

01 Paint Pigment Varnish & Lacquer Manufacturing
02 Paint Varnish Solvents & Coating Technology

EIRI PROCESS TECHNOLOGY BOOKS

03 Paint, Pigment, Solvent, Coating Emulsion, Paint Additives & Formulations
04 Technology Of Coating, Resins, Pigments & Inks Industries
05 Manufacturing Technology & Formulations Hand Book On Thinners, Putty, Wall & Industrial Finishes And Synthetic Resins
06 Technology of Synthetic Resins and Emulsion Polymers
07 Technology of Paints and Coatings with Formulations
08 Powder Coating Technology Hand Book

PLASTIC/POLYMER PROCESS, COMPOUNDING, INJECTION MOULDING, ROTATIONAL MOULDING, PLASTIC FILM, FIBRE GLASS, PLASTIC WASTE RECYCLING, MOULDS, PET & RESINS INDUSTRIES

01 Moulds Design & Processing Hand Book
02 Hand Book Of Plastic Materials & Processing Technology
03 Injection Moulding Of Plastics
04 Plastic Processing and Packaging Industries
05 Plastic Waste Recycling Technology
06 Technology Of Plastic Films
07 Rotational Moulding Technology Hand Book
08 Plastic Compounding, Master Batches, Pet & Other Plastics
09 Synthetic Resins Technology With Formulations
10 Technology Of PVC Compounding And Its Applications
11 Hand Book of Polymer And Plastic Technology
12 Hand Book of Fibre Glass Moulding
13 Technology of Reinforced Plastics
14 Plastic Additives Technology Hand Book
15 Technology of Pet Bottles, Preform and Pet Recycling
16 Modern Technology of Extrusion & Extruded Products
17 Technology of Synthetic Resins & Emulsion Polymers
18 Technology of Plastic Additives with Processes and Packaging
19 Technology of Disposable Medical Products

BAKERY, CONFECTIONERY & BREAKFAST, PASTA & CEREALS

01 Hand Book Of Bakery Industries
02 Hand Book Of Confectionery With Formulations
03 Breakfast, Dietary Food, Pasta & Cereal Products Technology
04 Hand Book Of Modern Bakery Products
05 Modern Bakery Technology & Fermented Cereal Products With Formulae
06 Technology of Confectionery, Chocolates, Toffee, Candy, Chewing & Bubble Gums, Lollipop and Jelly Products with Formulations

AGRO CULTIVATION, ANIMAL FARMING, AGRO PLANTATION & AGRO CHEMICAL/ PESTICIDES/ FLORICULTURE & BEE KEEPING

01 Poultry Farming & Feed Formulations
02 Hand Book of Pig Farming
03 Agro Based Hand Book of Plantation, Cultivation & Farming
04 Agro Based Plantation Cultivation & Farming
05 Agro Chemical Industries (Insecticide & Pesticides)
06 Modern Bee Keeping & Honey Processing

EIRI PROCESS TECHNOLOGY BOOKS

07 Technology Of Modern Rice Milling And Basmati Rice
08 Hand Book Of Goat Farming
09 Floriculture Hand Book (Flowers Growing Technology)
10 Aloe Vera Cultivation, Processings, Formulations And Manufacturing Technology
11 Poultry Farming, Hatchery and Broiler Production

DAIRY FARM, MILK PROCESSING AND ICE CREAM

01 Hand Book of Dairy Formulations, Processes & Milk Processing Industries
02 Milk Processing And Dairy Products Industries
03 Hand Book Of Dairy Farming to Produce Milk with Packaging
04 Hand Book Of Ice Cream Technology And Formulae
05 Hand Book Of Milk Processing Dairy Products And Packaging Technology
06 Dairy Farming for Milk Production Technology
07 Commercial Dairy Farming with Project Profiles

HERBS & MEDICINAL CULTIVATION

01 Herbs, Medicinal & Aromatic Plants Cultivation
02 Aushidhi And Sungndhit Paudho Ka Vaysayik (Hindi)
03 Hand Book Of Aromatic & Medicinal Plants And Biodiesel (Jatropha)
04 Hand Book Of Medicinal & Aromatic Plants (Cultivation, Utilisation & Extraction Processed)

FOOD & AGRO PROCESSING, PRESERVATION, SNACK FOOD, DEHYDRATION, FRUIT BEVERAGE, POTATO, MAIZE, MEAT PROCESS

01 Fruits & Vegetables Processing Hand Book
02 Fruit Beverage & Processing With Mango
03 Food Processing & Agro Based Industries
04 Manufacturing Technology of Snacks Food, Namkeen, Pappad & Potato Products
05 Preservation & Canning Of Fruits and Vegetables
06 Hand Book of Food Dehydration & Drying
07 Meat Processing & Meat Products Hand Book
08 Technology of Food Preservation & Processing
09 Hand Book of Food Packaging Technology
10 Agro Based & Processed Food Products
11 Potato & Potato Processing Technology
12 Technology of Maize & Allied Com. Products
13 Technology of Food Processing Industries
14 Technology of Sweets (Mithai) with Formulae
15 Modern Technology of Frozen Food Products
16 Technology of Water and Packaged Drinking water
17 Complete Hand Book on Frozen Food Processing and Freeze Drying Technology
18 Technology of Sweets (Mithai), Namkeen and Snacks food with Formulae
19 Complete Book on Banana Cultivation, Dehydration, Ripening, Processing, Products and Packaging Technology

EIRI PROCESS TECHNOLOGY BOOKS

20 Technology of Chicken Meat and Poultry Products
21 Agro Food Processing and Packaging Technology
22 Modern Technology of Tomato Processing and Dehydration (Ketchup, Juice, Paste, Puree, Soup and Drying)
23 Modern Technology of Agro Processing and Food Packaging Products with Project Profiles
24 Start Your Own Coffee and Coffee Processing

SOAPS & DETERGENTS

01 Household Soap. Toilet Soap & Other Soap
02 Profitable Small Scale Manufacture Of Soaps & Detergents
03 Synthetic Detergents With Formulations
04 Modern Technology of Acid Slurry, Surfactants, Soap and Detergents with Formulae

ORGANIC FARMING AND FOOD

01 Hand Book Of Organic Farming And Organic Foods With Vermi-Composting And Neem Products

FISH FARMING & FISHERY PRODUCTS

01 Hand Book Of Fish Farming And Fishery Products

COSMETIC (SYNTHETIC & HERBAL)

01 Cosmetics Processes & Formulations
02 Herbal Cosmetics & Beauty Products With Formulations
03 Profitable Small Scale Manufacture Of Cosmetics (Synthetic & Herbal)
04 Hand Book Of Synthetic & Herbal Cosmetics (How to Make Beauty Products)
05 Technology Of Herbal Cosmetics & Toiletries Products With Formulae

OIL AND FATS

01 Hand Book of Oils, Fats And Derivatives With Refining & Packaging Technology
02 Technology of Oilseeds Processing, Oils & Fats and Refining

ESSENTIAL OILS & AROMATIC

01 Essential Oils Manufacturing & Aromatic Plants
02 Modern Technology Of Essential Oils
03 Technology Of Perfumes, Flavours & Essential Oils
04 Essential Oils Processes & Formulations

PERFUMES AND FLAVOURS

01 Hand Book Of Flavours & Food Colourants Technology
02 Hand Book Of Perfumes & Flavours
03 Hand Book of Perfumes With Formulations
04 Technology Of Perfumes, Flavours & Essential Oils
05 Hand Book Of Flavours Technology

BUILDING MATERIALS & CHEMICALS

01 Technology Of Building Materials & Chemicals With Processes

CONTENTS CUM SUBJECT INDEX

EIRI PROCESS TECHNOLOGY BOOKS

TEXTILE, GARMENTS, DYEING

01 Modern Technology Of Bleaching, Dyeing, Printing & Finishing Of Textiles
02 Technology Of Textiles (Spinning & Weaving, Dyeing, Scouring, Drying, Printing And Bleaching)
03 Garments Manufacturing Technology

SPICES & COLD STORAGE

01 Spices & Packaging With Formulaes
02 Start Your Own Cold Storage Unit

PULP & PAPER TECHNOLOGY

01 Hand Book Of Pulp & Paper, Paper Board & Paper Based Technology

NON WOVEN TECHNOLOGY

01 Complete Technology of Nonwovens Fabrics, Carry Bags, Composite, Geotextiles, Medical Textiles, Fibres, Felts, Apparels, Spunlace and Absorbent Nonwoven

TEXTILE AUXILIARIES & CHEMICALS

01 Textile Auxiliaries And Chemicals With Processes And Formulations

PHARMACEUTICALS & DRUGS

01 Pharmaceuticals And Drugs Technology With Formulations

LEATHER & LEATHER PRODUCTS

01 Hand Book Of Leather & Leather Products Technology

BIOTECHNOLOGY

01 Hand Book Of Biotechnology

CERAMICS

01 Hand Book of Ceramics and Ceramics Processing Technology

TREE FARMING

01 Hand Book Of Tree Farming

MUSHROOM CULTIVATION/ PROCESS

01 Hand Book Of Mushroom Cultivation, Processing and Packaging

BIOFERTILIZERS & VERMICULTURE

01 Hand Book Of Biofertilizers & Vermiculture

MINERAL AND MINERALS

01 Hand Book Of Minerals And Minerals Based Industries

RUBBER CHEMICALS, COMPOUND & RUBBER INDUSTRIES

01 Rubber Chemicals & Processing Industries
02 Modern Rubber Chemicals, Compounds & Rubber Goods Technology
03 Technology Of Rubber & Rubber Goods Industries

AYURVEDIC MEDICINES

01 Ayurvedic & Herbal Medicines With Formulaes
02 Hand Book Of Ayurvedic Medicines With Formulations (A

1

Printing Technology

All Printing processes reproduce lines and/or dots that form an image. **Printing** is the process of manufacturing multiple copies of graphic images. Although most people think of printing as putting ink on paper, printing is not limited to any particular materials or inks. The embossing process uses no ink at all, and all shapes and sizes of metals, wood, and plastics are common receivers of printed messages.

Major Printing Processes

The following five major printing processes are used to reproduce graphic images:

- Relief printing
- Intaglio printing
- Screen printing
- Lithographic printing
- Electrostatic printing

Each of these processes is suited for specific applications, such as newspaper, book, package, or textile printing.

Relief Printing

The **relief printing** process includes letterpress printing, flexographic printing, and all other methods of transferring an image from a raised surface (figure 1.1a). Although it was once a major process in the printing industry, letterpress printing has been replaced largely by other printing processes. Most relief printing done today is done with flexography. Flexographic printing is used extensively in the packaging industry for printing on corrugated board, paper cartons, and plastic film. Flexography is also becoming a significant process for printing newspapers, newspaper inserts, catalogs, and directories.

Intaglio Printing

Intaglio printing is the reverse of relief printing. An intaglio image is transferred from a sunken surface (figure 1.1b). Copperplate etching and engraving are two intaglio processes. Industrial intaglio printing is called **gravure.** Gravure is used for extremely long press runs. Cellophane and aluminium-foil candy bar wrappers are two common packaging materials printed with gravure printing. *Reader's Digest* and *National Geographic* are but two of the many national magazines that are printed with gravure.

Screen Printing

Screen Printing transfers an image by allowing ink to pass through openings in a stencil that has been applied to a screen mesh (figure 1.1c). The screen process is sometimes called silk screen printing. Silk is rarely

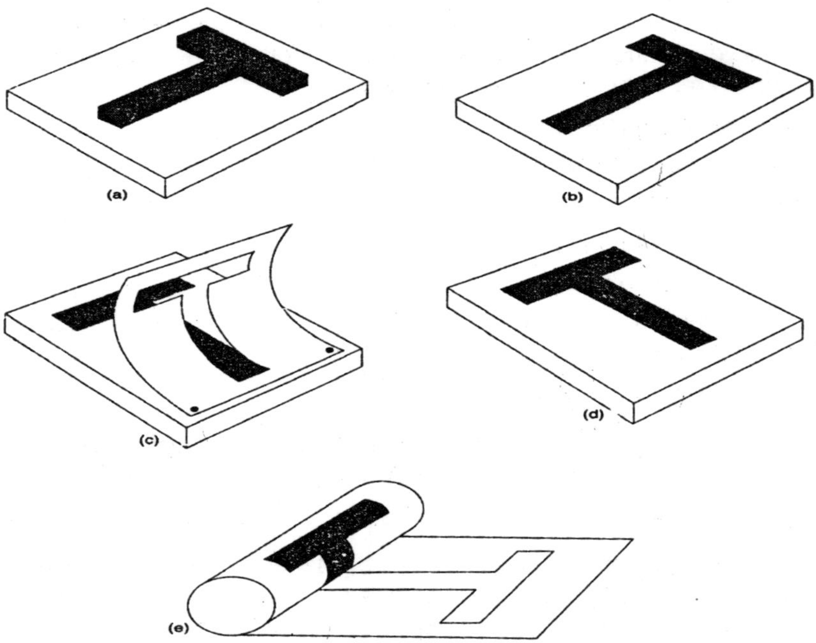

Figure 1.1 Five main printing processes

(a) Relief printing transfers an image from a raised surface.

(b) Intaglio printing transfers an image from a sunken surface.

(c) Screen printing transfers an image through a stencil.

(d) Lithographic printing transfers an image chemically from a flat surface.

(e) Electrostatic printing transfers an image using an electrostatic charge.

used to hold the stencil industrially, however, because silk is not as durable as industrial screen materials. Some of the industrial uses for screen printing include printing on plastic such as round plastic containers, printing large display signs and billboards, and printing on textiles. Another major use for screen printing is in the manufacture of printed circuit boards for electrical/ electronic equipment.

Lithography

Lithography as it is known today is a relatively new process, dating from around 1798. A lithographic image is transferred from a flat surface. Certain areas on the surface are chemically treated to accept ink while other areas are left untreated so that they will repel ink. When the surface is inked, the ink remains in the ink-receptive areas, but not in the untreated areas. When a material such as paper contacts the surface, ink is transferred to the paper (figure 1.1d). This process is sometimes called planography, offset lithography, offset, or photo-offset lithography.

Offset lithographic printing is the most widely used printing process in the commercial printing industry. Its major application is for printing on paper; thus it is ideal for printing newspapers, books, magazines, pamphlets, and all other forms of paper publications.

Electrostatic Printing

Electrostatic printing was invented in 1937 by Chester Carlson. It involves creating an image by electrostatically charging areas of a special drum. As a result, the drum attracts a dry or liquid toner. The toner is then transferred and fused to a sheet of paper (figure 1.1e).

Even a decade ago, electrostatic printing would not have been classified as one of the major printing processes. However, today electrostatic reproduction has become a standard part of digital printing. When a "copier" can deliver 6000 or more images that meet or exceed traditional ink density targets in an hour, the copier must be recognized as a printing press.

Although there are several older printing processes, such as collography, which prints from a fragile gelatin emulsion, these five major printing processes account for nearly 99 per cent of all work done in the contemporary printing industry.

Printing technology has long been a powerful tool for social change. Edward George Bulwer-Lytton wrote, "The pen is mightier than the sword". But his statement assumes that the ideas the pen recorded are distributed. Without printing, few would read the worlds, and the pen would be a very weak weapon.

Printers have long been the most influential individuals in the community. Early colonial printers helped to shape our country by reproducing, recording, and distributing the ideas and events of the period. Benjamin Franklin, an early American patriot, was proudest of his role as a printer. After being active in the Revolution, a signer of the Declaration of Independence, a member of the First and Second Continental Congresses, founder of the first American library, an author, an inventor, a publisher, and ambassador to France, he directed that his epitaph should read "B. Franklin, Printer."

Printing Cycle

Since the time of Franklin, the basic cycle of the printing industry has not changed much from the following procedures:

1. Identifying a need
2. Creating an image design
3. Reproducing the image design
4. Distributing the printed message

The printing cycle begins with an identified need. The need might be as simple as the reproduction of a form or as sophisticated as a poster intented to change human attitudes. It could be as ordinary as package designed to convince a consumer to buy one brand of cereal rather than another. Whatever the need, a graphic design evolves. Special design agencies are often set up whole sole purpose is to sell ideas to clients and work closely with the printer as the design is turned into print.

The function of printing management is to be responsible for creating and controlling the reproduction process. Skilled workers must be employed and an organization created that efficiently and effectively delivers printed products. Once a job is proposed, the most efficient printing process must be identified. Such variables as the type of material to be printed, length of run, number and types of colors, time requirements, desired quality, and customer's cost limitations must all be considered. An estimate must be made for each job. A profit must be made, and yet the estimate must be low enough to attract work in a very competitive market. If the customer approves the estimate, management must schedule the job, arrange to obtain all materials, ensure quality control, and keep track of all phases of production so the job is finished on schedule.

The final test of the cycle is the method of distributing the printed message. Without an audience for the graphic images created by the artist and printer, the printing cycle is useless. Printing is mailed; handed out on busy streets; sold on street corners; and shipped to department stores, corner drugstores, and local news stands. It is passed out in highway toll booths, filed in offices, pasted on bill boards, carried on placards, or even thumbtacked to poster boards. The purpose of all this activity is to place printed matter in the hands of consumers.

Sequence of Steps in the Printing Processes

The printing industry has historically consisted of shops identified with particular processes, such as relief printing or lithography. Craftsmen were trained and then bound by union or guild structure to a particular process. Recently it has become common for a printing establishment to use a variety of printing processes. Regardless of the printing process used, however, there is a sequence of production steps that all printing follows. This sequence consists of the following steps:

Image design

Image generation

Image conversion

Image assembly

Image carrier preparation

Image transfer

Finishing

The point of a printed product is to meet a need, such as to inform, pursuade, entertain, or convince. The design must fulfill the customer's need. In the **image design** step, sketches and final layouts are made. Design variables such as type style, visual position, type size, balance, and harmony are all considered. After the customer approves the design, the image must be generated and made into a final form. Specifications developed during the image design stage are carefully followed during image generation. Traditionally, image generation was performed by hand, perhaps using photographic techniques. Advances in computer technology, however, have moved most image generation to the computer monitor and keyboard.

Traditionally, printing processes relied to a large degree on photography. Images were converted to transparent film in the **image conversion** step and then photographically transferred to an image carrier during the **image carrier preparation** step. Computers have revolutionized image generation and forever changed the way that image conversion and assembly happen. It is now a digital process happening within software. But no matter how advanced the processes have become, images and pages must still be assembled and prepared for an image carrier, whether film, plate, direct to digital press or prepared for a digital venue like the Internet.

The image must be printed onto a receiver material during the **image transfer** step. **Finishing** is the last step, which combines the printed material into a final finished form that can be delivered to the customer. This may include cutting, performing, scoring, folding, inserting, stapling, binding, or packaging.

This book is designed to reflect traditional and contemporary printing technology. Letterpress printing is presented first to give a historical overview of the development of modern printing. The discussion of letterpress also introduces new terms and concepts needed to understand the other printing processes. For the same reason, traditional processes, many no longer done, are reviewed to develop a total understanding of the language and techniques that the digital revolution has incorporated. All the steps in the printing sequences for the lithographic and screen processes are examined in detail. The gravure process is discussed only briefly because the complexity of image carrier preparation and image transfer for gravure printing are beyond the scope of this work. Flexography is presented in some depth because it represents a contemporary relief process and is a significant and growing part of the printing industry. Digital printing, both direct imaging presses and toner-based printing, is also discussed briefly to introduce the concepts and technologies involved. This technology is rapidly gaining in importance and presents a challenge to traditional processes that will ultimately forever change the way we approach the craft of printing.

The printing industry is much larger than any one printing process. The printing industry reflects a manufacturing technology. A technology cannot be learned by examining tools or materials. A technology is mastered by understanding concepts. The first six chapters of this text review the traditional prepress procedures and fundamental concepts from which our current technology has evolved. The next four chapters cover the new digital workflow that has become the standard method for prepress and output today. The remaining chapters cover the concepts and technology behind each major printing process and explore the business of the graphic arts industry.

2

Organization of Printing Services

There are several ways to categorize printing companies. One traditional way is by the kinds of services or produces the companies deliver. With this view, printing organizations can be classified into eight distinct areas:

- Commercial
- Trade shops
- Special purpose
- Quick printing
- In-plant
- Publishing
- Packaging
- Related industries

Commercial Printing

Commercial printing is done by a company that is willing to take on nearly any sort of printing job. Commercial printers can usually handle a large variety of printing jobs, regardless of sheet size, number of ink colors, length of run, or even binding requirements. Typical products produced in a commercial printing shop include small business cards, letterhead stationery, posters, and four-color glossy advertising sheets for mailing. If a commercial printer does not have all of the equipment or skilled staff to perform a whole job, parts of the job, such as die cutting, foil stamping, or binding, may be subcontracted to trade shops.

Trade Shops

Some companies provide services only to the printing trade. These are called **trade shops**. Not all commercial, special purpose, in-plant, publishing, or packaging companies can afford to own and operate all the equipment necessary to meet their total production requirements. For example, a company might decide not to buy bindery equipment because only a small percentage of its work requires folding, collating, or binding. When the company receives a contract that requires binding, it sends it to a trade that specializes in binding.

Special Purpose Printing

Special purpose printing is defined by the limited type of jobs performed by a company. One printer might print only labels. The printer would purchase special equipment and accept orders for only labels. However, the printer would make labels to any size, shape, number of colors, length of run, or purpose. Another printer might specialize in business forms, such as order forms, estimate blanks, filing sheets, school notepaper, or duplicate sales slips. Forms printing is an important area of the printing industry in terms of size and yearly sales. Yet another example of special purpose printing is called *legal* printing. Legal printers are concerned with reproducing pieces such as corporate stock offerings, insurance policies, or financial reports.

Quick Printing

Within the last several years, a **quick printing** sector of the printing industry has grown around the use of electrostatic printing. More commonly known as photocopying, or simply copying, the electrostatic process allows copies to be reproduced without the use of a traditional printing plate press. The explosion of quick print shops has revolutionized public access to the reproduction process. Most copy centers offer walk-in services that include copies up to 11 inches x 17 inches or larger, spot (solid) color, full-color images, binding, computer rental time, direct digital copiers, and an array of supplies such as matching envelopes, binders, and presentation folders (figure 2.1). In addition, many quick print shops offer access through the Internet for online ordering and document creation.

In-Plant Printing

In-plant printing is defined as any printing operation that is owned by, and serves the needs of, a single manufacture a variety of products that must each be packaged with an instruction sheet. Management could decide it is more convenient and cost efficient to set up its own shop to print the instruction sheets rather than to send the job to a commercial printer. The company would then also be able to produce in-house forms, promotion pieces, company letterhead and stationery, time cards, and almost all of its printing needs.

Many in-plant printers use the lithographic process. Equipment manufactures are currently marketing systems with a platemaker "in line" with a press, collator, and binder. With the *systems* approach, the in-plant printer can enter original copy (such as a typed form) into an automatic direct image platemaking system, select the number of copies to be printed and the preferred binding method, and produce bound copies in a matter of minutes (Figure 2.2).

Figure 2.1 Electrostatic printing.
This unit has a high-resolution digital scanner capable of creating 600 dot-per-inch (dpi) digital masters, which can then be adjusted if needed. The system can print 135 prints per minute at 600 dpi and has optional add-ons capable of binding and outputing booklets.

Publishing

Another category of printing services that we use nearly every day is **publishing**. Within this category are the thousands of companies that produce daily or weekly newspapers, and the even larger group of companies that produces periodicals, such as *Time* and *Newsweek,* which sell to a national market. Consider also the group of businesses that produces and markets books. The publisher of this book is a private company that produces textbooks. It is important to understand that printers do not usually make decisions to publish a book, CD, or magazine. Printers are rarely the publisher. Publishers, however, require the skills of the printer

Package Printing

Hundreds of different containers we use every day are produced by **package printing**. The idea of impulse buying (buying a product on the visual appeal of the package) has skyrocketed the demand for high-quality, multicolor packages that attract the consumer's attention. Package printers decorate and form hundreds of millions of folded paperboard boxes, flexible packages, and corrugated boxes each year. Millions of printed plastic bags are used every day in grocery stores and companies that distribute or package food. Corrugated boxes and thin plastic film are both printed by flexography. Packaging, however, is not restricted to paper or plastic. Think of all the steel and aluminium soft drink and beer containers sold every day. These packages are produced by a special process called *metal decorating*.

Figure 2.2 Digital color press.

The DocuColor Digital Color Press allows a company to handle many of its printing needs in-house. This system can print four-color pages at 60 pages per minute (ppm) on a variety of substrates at a cost of less than 10 cents per page.

Related Industries

The last category of services in the printing industry is called **related industries**. The raw materials of the printer are such things as ink, paper, plates, chemicals, and many other supplies. Printers also use special purpose equipment, such as presses, paper cutters, platemakers, cameras, and light tables, to produce their products. Companies that provide services to printers by either producing or selling these supplies and equipment are called related industries. Other businesses, such as consulting firms and advertising agencies that prepare designs for reproduction, might also perform a service, but they do not make or sell a physical product.

3

Typographic Technology

The invention of typography has been called the beginning of the Industrial Revolution. It is the earliest mechanization of a handicraft: the handlettering of books Typographic design has been closely bound to the evolution of technology, for the capabilities and limitations of typesetting systems have passed constraints upon the design process. At the same time, typesetting has offered creative challenges as designers have sought to explore the limitations of the available systems and to define their aesthetic and communicative potential.

From hand composition to today's electronically generated typography, it is important for designers to comprehend the nature and capabilities of typographic technologies, for this understanding provides a basis for a thoughtful blending of design and production.

Hand Composition

The traditional method of setting foundry type by hand is similar to the method used by Gutenberg when he invented movable type in 1450. For centuries, hand composition was accomplished by assembling individual pieces of type into lines. A typographer would hold a composing stick (Fig.1) in one hand while the other hand placed type selected from a type case (Fig. 2) into the stick. Type was set letter by letter, line by line, until the desired setting was achieved. When it was necessary to justify a line, additional spaces were created in the line by inserting metal spacing material between words. Letterspacing was achieved by inserting to extent of 1/30 thin pieces of lead spaces between letters until words appeared to be evenly spaced. When additional space between lines was desired, strips of lead were inserted between lines until the type column was the proper depth. By adding lead, the exact proportion and size of the column could be formed, assuring readability through consistent spacing.

Once the matter was set for printing, it was "locked up" in a heavy rectangular steel frame.This was done on a table called a stone. The type was surrounded by wood or metal spacing material and the contents of the chase were made secure by tightening steel wedgelike devices called quoins. After the type was secured in the chase, it was ready to be transferred to a press for printing, and after printing, the individual pieces of type were again distributed back into the type case by hand. This process continues with one set of type fonts even for couple of years in a small letter press unit and lacs of impressions were taken.

Figure 3.1 Composing stick

Figure 3.2 Type case

Hand composition was tedious and time consuming. When typesetting became automated as a result of the invention of Linotype and Monotype machines, hand composition was used only for setting small amounts of type or for display type. Currently, hand composition is obsolete as a practical means of setting type, but as an art form there has been a revival. Private presses produce limited-edition books and a variety of experimental materials by hand. Many of our typographic conventions and traditions have their origins in the rich heritage of handset metal type.

Linotype

One of the most profound developments in typesetting technology was the invention of the Linotype machine (Fig 3.3) by Ottmar Mergenthaler in 1886. This machine represented the first great step toward typographic automation. Its name was coined because it.

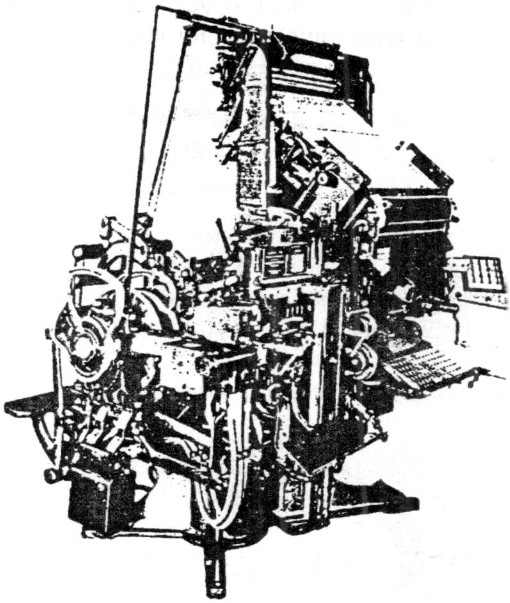

Figure 3.3 Linotype machine

Produced a single line of type to a predetermined length specified by the keyboard operator.

(The operation of the Linotype was based on the principle of a circulating matrix. Each time a key was pressed, a single brass matrix (Fig. 3.4) was released from an overhead magazine, divided into ninety vertical channels, each containing matrices for one character. The magazine was the character storage case for the machine. Once an entire line had been typed the matrices moved into an automatic casting mechanism where the line of type was cast from molten lead. As each line was being cast, the operator typed the next line. After the casting process was complete, cast lines of type called slugs (Fig. 3.5) were ejected from the mold, and the matrices were automatically returned to their appropriate slot in the magazine for reuse.)

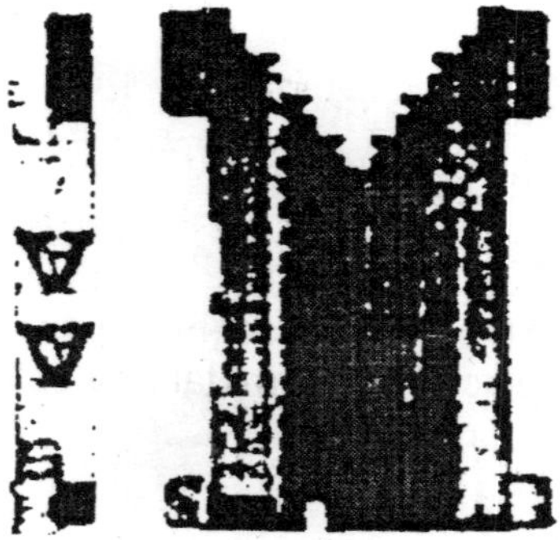

Figure 3.4 Linotype matrix

(The advantages of machine composition as compared to hand composition were obvious. It was faster and more accurate; the problem of type distribution (returning characters to the type case) was eliminated, for the cast lines of type were simply melted, and the lead was reused. Justification of type was automatic, eliminating the tedious process of inserting spaces between letters and words. A standard Linotype could cast lines up to thirty picas in length.)

An important development for linecasting type was Teletypesetter. This perforated tape-driven machine— an attachment to Linotype and INtertype — was introduced in 1928. Tape that was punched by a machines similar to a standard typewriter would be generated from a distant office and transmitted to the linecaster by wire, which made the machine invaluable to news services.

Figure 3.5 Linotype slug

Monotype

(Another significant achievement leading to fully automated typesetting was the Monotype machine, invented by Tolbert Lanston in1887. This machine case one character at a time rather than an entire line. It was composed of two parts: a keyboard and type caster (Fig. 3.6). When an operator typed at a keyboard, a perforated paper tape was generated. This coded tape was used to drive the second part of the system— the typecaster. Compresses air, blown through the punched holes of this revolving spool of coded paper, determined which characters would be cast by the typecaster. Actual casting of type occurred when hot metal was forced into matrices from the matrix case (Fig.3.7). Once the cast characters had cooled, they were placed into a metal tray called a galley, where the lines were assembled. Monotype lines could reach a maximum length of about sixty picas.)

(Monotype became an efficient way to set type for several reasons. Corrections could be made by changing individual letters instead of complete lines. Therefore, complex typesetting, such as scientific data and tabular information, was easier. The Monotype matrix case held many more

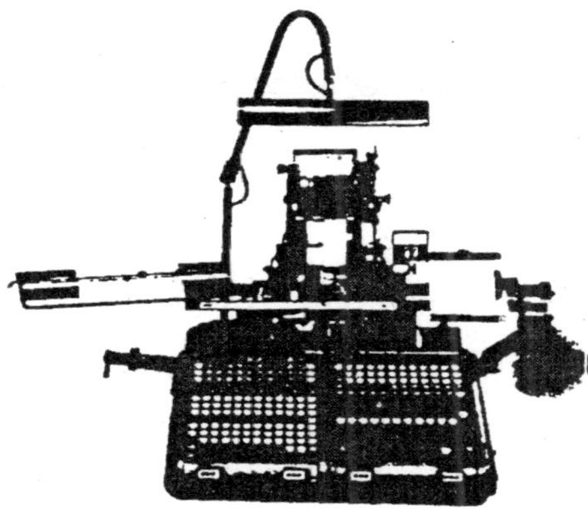

Figure 3.6 Monotype Keyboard

characters than a Linotype magazine, and the casting machine was relatively fast, casting one hundred fifty characters per minute. Since the system consisted of two seperate machines, an operators could keyboard information for later setting.)

Ludlow

Ludlow, a semiautomatic linencaster, is another machine that found a place in the development of automated typesetting (Fig.3 8). Unlike the Linotype and Monotype, the Ludlow did not have a keyboard but combined both hand and machine production. An operator took amtrices from a matrix case similar to a handset type case and placed them into a special composing stick, one by one. The stick would automatically justify or centre lines by inserting blank matrices where necessary. Once a line of matrices was assembled, it was placed into a casting device where it was automatically cast into slugs. If a correction was necessary, matrices were inserted into the stick, cast, locked up, and printed. Although partially automated, this process was time consuming. Distributing the matrices back into the type case by hand added to the production time.

Type produced by the Ludlow machine ranged from 6-to 144-point. Its major use was to produce display type for headlines and other purposes requiring larger typefaces. As was true in the case of handset composition, the Ludlow was neither practical nor efficient for setting large volumes of type.

Although some research in the area of phototypesetting had been done as early as the 1880s, the practicality of this new form of typesetting was not fully recognized until the close of World War II. Printing technology advanced from letterpress to the photographic process of offset lithograph; typesetting underwent a similar technological change.

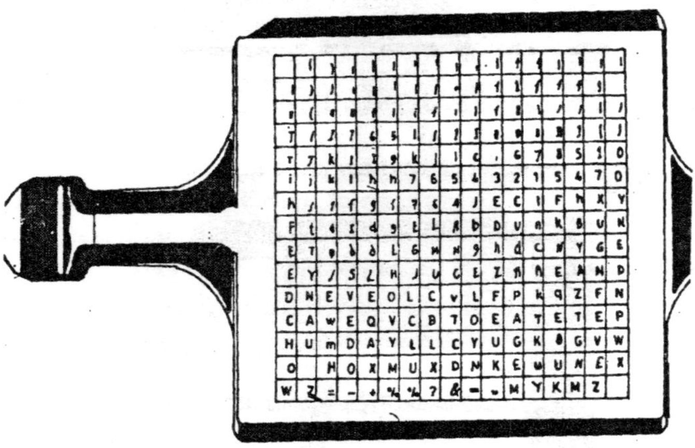

Figure 3.7 Monotype matrix case

Phototypesetting and digital typesetting are currently the primary methods of setting type in the graphic arts. Since the development of Intertype's Fotosetter and Mergenthaler's Linofilm, the first generation of keyboard phototypesetters, introduced in 1950, numerous other systems have been developed (Fig. 3.9). Typesetting speed, character definition, and ergonomics (the relationship between man and machine) have continued to improve. Despite varying degrees of electronic sophisticating among systems, photosetting can be divided into two basic classes: photo-optical systems and photo-scanning systems.

Figure 3.8 Ludlow linecaster

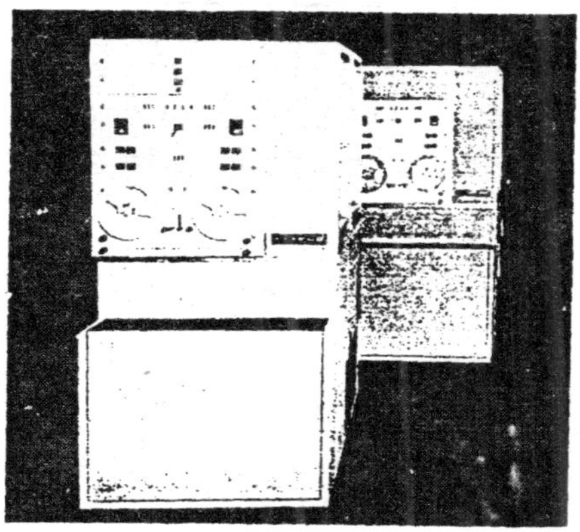

Figure 3.9 Linofilm machine

Photo-optical Systems

(Photo-optical systems store characters in the form of a master font on film, discs, grids, strips, of drums (fig. 3.10). These negative images are the "matrices" of phototypesetting systems. They are optically project onto photographic film or paper. A variety of type sizes can be obtained from a single master font in most systems. An operator enters text and specifications at an editing terminal. Advanced computer technology is used to control this typesetting process.)

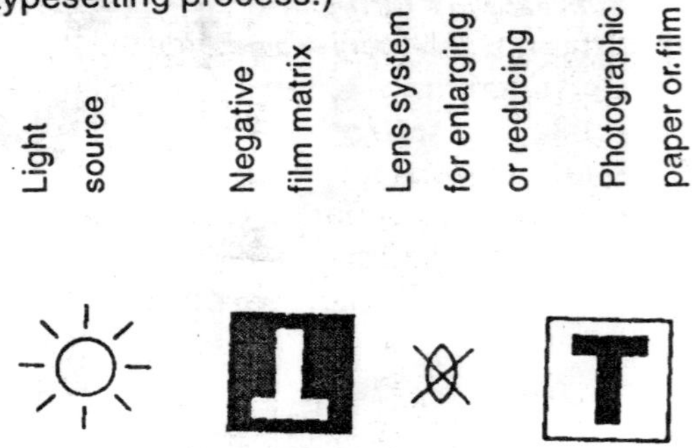

Figure 3.10 Photo-optical system

Photo-scanning Systems

Photo-scanning systems store characters in the form of a master font, not unlike those of photo-optical systems (Fig. 3.11). However, characters are not photographically projected onto film or paper; rather, they are scanned electronically and broken down into either dots or lines. These digitized characters are then projected onto a cathoderay tube from which they are optically projected onto photographic paper or film. Once, the characters have been digitally generated, their appearance can easily be altered. Weight, width, and slant can be changed automatically. Photo-scanning systems operate at much higher speeds than photo-optical systems.)

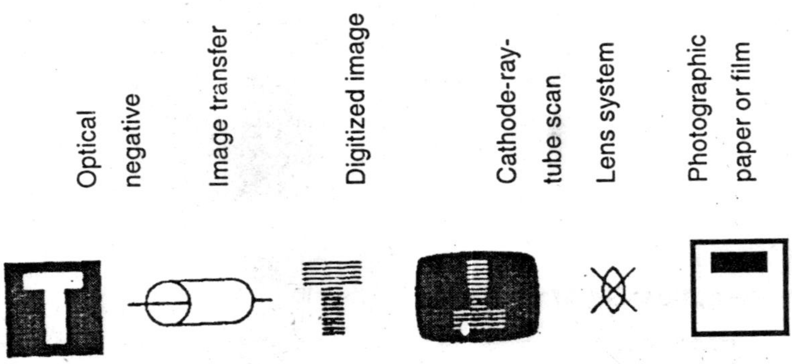

Figure 3.11 Photo-scanning system

Systems Components

(Typically, a phototypesetting system is composed of five parts, which perform input, output, editing, and storage function) (Fig. 3.12)

Keyboard and visual display terminal (VDT). Copy is typed at the keyboard and viewed on the VDT. Although keyboards vary from one system to another, their basic function is to enter and edit text. While a phototypesetter keyboard is similar to a typewriter's keyboard, additional keys are provided to perform special functions. The editing capabilities of terminal save considerable time and effort. Words, lines, and paragraphs can be added, deleted, and moved from one area of the screen to another with ease. Changes can be made at the keyboard before type is processed.

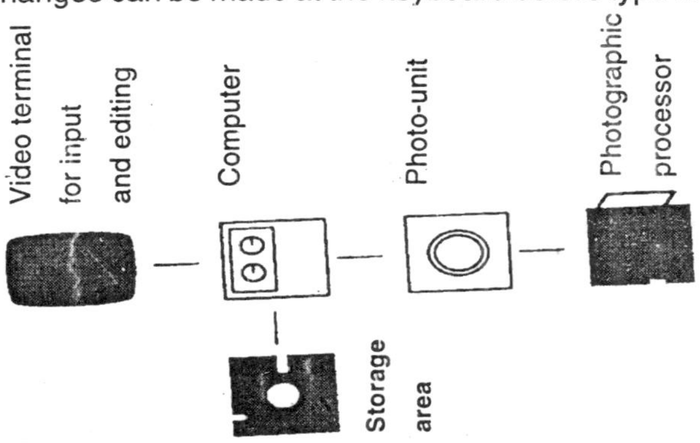

Figure 3.12 Typical Phototypesetting system

Storage area : Storage is very important part of the editing system of a phototypesetter and may or may not be a part of the VDT. Input is stored temporarily in the computer's memory or permanently on a magnetic disk or on tape. If a later date the type needs to be altered, the contents of the disk or type are simply loaded into the computer and changes are made at the keyboard.

Computer : This component, which is connected to the VDT, relays signals between the keyboard, the screen, memory, photo-unit, and processor.

Photo-unit : The photo-unit is the part of the system that actually generates type. A photo-optic system, for example, would optically exposed an entire character from a photographic negative onto paper or film.

Processor : After type has been set, the exposed film or paper is developed in a photographic processor. This may be a part of the typesetter, or it may be separate unit.

(A machine that combines all the above components into one unit is called a direct-entry photosetter. These machines are very popular for a number of reasons. They are small, affordable, easy to operate, and capable of handling difficult typesetting demands. Since input and output are shared by a single unit, a direct-entry machine can be used as word processor and editing terminal that produces high-quality type. Some units can connect with other word-processing terminals, enabling input from more than one operator. Other units have full-page display capabilities. Called area-composition terminals, these units enable complete page makeup and presentation on a visual display terminal.) (Phototypesetting systems, especially those of newspapers and other large publishers, can be part of complex and extensive networks, with links to word processors and mainframe computers. Text generated at one location can be transferred to another via telephone modern or satellite.)

(The advantages of phototypesetting over hotmetal composition are obvious. Phototypesetters are highly flexible and very fast. The typical photosetter can set as many as five hundred characters per second, while hot-metal machines may set only five characters per second. Hot metal machines are operated mechanically; phototypesetters are controlled and operate delectronically. Type generated from a phototypesetting system takes up very little physical space because its final form is a film or paper proof. In contrast, the space required to store lead slugs is enormous. Another major advantage of phototypesetting is that text input uses computerized editing capabilities. This speeds up the process of entering the text considerably, as corrections are made electronically at the keyboard.)

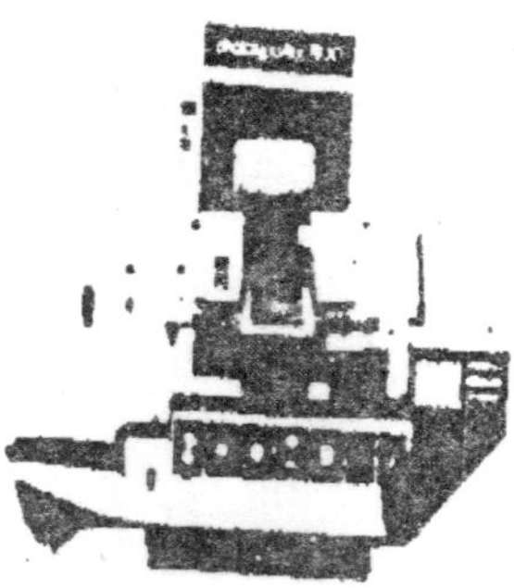

Figure 3.13 Display photographic typesetter

Typography created by phototypesetting is free of the physical restrictions inherent in lead type. There is flexibility in the spacing of typographic elements through kerning, letterspacing, overlapping, interline spacing, and special effects such as runarounds (type that runs around another element such as a photograph or illustration). A designer should understand the capabilities and limitations of typesetting technology so that it can be controlled and used to a greater advantage.

A rapid increase in the use of display photographic typesetting during the 1960's brought new design capabilities to the designer in display typography (Fig. 3.13). As with keyboard phototypesetters, a light source projects in the image of a letter form from the film font through a lens system onto photographic paper or film. There is no keyboard; each character from the film font is brought into position by the operator using hand controls. Because the operator is able to see the recently set characters as they develop photographically, letterspacing can be precisely controlled (Fig. 3.14).

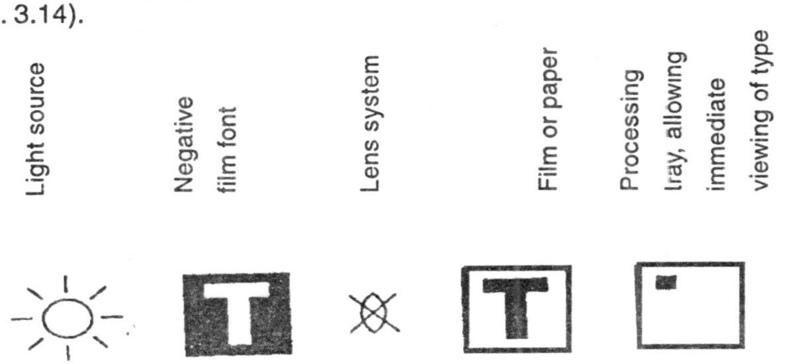

Figure 3.14 Typical display photographic typesetter

The numerous design advantages of display photographic systems led to their becoming the dominant method for heading typesetting within a few years. Instead of being bound by the sizes of handset composition, the designer could now specify any enlargement or reduction of the master font — which has capitals about one inch high — from twice up to four times down with perfect sharpness. Unlike metal display fonts, which have a limited number of characters, display phototype offers an unlimited supply. Spacing flexibility was a major innovation, for display type could now overleap, touch, and be set at any interletter-spacing interval specified by the designer. The constraints of blocks of metal yielded to the elasticity of photographic processes, and innovative designers rapidly explored new possibilities. The lens system enables letter form distortion. Characters can be expanded, condensed, italicized, and even backslanted (Fig. 3.15). The trmendous expense of introducing new metal typefaces, requiring punches, matrices, and cast letters in each size, was replaced by one economical film font. As a result, the introduction of new typefaces and revivals of earlier styles greatly increased.

Digital Typesetting

The digital computer, in combination with the high-resolution cathode-ray tube (CRT) and laser, is revolutionizing the communications industry. Because digital computers have no mechanical parts and are entirely composed of electronic components, they set and process type at speeds never thought possible. In addition, the text type from digital typesetters has now been developed to rival the quality of phototype.

Knowledge of digital-computer functions is critical to an understanding of digital typesetting . A digital computer is an electronic device that uses electricity to process information. It can perform repetitive logical arithmetic operations and store the results of those operations in memory. A computer system is composed of hardware, software, and firmware. Hardware consists of the physical components of a computer; software is the program data which controls the operation of the hardware; firmware is software in hardware form.

The computer component that controls all other parts, performs logical operations, and stores information in the central processing unit (CPU). All components that do not belong to the CPU are called peripherals. A typical digital-typesetting system is composed of a CPU and various peripherals that perform functions necessary to the setting of type— for example, editing and storing text, displaying text on a screen, and printing typeset copy.

A CPU consists of three interdependent components: arithmetic-logic unit(ALU), main, memory, and control unit. These three components work together to control the operations of the computer. The ALU performs both arithmetic and logical functions such as adding two numbers together and determining which of two numbers together and determining which of two numbers is the greatest. In the main memory, called the random-access memory (RAM), data is stored and retrieved by the control unit. This unit also governs the functions of ALU and RAM. Consisting of these three parts, the CPU is the brain of a computer. It controls all functions, including the generation and setting of type in a digital-typesetting system.

A digital-computer system is based on the biconditional state of electronic circuitry. An electronic line can exist in only one of two states: it is either on or off. Each on/off state represents one binary digit or bit, enabling a computer to operate within the laws of the binary-number system. The binary system is a base-2 numbering system using only two numbers, 0

Figure 3.15

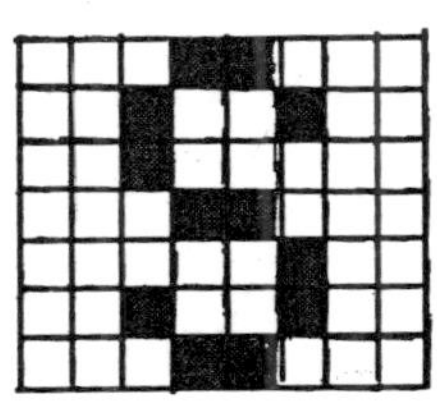

Figure 3.16

and 1, these numbers coincide with the biconditionals: off and on, respectively. The binary system is the exclusive language of any digital computer.

A computer communicates and processes information through the use of data structures. Thse are bits that have been grouped together into various configurations large enough to store significant information. The smallest bit structure is a byte, which consists of a group of bits linked together, such as the ASCII code (American Standard Code of Information Interchange, an information code in which the numbers zero to one hundred twenty-seven represent alphanumeric characters on the keyboard). These data structures are binary codes representing characters or numbers. Translating our alphanumeric characters into the binary system enables computers and people to communicate.

In digital type setting, when the operator punches a key to enter a letter or issue a command (such as lien length or paragraph indent), the computer receives it as a binary code. Once information has been entered, it can be stored, edited, and sent to a peripheral device for typesetting.

A digital-typesetting system encodes typographic characters digitally on a grid, defining the shape of each letter as a certain number of distinct points. Every detail of a letter is defined, including horizontal strokes, vertical strokes, and curves. The coded characters are stored electronically as digital instructions designating the x and y coordinates of the character on the grid. These instructions are then sent to a CRT, where the character is generated onto the screen.

Figure 3.17

Example of digital letter-forms demonstrating decreasing resolution from top to bottoms the number of elements is reduced.

A CRT is much like a television set. It has a vacuum tube with a cathode at one end and a plate of phosphorus and aluminium at the other. When the CRT receives the digital instructions from the computer, defining the shape of the characters, the cathode emits a beam, which scans the tube in a series of parallel, back-and-forth sweeps. The cathode beam is programmed to be either on or off, depending upon the design of the letterforms that have been digitally encoded into the computer. When the beam is on, it excites the phosphorous and aluminium plate. The light emitted by the plate defines each character being typeset. The type is then digitally exposed to photographic paper.

The level of resolution in digital letterforms is an important considerations. Basically, the more dots or lines used to describe a letter form the higher the resolution becomes. Because letters are constructed on a grid, the curved lines consist of a series of stair-stepped contours (Fig. 3.16). when more dots are used to represent a curve, the curve appears smoother to the eye. Large characters require more dots than do small characters to achieve a refined appearance. The quality of a letter form is determined not only by its original design, but also by its digital resolution (Fig. 3.17). The designer of digital type must consider digital technology and its effect upon the resolution of letterforms.

One major difference between digital type and phototype is the maner in which type is stored. Rather than storing master fonts on photographic disks, drums, grids or strips, digital master fonts can be stored electronically as bit patterns on a magnetic disk. Some machines are capable of storing hundreds of fonts,with each size stored independently.

Scanning and Laser Systems

There are two classes of digital typesetters: digital-scanning systems and digital-laser systems. In digital-scanning system (Fig. 3.18), photographic characters are digitally scanned and recorded electronically on a magnetic disk or type. The characters are transferred into grid of extremely high resolution and are transmitted as a set of instructions to a CRT. Next, the characters are generated onto CRT by a series of scan lines. The letter form images then projected from the CRT onto paper, film, or an electrostatic drum. Because the output type is digital, it can be modified automatically to reflect a number of typographic variations. For example, it can be made heavier, lighter, slanted, condensed, or expanded at the command of the operator.

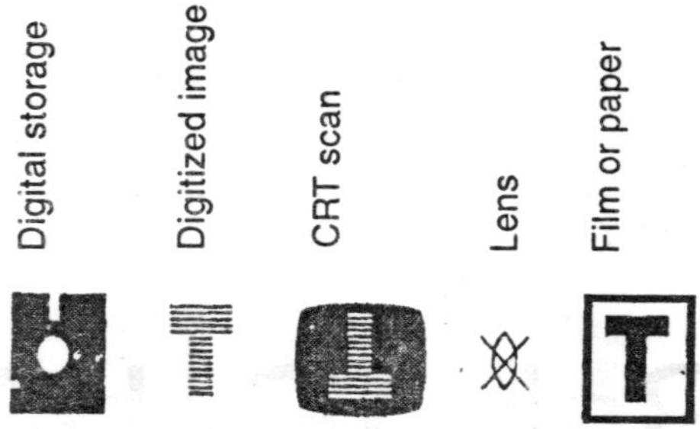

Figure 3.18
Digital scanning typestter

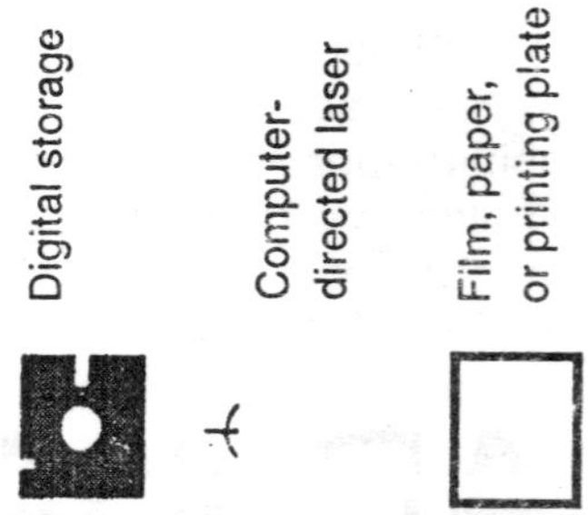

Figure 3.19
Digital laser typestter

Digital-laser systems (fig. 3.19) also store characters digitally, but do not employ a CRT to generate characters. A laser beam scans photographic paper as it reads digital information stored in the typesetter. As the paper is scanned, a series of dots forming the characters are exposed to the paper. The information controlling the laser includes the typographic font and spacing, such as hyphenation, justification, kerning, and letterspacing.

Because digital typesetting is so fast, it is particularly suited to industries requiring the processing of enormous amounts of information, such as news services and publishing companies. However, smaller offices and type houses are also using digital type because of its efficiency.

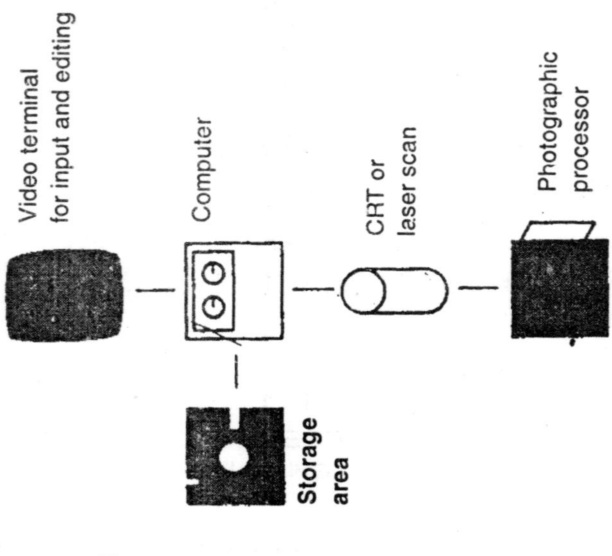

Figure 3.20 Typical digital typesetter

Direct-entry digital typesetters (Fig. 20) are similar to direct-entry phototypesetters, for they are both-self contained. However, direct-entry digital typesetters are much faster and more versatile. Because they generate modified characters and wider range of sizes and spatial intervals, direct-entry digital typesetters bring greater flexibility to the typesetting process (Fig. 3.21).

ABabABabABabA

ABabABabABabAB

ABCabcABCabcABCab

ABCabcABCabcABCab

ABCabcABCabcABCabcAB

ABCDabcdABCDabcdABCDabc

ABCDEabcdeABCDEabcdeABCDEa

ABCDEFabcdefABCDEFabcdefABCDEF

ABCDEFGabcdefgABCDEFGabcdefgABC

ABCDEFGHabcdefghABCDEFGHabcdefghA

ABCDEFGHIabcdefghiABCDEFGHIabcdefghiABCD

ABCDEFGHIJabcdefghijABCDEFGHIJabcdefghijABCD

ABCDEFGHIJabcdefghijABCDEFGHIJabcdefghijABCD

ABCDEFGHIJKabcdefghijkABCDEFGHIJkabcdefghijkABCDEFG

ABCDEFGHIJKabcdefghijkABCDEFGHIJkabcdefghijkABCDEFGHI

ABCDEFGHIJKLabcdefghijklABCDEFGHIJKLabcdefghijklABCDEFGHIJKLabcdefg

Figure 3.21

An Infinite variety of letterform alteration is possible with digital typesetter.

Basic classification of typefaces

(An infinite variety of typestyles are available today. Phototypography, with its simple and economical introduction of new typefaces, has made the entire history of typography accessible. Numerous efforts have been made to classify typefaces, with most falling into the following major categories. Some classification systems add a decorative or novelty category for the wide range of fanciful typestyles that defy categorization).

Old Style

Old Style type began with designs of the punchcutter Francesco Griffo, who worked for the famous Venetain scholar-printer Aldus Manutius during the 1490s. Griffo's designs evolved from earlier Italian type designs. His Old Style capitals were influenced by carved Roman capitals; lowercase letter were inspired by fifteenth-century bumanistic writing styles, based on the earlier Carolingian minuscules. Old Style letterforms have the weight stress of rounded forms at an angle, as in handwriting. The serifs are bracketed (that is, unified with the stroke by a tapered, curved line). Also, the top serifs on the lowercase letters are at an angle.

&

Old Style

Italic

Italic letterforms slant to the right. Today, we use them primarily for emphasis and differentiation. When the first italic appeared in the earliest "pocket book," printed by Aldus Manutius in 1501, it was used as an independent typestyle. The first italic characters were close-set and condensed; therefore, Manutius was able to get more words on each line. Some italic styles are based on handwriting with connected strokes and are called scripts.

&

Italic

Transitional

During the 1700s, typestyles gradually evolved from Old Style to Modern. Typefaces from the middle of the eighteenth centuary, including those by John Baskerville, are called Transitional. The contrast between tick and thin strokes is greater than in Old Style faces. Lowercase serifs are more horizontal, and the stress within the rounded forms shifts to a less diagonal axis. Transitional characters are usually wider than Old Style characters.

Transitional

Modern

Late in the 1700s, typefaces termed Modern evolved from Transitional styles. These typefaces have extreme contrasts between thick and thin strokes. Thin strokes are reduced to harilines. The weight stress of rounded characters is vertical. Serifs are horizontal hairlines that join the stems at a right angle without bracketing. The uppercase width is regularized; wide letters such as M and N are condensed and other letters, including P and T, are expanded. Modernstyle typefaces have a strong geometric quality projected by regorous horizontal, vertical and circular forms.

Egyptian

In 1815, the English typefounder Vincent Figgins introduced slab-serif typestyles under the name Antique. At the time, there was a mania for ancient Egyptian artifacts, and other typefounders adopted the name Egyptian for their slab-serif designs. These typestyles have heavy square or rectangular serifs that are usually unbracketed. The stress of curved strokes is often minimal. In some slab-serif typefaces, all strokes are the same weight.

Sans Serif

The first sans serif typestyle appeared in an 1816 specimen book of the English typefounder William Caslon IV.the most obvious characteristic of these styles is, as the name implies, the absence of serifs. In many sans serif typefaces, strokes are uniform, with little or no contrast between thick and thin strokes. Stress is almost always vertical. Many sans serif typefaces are geomatric in their construction; others combine both organic and geomatric qualities.

&

Sans Serif

Typographic Measurement

Our measurement system for typography was originally developed for the handset metal type invented by Johann Gutenberg around A.D. 1450. The rectangular, metal block of type (Fig. 3.22) has a raise letter form on top, which was inked to print the image.

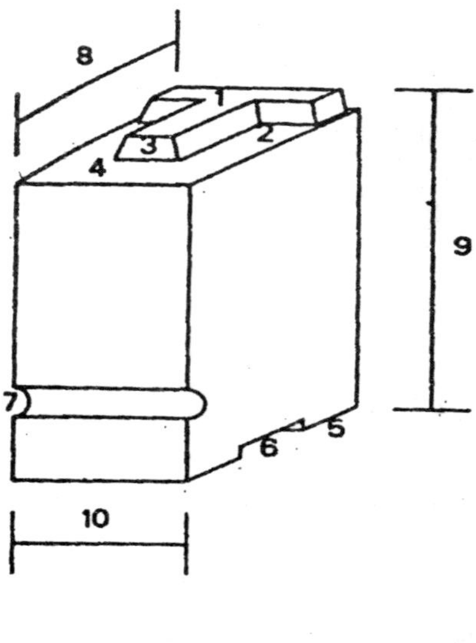

1. Face (Printing surface)
2. Counter
3. Beared
4. Shoulder
5. Feet
6. Groove
7. Nick
8. Point size (body size)
9. Type-high (918" height)
10. Set-width

Figure 3.22

Metal Type Measurement

The small sizes of text type necessitated the development of a measuring system with extremely fine increments. There were no standards for typographic measurements until the French type designer and founder Pierre Simon Fournier leJenue introduced his point system of measurement in 1737. The contemporary American measurement system, which was adopted during the 1870s, has two basic units: the point and the pica(Fig. 3.23). There are approximately seventy-two points in an inch (each point is 0.138 inches) and twelve points in a pica. There are about six picas in an inch.

Metal type exists in three dimensions, and understanding of typographic measurement begins with this early technology. All metal type must be the exact same height, which is called typehigh (0.918 inch). This uniform height enabled all types to print a uniform impression upon the paper, the depth of the type, which is called the point size or body size, is measured in points. Before the development of the point and pica system, various sizes of type were identified by names, such as brevier, long primer, and pica, these became 8-point, 10-point, and 12-point type. The chart (Fig. 3.24), reproduced from a nineteenth century printers' magazine, shows the major point sizes of type with their old names. Types that are 12-point and under are called text type and are primarily used for body copy. Sizes above 12-point are called disply type, and they are used for titles, headlines, signage, and the like. Traditional metal type had a range of text and display sizes in increments from 5-point to 72-point (Fig. 3.25).

As shown in Figure 3.22, the width of type is called the set width and varies with the design of each individual letter. The letter M and W have the widest set width; i has the narrowest. The length of a line of type is the sum of the set width of all the characters and spaces in the line. It is measured in picas.

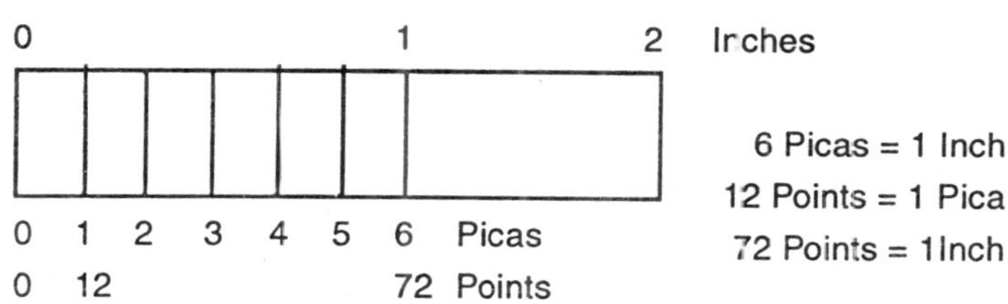

Figure 3.23

AMERICAN SYSTEM OF INTERCHANGEABLE TYPE BODIES.

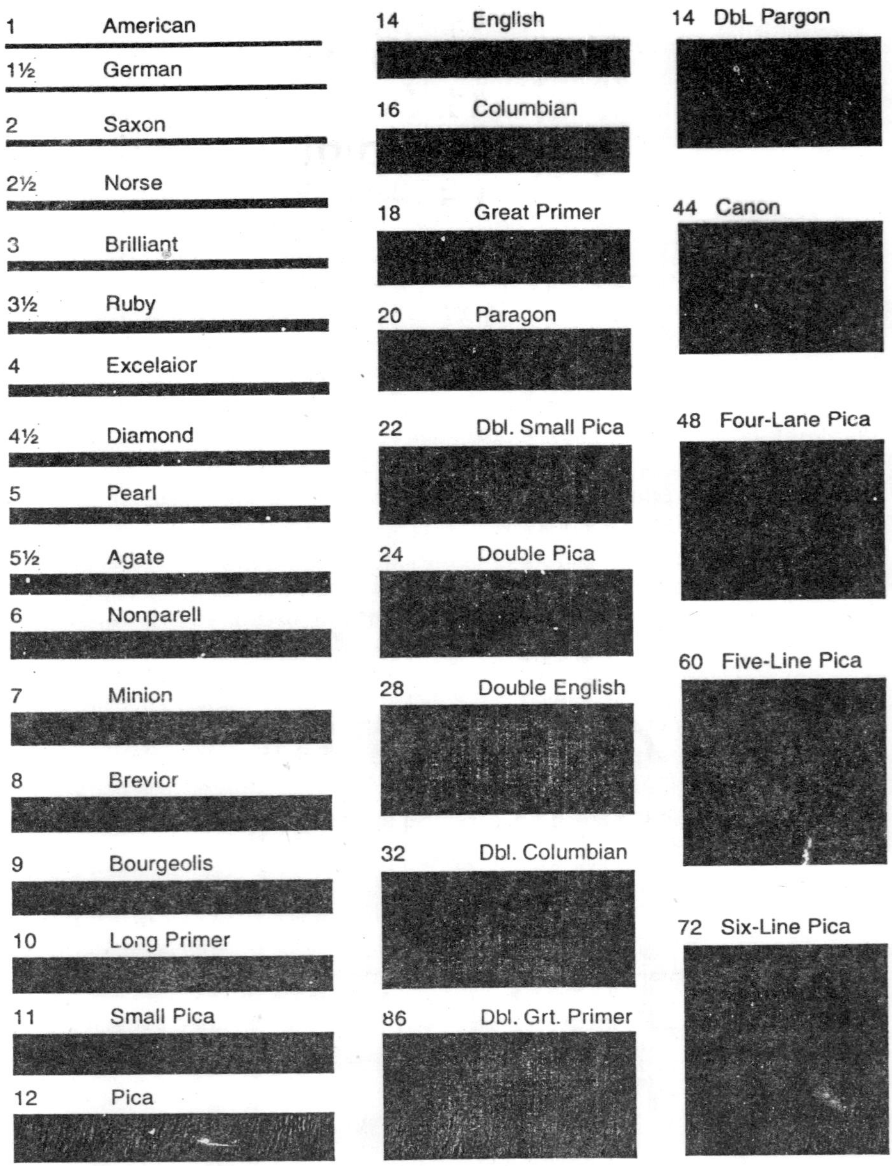

Figure 3.24

Reproduced actual size from the Inland Printer, April 1885

5 Point
6 Point
7 Point
8 Point
9 Point
10 Point
11 Point
12 Point
13 Point
14 Point
18 Point
24 Point
30 Point
36 Point
42 Point
48 Point
54 Point
60 Point
72 Point

Figure 3.25

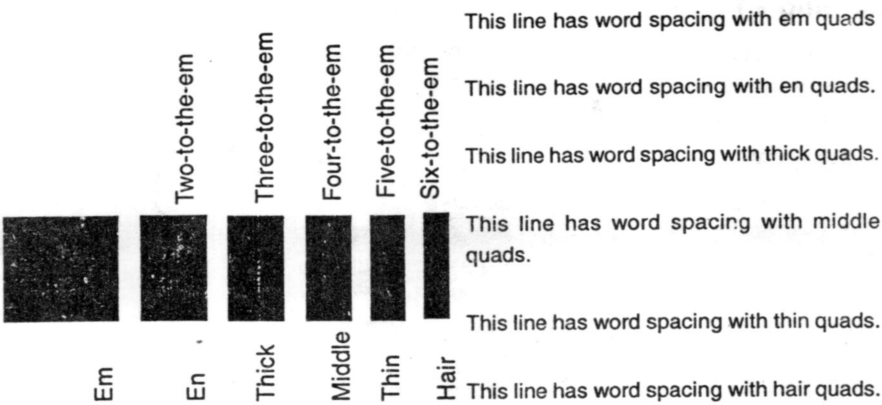

Figure 3.26

Spatial Measurement

The designer measured and specifies the spatial intervals between typographic elements. These intervals are: interletter spacing, traditionally called letterspacing, which is the interval between letters; interword spacing, also called wordspacing, which is the interval between words; and interline spacing, traditionally called leading (because thin strips of lead are placed between lines of metal type to increase the spatial interval between them), which is the interval between two lines to type.

In traditional metal typography, interline and interword spacings are achieved by inserting metal blocks called quads between the pieces of type. Because these are not as tall as the type itself, they do not print. A quad that is a square of the point size is called an em; one-half an em quad is called an en. In metal type, other smaller divisions of space are fractions of the (Fig. 3.26). These metal spacers are used for letter and wordspacing, paragraph indentations, and centering or justifying lines of type.

For design considerations, the em of a condensed typestyle can be narrower than a square, and the em of an expanded typestyle can be wider than a square. This is demonstrated by the em quads from four styles in the Univers family of typefaces (Fig. 3.27)

Univers

Figure 3.27

Planning a Page

The layout reflects the visual appearance of the printed job and also serves to guide its production. A will planned layout is one that integrates the principles of a good typography with those of good design.

Typography

Typography refers to the selection and arrangement of type elements that are included in the layout. Good typography requires an understanding of type, typefaces, type measurement, and the various formats that are available for presenting a printed message.

Type

Type provides the means for including symbols (letters, numbers and punctuation marks) in the printed message. Two basic kinds of type are available for this purpose-hot type and cold type.

Most hot-type composition methods employ type made from molten metal. Each piece of hot type consists of one or more raised characters on a metal body. Hot type can be set by hand (foundary type) or by machine (Monotype, Linotype, Ludlow). After the desired words and sentences have been set, the type is inked and then printed on dull, coated white paper. The resulting print (reproduction proof) serves as the copy used in the photo-offset reproduction process.

Metal type is not used in cold-type composition methods. Instead, the copy required for photo-offset reproduction is produced using one or more of the following techniques : It can be hand assembled from preprinted paper and plastic letters; it can be set with a typewriter and it can be generated by photographic methods.

Type Faces

The term typefaces refers to the unique shape or design of the characters included in a type font, type series, or type family. A font is a assortment of characters of a single size and design. It includes upper(capital) letters lowercase (small) letters, numbers and punctuation marks. The word series is used when referring to a particular type font that is available in more than one size and a family is a group of related, but slightly different type fonts and type series. Although family members all possess the same general design characteristics, they may differ with regard to the spacing weight, and/or slope of their individual characters.

Although thousands of typefaces exist, all can be categorized into six major classes or styles. Understanding the characteristics of each style of type will help customer when selecting type for a particular job. The six major type styles are text, roman, sans serif, square serif, script and novelty.

Text (also known as Back letter and old English) typefaces resemble and the hand-drawn letters of the early scribes text faces are generally utilized when it is necessary to cannot a feeling of age, reverence and formality. However, faces in this category are difficult to read and should not be used for large masses of type. Also never set text type in all capital letters. A message prepared in this manner would be almost impossible to decibpher.

TIMES ROMAN 327

14 Point

ABCDEFGHIJKLMNOPQRSTUVWXYZ &
abcdefghijklmnopqrstuvwxyz 1234567890

18 Point

ABCDEFGHIJKLMNOPQRSTUVWXYZ &
abcdefghijklmnopqrstuvwxyz 123456789

24 Point

ABCDEFGHIJKLMNOPQR &
abcdefghijklmnpqrstuv 123

30 Point

ABCDEFGHIJKLM &
abcdefghijklmnpq 123

36 Point

ABCDEFGHIJK &
abcdefghijklmn 123

TIMES BOLD 334

14 Point

ABCDEFGHIJKLMNOPQRSTUVWXYZÆŒ &

abcdefghijklmnopqrstuvwxyzfiffflffiffl æœ 1234567890

18 Point

ABCDEFGHIJKLMNOPQRSTUV &

abcdefghijklmnopqrstuvwxyzffififlfiffffl 123

24 Point

ABCDEFGHIJKLMNOPQ&

abcdefghijklmnoprstuvw 12345

30 Point

ABCDEFGHIJKLMN

abcdefghijklmnopqrst 23

36 Point

ABCDEFGHIJK M

abcdefghijkm 12345

MODERN SANS LIGHT

14 Point

ABCDEFGHIJKLMNOPQRSTUVWXYZ &£

abcdefghijklmnopqrstuvwxyz 1234567890

18 Point

ABCDEFGHIJKLMNOPQRSTUVWXYZ £)

aabcdefghijklmnopqrstuvwxyz & :;?!] 11234567890

24 Point

ABCDEFGHIJKLMNOPQRSTUV&

abcdefghijklmnopqrst 1234567890

36 Point

ABCDEFGHIJKLMNOPQ

abcdefghijklmnopqrstuvwxy:

OLD ENGLISH

10 Point

ABCDEFGHIJKLMNOPQRSTUVWXYZ &(-';:,!?

abcdefghijklmnopqrstuvwxyz [fiffflffiffl] 1234567890

12 Point

ABCDEFGHIJKLMNOPQRSTUVWXYZ &,.;:!';-

abcdefghijklmnopqrstuvwxyz 1234567890

14 Point

ABCDEFGHIJKLMNOPQRSTUVVWWX &

abcdefghijklmnopqrstuvwxyz fiſſflſtchck 1234567890

18 Point

ABCDEFGHIJKLMNOPQRSTUV

abcdefghijklmnopqrstuvwxyz 1234567890

24 Point

ABCDEFGHIJKLMNOP

abcdefghijklmnopqrstuvwxyz 1234

36 Point

ABCDEFGHIJ K

abcdefghijklmnopq 190

PALACE SCRIPT

14 Point

A B C D E F G H I J K L M N O P Q R S T

abcdefghijklmnopqrstuvwxyz *1234567890*

18 Point

A B C D E F G H I K L M N P W R X

abcdefghijklmnopqrstuvwxyz *1 2 3 4 5 6 7 8 9 0*

24 Point

A B C D E G F H I K J L M N

abcdef ghijklmnopqrstuvwxyz12345678

36 Point

A B C D E F G H I

abcdefghijklmnopqrstuvwxyz1

Roman typefaces are easy to read. They are, therefore used widely in books, magazines, newspapers and other printed items that contain large masses of type matter.

Roman faces are classified according to the width of the strokes that form the characters and by the shape of the serifs attached to these strokes Serifs are the short crosslines used to end each main stroke of the letter or number. These are old style, modern and transitional.

Old style roman faces are open, rounded and very readable. There are only moderate contrast between the thin and thick strokes used to form each symbol. Serifs are pointed.

Modern roman faces are also quite readable. There is considerable contrast between the thin & thick strokes. Serifs on modern roman faces are generally thin and delicate.

Transitional faces provide a compromise between old style and modern roman faces. Variations in the width of transitional letter strokes are greater than in old style but less than in modern. The serifs on transitional typefaces are not so heavy as old style serifs and not so delicate as the serifs on a modern typeface.

Sans serif typefaces connote a "modern" feeling as well as one of simplicity. Sans serif faces are characterized by their lack of serifs and by the uniformity of the strokes that form the symbols. Headlines and captions are often set in sans serif type. Although not quite so readable as roman, sans serif faces are often used for the text matter in books & magazines.

Square serif typefaces are geometric in shape. Characters are clean & open. A uniform stroke is used to form both the letters & their serifs. Square serif typefaces are appropriate for headlines and for short pieces of reading matter. They should not be used to set large amounts of copy.

Script (also called cursive) typefaces are designed to resemble handwriting or hand lettering. Script faces are characterised by variety in the thickness of letter strokes and by their graceful, flowing design. Script faces are difficult to read and should not be used to set large amounts of copy. Also, never set script type in all capital letters. Script typefaces are appropriate for use in invitations & formal announcements.

Novelty (also known as decorative) typefaces are designed to attract attention or to establish a mood. Some novelty faces attract attention because of their size, their boldness, or some other unique feature. Other novelty faces are designed to affect the mood of the receiver so that he or she will be more receptive to the printed message. These are mostly used for advertising purposes.

Type Measurement

The printer's point system is used to designate type size and for measurements that relate to type composition. The basic units of the point system are the point and the pica. A point is equal to about 1/72 inch and 12 points equal to pica. Six picas (72 points) equal 0.996 inch or 0.004 less than a full inch.

A line gage is used to measure points and picas some line gages contain an agate scale as well.

Type is measured in points. Generally sizes range from 6 through 72 points. Larger type is sized by the line, which equals 12 points. Therefore, a piece of 10-line type used for headlines measures 10 picas or 120 points in height.

The size of a typeface is always smaller than its point-size designation. This is necessary in order to provide sufficient white space between lines when type is set solid without extra space added.

Cold-type sizes are arrived at by measuring from the top of an ascender letter to the bottom of a descender letter. Ascender letters are b,d,f,h,k,l, and t. Descender letter are g,j,p and q.

Inserting additional space between lines of type is termed leading (pronounced leading). The purpose of leading is make type matter easier to read. Long lines and large type faces generally require more leading than to short lines and small typefaces.

1 Point	= 1/72 inch
1 Inch	= 72 points
12 Points	= 1 Pica
6 Pica	= 1 Inch
6 Points	= 1 Nonpariel
(½ pica)	
5 ½ Points	= 1 Agate

Type Formats

A variety of formats are available for presenting the printed message in type. The most common formats are flush left, flush right, centered and justified.

Copy Fitting

Copyfitting is the process of determining, before typesetting, the exact space needed for a given amount of copy copyfitting can also be used to determine the amount of copy that must be writing in order to full a given amount of space. If copyfitting indicates that too much or too little space is required then adjustment can be made in several ways, by cutting or adding to the manuscript, by selecting an alternate typeface, by using a different type size, or by changing the amount of leading between lines type.

Different procedures are followed to copyfit text type and display type. Text type is generally used for the body copy of general reading matter such as a newspaper or a book. Type face 12 points and smaller are termed text or body types. Display type is most often used for headlines and for advertisements. Typefaces 14 points and larger are called display types.

4

Composing The Type

Copy for photo- offset reproduction may be set in either hot or cold type. In general, hot type composition methods employ Three-dimensional type made from molten metal. Examples include foundry type, Monotype, Linotype and Ludlow. Wood type is also classified as a kind of hot type even though it is not made from metal.

Cold-type composition methods do not employ metal type to form the desired copy. Instead, two dimensional images of the required numbers, letters and punctuation marks are used. Cold type composition methods include hand-lettering with pen and ink, hand assembling preprinted paper and plastic letters, typing as a typewriter (strike-on composition), and setting type, photographically (phototypesetting).

Hot-type composition

Hand -set Hot Type

foundry type and wood type are both set by hand. Foundry type is named for the fact that it is produced in a foundry. Here, molten lead with some antimony and tin added is poured into type face moulds.

Wood type is much lighter than foundry type and is generally used to reproduce very large head lines. Each piece of wood type contains a single relief character that is carvel in reverse into the end grain of specially treated hardwood. Like foundry type, each piece of wood type must also be 0.918 inch high. This is called type high.

Foundry and wood type characters are stored in special drawers called job cases. Each case contains a complete font of type. A font is an assortment of type of one size and style including uppercase and lowercase letters, numbers, and punctuation marks.

HOW TO SET FOUNDRY TYPE

1. Select the type

Place the job case that contains the style and size of type wish to use on top of type bank (cabinet). The bottom edge of the case is behind the lip of the slanted top. Otherwise the case may slide off.

2. Adjust the composing stick to the desired measure

Foundry type is assembled in a composing stick, the stick is adjustable in pica and half-pica graduations to the desired measure (width) of the line. To adjust the composing stick, lift the clamp and move the knee to the appropriate mark as the throat. Then press the clamp down to lock the knee in position.

3. Cut leads and slugs to length

Leads and slugs are this strips of type metal used to provide white space between lines of type Leads are generally l or 2 points thick. Slugs are usually 6 points Thick. The strips may be cut to any desired length with a lead and slug cutter. Set the stop for the desired length of the lead or slug. Insert the strip and push down on the lever. Cut are more lead or slug than there are lines of type to be set.

4. Set the type

Insert a slug in the composing stick and begin to set the type. Hold the stick in the left hand and use right hand to pick out the type from the case and transfer it to the composing stick. Always place the first letter in the line at the knee and work toward the solid end of the stick. The nick on each piece of type must face up toward the open side of the composing stick. Place a 3-cm space after each word that is set. Try to fit as many words as possible into the line. Try various combinations of spaces and quads to fill completely the measure to which the stick is set. Should never have to force a space into the composing stick, yet the line should be tight enough to stand alone in the stick when the entire line is tilted forward. When the line is full, rearrange the spaces so that the smaller ones are next to the last word set and larger ones are at the end of the line. This is done to prevent the spaces from falling when the type is removed from the stick. After the line is completed, insert a lead or slug and repeat the procedure until the stick is full.

Prepare a Reproduction Proof

Hot type cannot be reproduced directly by the photo-offset process. Instead, a proof of the type called a reproduction proof or "repro" is used the repro is prepared by inking, and then printing the type form as a dull, coated white paper.

Machine -set Hot Type

Several different typesetting machines are available for casting and setting hot type. Monotype, Linotype, and Ludlow are the names of the important of these machines.

Monotype typesetting machines are used to cast and set individual pieces of type from molds or matrices. Two basic units comprise the monotype system a keyboard and a caster.

Input to the Monotype keyboard unit is controlled by compositor as he or she retypes the manuscript. Keys are provided for all uppercase and lowercase letters, numbers and punctuation marks. The keyboard also has additional keys for inserting special instructions as the copy is typed. The output of the Monotype keyboard unit is a perforated paper tape. Each character or instruction that was typed is represented as the tape by a particular performation code.

After the tape is completed, it is removed from the keyboard unit and placed on the monotype caster. Here the code for each inputted character is read in sequence by means of compressed air that is forced through the perforations. As each character code is read, a corresponding matrix is automatically positioned is the caster. Molten metal is then injected into the matrix where it immediately solidifies. Each of the required type characters are formed in this manner. It should be noted that all the type characters delivered by the monotype machine are is correct sequence and properly spaced and justified.

Linotype typesetting machines produce a complete line of type with all characters cast on a single body. Each line of type is called a slug.

Input to the linotype can be by keyboard or by punched tape. Each time a key is depressed or the tape advances, a matrix is released from its storage case (magazine) and falls onto a moving belt. The belt also carried space bands, which are wedge shaped devices used to space between words. Matrices and space bands are transferred by the belt to an assembler stick where they are formed into a line and justified.

From the assembler stick the line of matrices and space bands is sent to the casting section of the machine. Here the assembled line is positioned over a mold, sealed in place and then filled with molten metal. After the metal solidifies, the Linotype slug is trimmed to a preset point size. Finally, the completed slug is delivered to a receiving that is located next to the operator. Meanwhile all matrices and space bands are automatically returned to their proper compartments in the magazine in preparation for reuse.

Ludlow typesetting machines also produce a complete line of type characters on a single body (slug). However, both hand and machine operations are utilized in the Ludlow process. A special composing stick is used for this purpose. Then the line of assembled matrices is placed in the Ludlow casting machine where the slug is automatically cast.

Ludlow is used mostly for display composition. Any number of duplicate slug can be cast from a single matrix assembly. Also several sizes of display type may be combined a single slug.

HOW TO PREPARE A REPRODUCTION PROOF

Hot-type forms (the assembled type) are not used directly in the photo-offset process.Instead, printed proofs made from the type forms are required. These proofs are called reproduction proofs or repros. Similar procedures are followed to prepare reproduction proofs from both hand-set and machine-set hot type forms.

1. Dump The Type

Dumping is the process of removing the type from the composing stick and transferring it to a three-sided steel tray called a galley. To dump the type form, place stick in the galley and grasp the type. Then slide the form out of the stick and into one corner the galley. The type form is now supported by two sides of the galley.

2. Space Out The Form

If necessary, leads and slug can be added between any or all of the lines of type in order to space out the form. Remember, leads are usually I or 2 points thick -and slugs measure 6 points in thickness. Add the leads and slugs so as not to knock over the type. Use a line gage to check the length of the spaced out form.

3. Tie up The Form

String is used to hold the type form together. The string should be long enough to wrap around the form five or six times. Tie a knot at one end of the string and hold the knotted end in left hand. Position this knotted end at the upper left corner of the form and begin to wind the string with your right hand around the form in a clockwise direction. Cross over the knot each time come to it. Overlapping the knot in this manner helps to bind the string in position. After making five or six turns, secure the loose end of the string by tucking it under the several layers of string that now surround the form. Use a lead or other this strip of metal to tuck the end of the string in place.

4. Proof The Type for errors

To make a proof, place the galley as the flat surface bed of the proof press. The open end of the galley must face toward the press's impression cylinder. Then proceed to ink the form. First, place a small amount of ink as the ink plate. Next, run a inking roller back and forth over the ink until the ink is evenly distributed across the plate. Finally, roll the inked brayer (inking roller) lightly over the type form. Make sure that the entire surface of the form is inked.

Now carefully cover the inked form with a clean sheet of paper and pull the proof. This is done by rolling the impression cylinder over the paper covered form. Pressure created by the impression cylinder causes the ink to transfer from the form to the paper. To avoid smudging the resulting print, take care as strip the proof from the form.

Any number of proofs can be pulled from a single type form. However, the form must be re-inked prior to each printing. Also the open end of the galley must be turned to face the cylinder each time an impression is made.

After the required number of proofs have been pulled, clean the remaining ink from the form with a cloth pad moistened with type wash. Remember type wash is a flammable solvent. Therefore not to use near an open five.

The printed proof is now checked for errors. If any are found, they must corrected by making the necessary changes in the type form.

5. Pull The Reproduction Proof

Reproduction proofs are prepared in the same manner as mentioned just above. Reproduction prepared in this way may be pasted directly as the mechanical or they may be photographed in order to reduce or enlarge the type.

Sometimes cellophane or acetate is substituted for the dull, coated white paper. The resulting transparent proof can then be contact printed on photographic film to produce a negative. Transparent repros may also be used to prepare offset plates that must be exposed through a film positive.

6. Distributed The Type

Once the reproduction proof has been pulled, the type form is no longer needed. If the form was set in the foundary type, keep the characters to their original locations.

Machine-set type is generally melted down and the metal refused.

COLD-TYPE COMPOSITION

Hand Lettering with Pen and Ink : Hand lettering with pen and ink is a cold type composition method. Special inking pens, stencils and lettering devices are available for this purpose.

Lettering devices are also used to create letters with pen and ink. It consists of plastic templates, each with recessed letters a scriber, an inking pen and interchangeable points of varying widths.

To letter with the Leryo set, faster the pen and point in the scriber and fill it with ink. Then place the pin end of the scriber on the template and begin to trace one of the second letters. Because the pen point follows the same movement as the pin, it will create and inked duplicate of whatever letter the pin traces.

One of the most versatile of the available lettering devices is the varigraph Head writer. It works very much like the Leroy set in that a pointed styles is used to trace characters engraved in template. As the letters are traced, an inked duplicate is generated on paper.

Hand-Assembling Preprinted Letters

Several kinds of preprinted letter are available for setting small amounts of cold type by hand. These include dry-transfer type, pressure-sensitive type and paper-type pads.

How to Set Dry-Transfer Type

Dry-transfer type comes in sheet form . It consists of preprinted characters attached to a carrier. The characters are printed in reverse on the back of the carrier that is made from transparent plastic or translucent paper.

Each carrier sheet usually contains a single size and style of type. Following are steps to set dry-transfer type.

1. Position the first letter

Draw a baseline as the paper with a light-blue pencil. Then position the base of first letter to be set on this baseline. Some brands of dry-transfer type have preprinted guide lines to aid in positioning the letter on the baseline.

2. Transfer the first letter

The back of the characters that are printed on the carrier sheet are coated with waxlike adhesive. To transfer, hold the sheet firmly in position. The carrier side of the sheet must face up so that all characters are right-reading. Now rub tightly over entire letter with a burnisher. Burnishing transfers the letter to the paper below.

3. Transfer the remaining letters

Carefully lift away the carrier sheet and position the next letter on the baseline. Then repeat to the process described just above (transfer the First Letter).

4. Make corrections as required

Letter that are improperly transferred or damaged during the transfer process may be removed in several ways. They can be scraped away with the edge of razor blade,picked off the surface with the sticky side of adhesive tape. Or erased. After removal, replace the letter as described previously.

5. Burnish the type

Place the protective sheet that was supplied with the carrier over the lettering. Then rub firmly over the type with the burnisher. Burnishing serves to improve adhesion between the letters and the paper.

6. Protect the copy

Several light coatings of a plastic spray or fixative serve to protect the type from dust, dirt and moisture.

7. How to set pressure – sensitive type

Pressure-sensitive type letters are printed on the top surface of a clear plastic sheet. The back of the sheet is coated with an adhesive and protected by a paper backing. Each sheet of pressure sensitive type usually contains a single size and style of letters. Following are steps to set pressure-sensitive type.

1. **Construct a baseline**

   Draw a baseline as the paper with a light-blue pencil. This baseline is needed position the letters correctly as they are set.

2. **Cut out the first letter**

   Use a sharp knife to cut around the selected letter. Cut through the plastic sheet only. Do not cut into the protective paper backing. Then use the point of the knife to lift the cutout letter from the protective paper backing.

   Then use the point of the knife to lift the cutout letter from the protective paper backing.

3. **Adhere the first letter**

   Place the guideline located at the bottom of the letter on the baseline that was constructed in step 1. Then rub lightly over the surface of the letter with a blunt tool such as a burnisher or the handle of a knife. This adheres the letter to paper below.

4. **Set the remaining letters** by repeating step 2 and 3

5. **Make corrections as required**

6. **Burnish the type**

HOW TO SET TYPE FROM PAPER - TYPE PADS

Preprinted pads of paper type are sometimes used to set small quantities of display type. Each of the padded sheets contains a single type character. Blank pad sheets are also supplied for spacing between words. Preprinted paper-type pads are available in a wide variety of type styles and sizes. Following are steps to set type from paper-type pads.

1. Assemble characters in the composing stick

Preprinted paper - type characters are assembled in a special composing stick provided for this purpose. Remove the first required letter from its pad. The characters are printed on both sides of the paper tab. On one side it is printed is blue ink and on the other in blank ink. Slide the paper tab, blue side up into the composing stick until it rests against the solid end. Repeat this process pushing each successive letter against the one before it, until all the characters have been set. Use blank tabs to space between words.

2.Remove Type from The composing Stick

The assembled type is removed from the stick with masking tape. Cut a piece of tape a little longer than the line and lay it over the assembled letters. Then open the stick slightly and lift up the tape. The type is adhered to the tape.

3. Trim Type To Size

Turn the line of type over so that the black printed characters face up. Then use a sharp knife or pair of shears to trim away excessive material. Trim the line to within 1/8 inch of the printed characters.

TYPING WITH A TYPEWRITER

Typewriters are often used to set type for photo-offset reproduction.

Phototypesetting

Typesetting by means of photography is termed phototypesetting or photocomposition. Characters are generated by exposing photographic paper or film through a series of negatives or to an image that is formed electronically on the face of a cathode-ray tube (CRT). Machines used for photocomposition are called phototypesetters. Some photo typesetters produce display type. Others generate text type.

Display phototypesetters are used to set display type (type 14 points and larger). A number of different machines have been developed for this purpose. Some produce their copy by the contact printing process, whereas others make use of projection printing techniques to generate the desired material.

The contact printing process is shown in Fig. 4-1. A film negative of the desired character is placed in tight contact with the light-sensitive (emulsion) side of film or paper. The photographic film or paper is then exposed through the negative to a source of light. After exposure, the film or paper is advanced, the next negative positioned and a new exposure made. This process is repeated until all the desired characters have been set. The exposed film or paper is then processed. Processing causes the exposed areas of the film or paper to turn black. Because these black areas duplicate the clear areas of the negatives, a positive print of the required copy is obtained.

(Detail about PTS is given under the heading "Text Phototypesetting Machines")

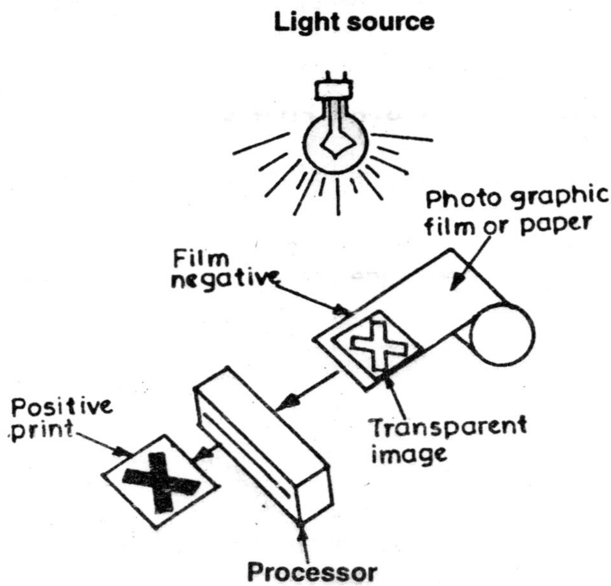

Fig. 4.1 The Contact Printing Press

Projection printing differs from the contact printing process in that the negative and light sensitive film or paper are separated during exposure (Fig 4.2) exposure is made by projecting light through the negative onto the film or paper. This makes it possible to vary the size of the size of the image that is formed from a single negative. Type can be enlarged or reduced on phototypesetting machines that utilize the projection printing process.

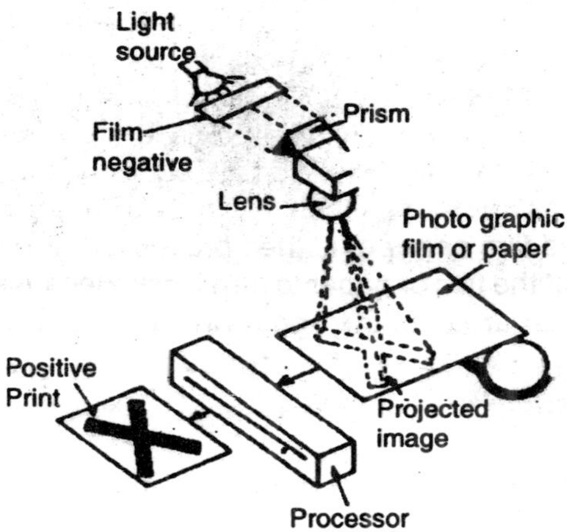

Fig. 4.2 The Projection Printing

HOW TO SET TYPE WITH A STRIP PRINTER

The strip printer is a display phototypesetter that utilizes the contact printing process to produce an image . A film strip containing one size and style of type serves as the negative. A roll of 35 mm photographic film or paper, positioned within the machine's housing and directly below the lamp, is used to record and exposed image.

1. Insert a Filmstrip

Select a filmstrip having the desired typeface and size and insert it under the exposure window and transport block.

2. Expose The First Character

A left and right registration mark provided below each character on the filmstrip. The distance between these marks is slightly larger than the width of the actual character. To set the first character, move the filmstrip until the appropriate negative appears in the window below the lamp. Then alight the left registration mark of the negative with the zero mark on the registration scale. Once aligned, lock the transport block with the transport lever. Next move the transport block slowly to the left until the registration mark to the right of the negative aligns with the zero mark on the registration scale. The first character can now be exposed by pressing the exposure button. After exposure return the transport block to its original position.

3. Expose The Remaining Characters

Spacing between characters on the strpprinter is accomplished by advancing the film or paper after each exposure. The film or paper advances each time the transport block is locked and moved to the left. The distance that the block moves should correspond to the distance between the registration marks that are provided on the filmstrip for each character. Aligning these marks with the registration scale as described just above will result in correct letter spacing.

To set the remaining characters, repeat the positioning and exposure process as described above step 2. spacing between words is accomplished visually by locking and moving the transport block to the left and then returning it to its original position. But do not press the exposure button when adding space between words.

4. Process The Film or Paper

Processing of the exposed strip printer film or paper can take place under subdued light (no bright sunlight or direct fluorescent light) or in a photographic darkroom. A darkroom is best. Process the exposed film or paper.

HOW TO SET TYPE WITH A HEADLINER

The headliner also produces images by means of the contact printing process. Unlike the stripprinter, however, this phototypesetter automatically provides the right amount of space between characters.

A plastic disc named a type master carries the needed negatives. Each plastic disc contains one size and style of type. The slots around the circumference of the disc. These slots vary in depth. The depth of any one slot will determine the spacing of the character that is located exactly 90 degree away.

A roll of 35 mm film or paper, housed in a plastic cassette, is used to record the characters that are set on the headliner. The cassette is positioned within the machine's housing and to the right of the light house control. The leading end of the roll passes under the lighthouse and is attached to a sprocket that advances the film or paper after each exposure is made.

1. Position The Type Master and set the controls

Select a type master having the desired typeface and size and position it in the machine. Then set the word spacing dial, the lighthouse setting and the front control switch according to the information printed on the type-master. Set the letter spacing dial to zero if normal spacing is desired. This setting can be increased or decreased when copy fitting or to obtain special effects.

2. **Set The Type**

To ensure that the leading end of the film or paper has not been exposed to light, removed it from the lighthouse area. To do so, turn the paper feed knob clockwise until 4 inches of the film or paper are fed into the machine.

After advancing the film or paper, turn the type-master until the first character to be set appears in the indexing window. Move the printspace lever to the print position (left) to make the exposure. Repeat this procedure for each of the letters in the first word. Now add a space by moving the print-space lever to the space position (right)

Then continue to set the rest of the type.

To process, turn the paper feed knob until 7 inches of unexposed film or paper have been fed into the machine. This is necessary in order to advance the last character that was set beyond the cutting knife mechanism. Now raise the cutoff lever and hold it up until the amber developing light goes on, Then release the lever.

Processing is completely automatic. Therefore, a second strip of film or paper may be exposed as the first is being processed. It takes about 90 seconds for a completely processed piece of film or paper to emerge from the side of the Headliner.

5

Desktop Publishing (Software)

As was once the case with filmsetting. DTP is proliferating. However, many people have not yet taken much notice of it. 'Desktop publishing', also known as computer publishing 'or' electronic publishing, is a revolution for everyone and not just the printing industry.

What is so revolutionary about DTP? Is it the new things that printers and graphic designers are able to do using it? No! What is new is that anyone can afford a 'printing firm' for their own desk and that everything is very simple to use. Many issues which were addressed either not at all or only insufficiently, because they were too expensive or required too much efforts, are no longer a problem. Furthermore, the way in which we work with graphic designers and printers has changed.

What is DTP?

In simple terms, it is no longer necessary to cut and paste at a light table. Instead anyone can create their own camera-ready copy on a personal computer. But just pressing a button, high-quality finished documents can be produced from a printer.

DTP only describes the function of arranging texts and pictures on a screen. Texts and pictures are shown on the screen at a size of almost 1:1 just as they will be printed later. Text may be created using a scanner. However, pictures may also be obtained by means of a video camera or produced using special programs, such as presentation, calculation or drawing programs of professional quality may be produced using a raster of 150 lpi. Using the new generation of photosetters they may even be produced with a raster of 200 lpi (80 lines per centimetre (L/cm).

If you desire perfect quality, you no longer send the printing firm text and pictures on paper, but instead you send the information by means of a data storage medium. The printer is no longer responsible for the creation and setting of work, but now only reproduces it. The advantage for the customer is that typing errors do not creep in during the printing process. It is cheaper and simpler, Even high-quality pictures may be entered into computer and processed.

An unusual application

By pressing a button a catalogue with more than 800 pages created literally within a few minutes. Text, drawings and photographs were integrated automatically.

The creation of this catalogue used to be laborious and required considerable effort and expense. Now it can be created just by pressing a button and without any manual work being carried out. It is used to be necessary to plan ahead as the preparation required 3 months, but now the catalogue may be created at the last minute.

Many people were of the opinion that it should be left well alone:'it won't work with DTP' was the general feeling. However, a few did not give up so easily. Their courage was justified since within a few weeks a functional concept had been produced.

It is possible with desktop publishing!

In this application, at regular intervals a comprehensive catalogue/ price list with more than 800 pages was published. The conventional process via a printing firm was very time consuming and required considerable effort. Whether it was a picture that was placed incorrectly or the infamous last-minute changes, the time and effort required to create an error-free catalogue was very expensive. It made more sense to jump on to the 'desktop publishing' band wagon and create the complete catalogue using a personal computer. This was especially so since all the prices had already been calculated using a computer and hence were already available on a data stoarage medium.

At the press of a button...

The task now existed of creating the catalogue without manual intervention simply by 'pressing a button'. Only in this way could the camera-ready copy for an error-free and comprehensive catalogue be produced within a few days.

Therefore, existing data needed to be prepared and created using DTP software. But, as often happens, it is the little things that cause problems. For example, one problem was that the layout (text and pictures) for the left and right hand pages was different. If one new page were to be inserted, all the following pages would needed to be altered. It is of course obvious that this cannot be done manually for a catalogue of several hundred pages.

The problem was solved by using a control program which, acessed from a user menu, combined all the required steps. The most important function of this program was the step 'Organise complete catalogue'. By pressing a key on the personal computer, all the text was organised, files for the DTP system (catalogue pages) were automatically created and the pictures and text were inserted accordingly. The formatting for the left and right hand pages was carried out at the same time.

The pictures required for the catalogue were stored before they were needed in a so-called picture database in the personal computer. They had been obtained either by means of a scanner or from a drawing program. The correct pictures were then automatically placed at the correct positions in the catalogue. Those pages on which items were to be placed could be determined very easily. A list was produced, also on the computer, where it was noted which product should be on which page with which picture.

...and within a few minutes 800 catalogue pages !

It thus took minutes to prduce a catalogue with over 800 pages on the personal computer. A sample was then printed using a laser printer and then information sent to the printing firm via a data storage medium. Next the printing firm photoset this data on to film and the complete camera-ready copy (films) was produced. Text and pictures were of course completely arranged on this camera-ready copy. It took only a few days from the initial work through the last change in price to the completed camera-ready copy. By using this process, it was primarly ensured that errors didnot creep in and that the document did not require manual creation.

Problems exist with hardware too !

No one considered that problems would arise in using standard hardware and software. For example, the selection of a high-performance scanner caused problems, even though according to the brochures, it should have been a really simple matter. It was the scanner software which actually proved to be the real problem.

For DTP it is important to have a powerful computer system. Graphical applications should not require patience. Above all, the performance and smooth operation between individual system components is extremely important. A 386 Mbyte SyQuest cartridges were used for backing up the data and transferring it to the photosetter. As in all components, but especially in the case of the monitor (the connection between man and machine), care was taken to have an ergonomically perfect model. An anti-dazzle, flicker-free ETAP double-page monitor with 75 Hz image scanning frequency was used. This system is now used not only to create catalogue, but also more and more for a diverse range of brochures and other documents.

HOW THIS BOOK WAS CREATED

Next, we will briefly describe how this book was produced using a DTP system. First the text was typed into a personal computer. We used the word processing programe WORD from Microsoft. WORD has many advantages, especially for the creation of longer texts: the more text created, more important the classification function of WORD became. At any time the headings alone could be viewed and hence an overall perspective was always maintained. There was also no problem in moving chapters, since moving a heading moved the corresponding text with it. When most of the text had been created it was reproduced on paper by means of a postscript laser printer to gain a better overview, to monitor the text and to carry out corrections.

The layout was then created using the Ventura Publisher DTP package. First basic layout such as the book size and page margins was set. Naturally it was worth while trying a few different ideas and printing a few samples. Finally the text typed using WORD was loaded into Ventura. For reasons of simplification and speed, the text was divided into separate pieces for each chapter. During the loading of text ventura automatically created all the required pages with the set format. As several formatting features had already been included when creating the text (e.g. for headings), the basic structure could be created within a few minutes.

Next, creativity and hard wok were required. The fonts, text sizes, headings and subheadings to be used needed to be set. Here, for example, a parameter in a menu was simply changed and all the headings in the book acquired a new format. The result could immediately be seen on the screen. Initially several samples of a few pages were printed in order to evaluate the final result. Once all the details had been decided upon the actual creation could be carried out.

The pictures for individual chapters were entered into the computer by means of scanner. As the final sizes of pictures were not be determined until later, all pictures were provisionally scanned at a very low resolution. They were then positioned and printed by a laser so that they could be checked. Next we worked from page to page and entered the pictures at the correct size. Text and pictures were shown at a size of 1:1 on the screen in addition, pieces of text were changes, deleted or inserted. Finally, it was time for a complete sample print. For this purpose a complete publication was created, i.e. all the individual chapters were put together. Before a well-earned break, the table of contents was created automatically. Left completely alone, the DTP program now had to print the complete book using a postscript laser printer.

By the next day we had recovered-fortunately perhaps, as we did not really like several aspects of the book. But thanks to DTP it did not really matter. Changes could be carried out relatively simply. All the pictures used were scanned again. As the sizes for all the pictures were set, it was now possible to scan using accurately calculated resolutions.

Once everyone concerned had given a general OK to the book, the glossary was created, then came the big moment; instead of sending the data to the printer, the data was copied on to 40 Mbyte magnetic tape cartridges. These cartridges were sent to the photosetter. Within a few days the film came back. After a quick check on the quality of the pictures the film was sent to the printing firm.

Printing-up to now

Everything is done the wrong way round, i.e.with the pages back to front. Even in modern filmsetting, work is carried out at a light table with the text and pictures being cut and pasted with the pages back to front (mirror image). What are the individual steps in this classical process?

Text creation by the author

First, as always, the text is created. Often the text is still presented on paper by the author.

Layout planning by the designer

The layout designer then plans format for the publication. The designer sets the so-called type area, determines the space required for pictures and text and marks the text with typesetting instructions. The typesetter is provided with information regarding which font and size of text should be used for typesetting.

The typesetting studio recreates the text

The processed text is sent to the typesetting studio. Here the typesetter reproduces the complete text and at the same time provides control commands for the typesetting system. So-called galley proofs are created. Galley proof are phototypeset lines of text on paper.

The galley proofs are corrected by the author

The galley proofs read and corrected either by the author or by a copy editor and errors which the typesetter missed are marked.

The layout designer pastes up the proofs

The layout designer pastes up the galley proofs and determines whether the text needs to be shortened or lengthened.

The author has to shorten texts which are too long

Such changes to the text are once again carried out by the author. After the author has confirmed that the text is correct the finished layout is returned once again to the typesetting studio. The typesetter carries out the specified changes to the text.

Imposition planning

The imposition planner converts the sketched layout into an exact reproduction copy. This includes cutting marks for the format size, text proofs (made up correctly), colour specifications. status and section specification for pictures.

Final control by author

Before everything is finally sent to the printer, the author once again determines whether all the material is correct and, if not, provides corrections.

Any other requests?

Now just imagine what would happen if the author wanted further changes to be made and the whole process had to be started from the beginning again...

The litho is created

The lithographers now require all the material; page layout pictures, slides, etc., in order to carry out yet another stage on the way to the end product. So that the individual parts are available on film, the page layout is photographed using a process camera and pictures and slides are scanned or, using a process camera, rasterised. These individual parts, pictures and photographed page layouts are assembled manually accordance with the layout specifications. This results in the final litho; page reversed positive films.

Printing preparation

At the printing firm the various processes merge again. Whether the processes were carried out conventionally or by using DTP, the printing firm then produces printing plates from the film and printing may commence.

Printing today-with DTP

Everything is so simple with DTP; number of columns, size of margins and line spacing are easily changed and the result is immediately visible on the screen. Another button is then pressed and the work is complete. Whereas before you had to busy yourself with scissors, scalpel and glue, now all you have to do is place texts and pictures on the screen using a mouse. The design may be changed repeatedly until it is acceptable . Imposition planning is no longer needed, as each design is an original and the level of quality may be controlled. With DTP both the time-consuming and decision-intensive steps are concentrated within your own firm or department. Text is created once only and so possible sources of introducing new errors are removed. The layout on the screen provides immediate information as to whether a text needs to be shortened or lengthened. You become more flexible and the infamous last-minutes changes are no longer a problem.

Using DTP you can maintain complete control over the entire production. As manufacture of the camera-ready copy you are in charge of the complete process of typesetting, producing graphics and carrying out the page layout. With this complete control you become extremely flexible and able to react quickly to changes, savings in time and cost are enormous. All the processes involved in creating camera-ready copy may be carried out at your own desk. There is no time-consuming postage to and fro, no need to proof-read repeatedly and no need to check the layout several times. Many tasks which had to be carried out elsewhere are either not required at all or may be carried out by yourself. Talking all these advantages into account, the cost of investment is relatively low. In addition, if you consider, for example, that the result obtained for smaller publications is considerably improved, or that an advertisement for a daily newspaper may be completed 5 minutes before the deadline, then it is difficult to imagine working without DTP.

DTP - simple?

Yes and no. The technical capabilities of DTP lead people to believe that nothing could be easier. However, DTP is not as simple as the advertisements in colourful brochures would like us to believe. Let us examine the issues from a critical point of view for a moment. DTP is very useful computer application, but also one which leaves you with the responsibility for what you print. You alone produce the layout and make the decisions. DTP often combines the activities of author, publisher, copy editor, editor, typographer, graphic designer and computer expert into one.

Especially when you begin using DTP do not overestimate your abilities. It is not just a case of learning how to produce page layouts. Though DTP is simple and perfect to use, it still needs to be learned. A computer novice without expert support will experience more frustration than joy.

However, an experienced computer operator should also ensure that he or she receives sounds advice. It is often minor revelations or even just a sensible introduction to it that will make DTP a successful experience.

WHAT DOES A DTP PROCESS LOOK LIKE?

Creation of text

First the text is required. For example, for a leaflet advertising special offers you would need product descriptions and prices. If these product descriptions and prices are already stored in you computer, you can transfer them and save having to retype the information. Otherwise you will need to type the required text as before, not by using a typewriter, but instead much more conveniently by using the computer. The text may be changed, corrected and moved without difficulty. If you are using a professional word processing program, many functions for the checking of spelling, automatic creation of the contents, etc., are available. You may then either send this text to a DTP service/printing firm for further formatting, safe in the knowledge that no more errors can creep in, or carry out the layout of the leaflet yourself.

Obtaining pictures

You may either create pictures yourself (e.g with a drawing program, table/calculation program, CAD system, etc.) or input them into the computer by means of scanning original pictures.

Layout

Now you may begin immediately! Using a DTP program the text and pictures which have already been created may be combined and laid out directly on the computer. On the screen you can view the result immediately, exactly as it will be printed. Pictures may be moved, reduced or enlarged and the text may be rearranged, with changes to size and line spacing. These changes may be applied repeatedly until the result is acceptable and the text fits within required number of pages.

Sample print

Once the layout has been completed a sample may be printed. Even printing experts are astounded with the quality of the printout from a PostScript laser printer. There is only one task which the laser printer cannot do, namely print high-quality photograph. It is not possible to attain more than newspaper-quality photographs with a laser printer. However, the layout may always be checked using a laser printer.

Inexpensive printing

If the printout from the laser printer meets the quality requirements sufficiently, it may be inexpensively copied or reproduced by the small offset process. In this case not film is produced, but printing is carried out instead using the paper camera-ready copy from the laser printer. The quality of photographs may be improved by requesting a copy shop or printing firm to rasterise and assemble the pictures. A true DTP user, however, will send the text and pictures to a photosetting studio in order to have them photoset.

Professional printing

Once a sample has been printed on the laser printer and the layout meets the requirements, a so-called PostScript printing file is created. This file is sent via diskettes (for example) to photosetting studio or printing firm and then photoset on to a professional output device (photosetter). Whilst a laser printer has a resolution of only 300 dpi, more than 2540 dpi can be attained using a photosetter. As a perfect result you will receive the camera-ready copy on film or paper. The pictures are of course included in this camera-ready copy.

The advantage

The advantage of producing documents using a DTP system is that from the initial creation of text right through to the complete layout, everything can be achieved by yourself. Errors due to the incorrect typing of text or incorrect arrangement of pictures, for example, will be avoided. Subsequent changes are simple with a DTP system, as all necessary information is stored within a data storage medium and only the required changes need to be applied. Once the new changes have been made photosetting will need to be carried out again and the new camera-ready copy will then be complete.

Comparison of the Conventional system with DTP

Creation of final print

Conventional printing		DTP system	
1.	Text is handwritten or typed using a type writer or word processor	1.	Text is typed into a personal computer.
2.	Layout planning changes are time consuming and expensive.	2.	Layout planning changes can be made at any time.
3.	Text is marked on the paper for the typesetter (labels, font,text size etc.)		N/A
4.	The text is sent with setting details to the typesetter. The text is copied taking into account the setting parameters (typesetting) and then photoset (galley proofs). It is then proof-read and if necessary photoset again.		N/A
5.	The author receives the galley proofs to check		N/A
6.	The printer receive the corrected galley proofs, carries out the corrections and photoset again.		N/A
7.	The author then receives the galley proofs again to re-check them.		N/A
8.	Pictures and drawing are selected.	3.	Pictures and drawings are selected and are then either entered by means of a scanner or created by of a computer program.
9.	Pictures and drawings are reproduced by a lithographer		N/A
10.	Galley proofs and reproductions (pictures) must be cut to the right size for each page and then individually mounted (pasted) on to each page.	4.	Inserting page breaks and arranging pictures are carried out on the screen. Changes may be undertaken immediately.

6

The Postscript Language

The postscript page description language dominates the world of desk top prepress. It has become the "defacto" standard method of communicating pages created on the desktop to an output device and is increasingly being used as a way of transferring graphics and documents between different applications and platforms.

Postscript is essentially a graphical programming language used to create precise descriptions of graphical objects for accurate transfer to an output device.

A fundamental strength of the language, and the main reason that it has elapsed alternative page description languages is its concept of device independence. Graphical objects are defined within a limitless coordinate system, and it is not until the page is output that it takes on the finite resolution of the specific output device. Device dependence allows pages to be printed at the highest resolution available on the printer or imagesetter that is used to output them.

Code that is written in a programming language is turned into a series of low-level instructions and the resulting programs can be executed on any compatible computer platform.

An interpreter is normally part of the environment in which the programm is executed. It accepts the code produced in a programming language and translates it line by line into low-level instructions. As the interpreter comes across a valid instruction that it has all the necessary information for, the instruction is executed.

Compiled programs execute slightly faster, since there is no need for line-by-line interpretation, but they are less able to exploit device specific features when they are executed.

Postscript language code is written not by human programmer but by another program (the user application). When the application's print function is invoked, postscript code in generated and sent to an output device. Here the device's interpreter translates the code into low-level instructions which it executes, producing a raster image of the page. Compared with more general-purpose programming languages such as C++, Postscript is a relatively simple language.

Postscript was developed in 1985 by Adobe systems, as one of several languages then available for controlling printing devices. Other printer control languages, such as Hewlett Packard's Printer control language (PCL), still exist, though they are of little interest to the graphic arts as they are unable to support high-resolution output devices.

Postscript was widely adopted by the nascent desktop publishing market, but as DTP matured the weakness of the original postScript language in controlling color and halftone screening became a barrier to further adoption, until language extensions were developed for these purposes. In 1990, adobe published Post Script Level 2, which incorporated these extentions and added a number of sophisticated functions. Implementing Postscript Level 2 is not an easy task for hardware and software developers, and so far not all applications exploit the full functionality that it offers. The number of applications and output devices that support the features of Level 2 is gradually increasing.

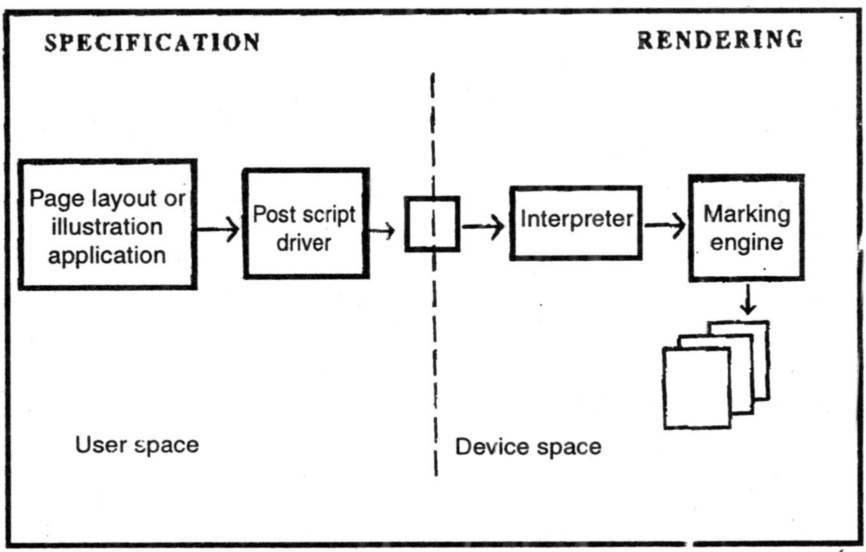

The main features of Level 2 are :

1. Better memory management
2. Better color handling
3. The introduction of forms caches and patterns.
4. Increased use of dictionaries.
5. The ability to handle file decompression during output.
6. Improved control of the output device from the front end.

Memory management reduces the chance of a memory limit being encountered when a file is being output, a frequent problem with Level 1 interpreters.

Color handling includes color conversions based on color lookup tables and CIE-based color space.

A forms cache stores repeated page elements (such as borders, logos, tints, or other graphics that are common to a number of pages) in memory, so that they do not need to be reprocessed for each page.

A dictionary is a looup table that holds a set of parameters used in defining a graphical object. An example is halftone dictionary that holds the screening parameters (screen frequency and angle and dot shape).

Post achieves device independence for page descriptions (and all the separate elements that comprise them, such as graphics, fonts, colours and so on) by maintaining a clear distinction between user space until they are actually output.

Creating and printing a document using Post Script can be divided into four separate phases as under conceptual postscript model.

First, the user creates a document (usually in a page layout application) and saves it in the native format of the application or the host computer. Next the document is "printed" from the application (either directly to an output device or to a file on disk) when the application's Post Script driver translates the document into Postscript code.

The resulting code is then sent to the output device, where the interpreter executes the code and translates. The page into a raster image in the coordinate system of the output device. Finally the raster images is sent to the marking engine, which reproduces only film or paper.

Creating the page and translating it to PostScript code both take place in device space. This distinction between user space and device space. This distinction between user space and device space is fundamental to PostScript's device independence.

Although a postscript interpreter will process all of the valid instructions it receives, a page description is more device-independent if it confirms to the document structuring conventions published by Adobe.

These conventions specify an overall structure for a PostScript page description and define structure comments (prefaced by percentages) that are incorporated into the page description. These comments are not actually processed by the interpreter but allow other programs to read and display the file and to carry out operations such as imposing the individual pages in a publication, managing the resources it needs, and handling any errors.

A Postscript program that confirms to the document structuring conventions has two parts : a prolog and a script. The main purpose of the prolog is to identify any procedures that will be used when the program is executed. It will also include a header that contains comments naming the document's creator and any resources, such as fonts that are used.

The script consists of a document setup section, followed by the individual page descriptions and a trailer that announces the end of the document. The trailer may also include some comments deferred from the prolog. Each page in the publication is defined in turn through a series of instructions to the interpreter to perform graphics operations. At the end of each page in the script, the instruction show page tells the output device to output the page. Each page may also have its setup section.

Postscript instructions are known as operators, and they are used in conjunction with the procedures defined in the prologue and with specific data or operands, operators carry out task like:

1. Constructing path
2. Controlling the specification of type
3. Controlling the specification of images
4. Drawing lines and objects.
5. Controlling the way that colours are output
6. Controlling the way that the page is output
7. Defining the graphics state (such as the current color or line width)
8. Defining transformations in user coordinates (allowing sealing and rotation to take place).

Most programming languages have difficulty in imaging both text and graphics together, since the methods used to draw characters are very different from. Those used for graphics the only way of combining them successfully is to render type as a graphic object, by turning it either into a bitmap or into a vector drawing procedure. Postscript describes everything (with the exception of scanned images) as vector graphics.

Image – Postscript describes the position and color of scanned images using the image operator. Image defines the following image properties:

(a) The image format (its height and width, its spatial resolution, and the number of colors).

(b) The coordinates of the user space where the image will be placed.

(c) The source data for the image (in other words the color values for each pixel).

(d) A description of how the source color values are to be translated into the current color space.

The Printer Driver : The role of a Postscript printer driver is to generate Postscript code when a document is printed. The driver creates code by translating from the host computer's native graphics format. The driver is also responsible for downloading any resources, such as fonts and dictionaries.

Unlike other kinds of printer drivers, a Postscript driver is not specific to the output device, but to the host platform or application. Postscript printer drivers are often written by the developers of the host application, and this tends vary in how they handle the creation of Postscript code.

Although Postscript code is created in user space and is therefore independent of the characteristics of the output device, it must be appropriate for the target output device. The driver does this by using printer description files that supply information specific to the output device, including the paper size, the maximum image size, and the available resolutions.

Postscript Errors : The most common Postscript errors are caused by:

(a) The correct fonts not being available to the interpreter.

(b) Interpreter memory limits being exceeded

Font errors can be fixed easily enough by making them available to the interpreter. Memory limits are much harder to predict or resolve and can be the most difficult to decipher from the error information provided by the interpreter.

To minimise the possibility of errors, the printer driver should ideally generate code that conforms to adobe's Document Structuring Convention. It should also make optimum use of memory resources during processing by bringing into memory only those items that are required by the document and removing them when they are no longer needed.

Output problems can of course be caused by user errors, such as supplying incorrect fonts or specifying composite output for a publication containing special colors. In such cases the file may be printed without Postscript errors but does not produce the output wanted.

Postscript Limitations : Postscript is an extremely powerful graphical imaging method and its dominance is unlikely to be challenged by any other page description language in short future.

The lack of transparency, are due to basic language characteristics that are unlikely to be modified is a limitation.

7

Proof-Reading

Proof-reading marks were developed to indicate mistakes with the least possible writing on the proofs. Earlier, proof-readers in India and many other countries followed proof-reading marks used by the British.

In the British system, a number of English terms were used such as caps for capital letters, s.c. for small capitals, l.c. for lower case, bold for bold types, n.p. for new paragraph, run on for no paragraph, stet for leave it as it is.

However, the existence of a large number of languages, especially in India, made it difficult to develop corresponding proof-reading words in different scripts and at the same time, most of the new typesetters of regional languages were unable to understand the proof-reading words used in the British system. It, therefore, became essential to develop standard proof-reading marks which could be used for any language in the world. This work has been done by BIS (Bureau of Indian standards) by introducing non-linguistic proof-reading marks, which has helped in achieving worldwide co-ordination in the proof-reading practice.

The following rules should be adopted while marking corrections on proofs;

(a) Only the position with the appropriate symbol should be indicated in the text of the proof and the corrections should be written in the margin.

(b) When more than two corrections occur in a line, the corrections appearing in the left half may be marked in the left margin and the corrections in the right half may be marked in the right margin of the proof.

(c) The corrections in the margin should be strictly marked opposite the line to which they belong.

8

Pre-Press Processes

The technological developments in the pre-press areas of printing production have had quite a profound effect already but one must expect further developments that will lead to :

(i) Increased accuracy
(ii) Increased speed/productivity
(iii) Less labour-intensive processes
(iv) Not much of skill required
(v) Lower consumable materials costs

In other words, systems that are more predictable, quieter and cheaper in application. One must look "all" areas of pre-press, even the printing processes and alternative technologies must be scrutinised to see what are implications for the future.

Some of the developments that have been emerging in pre-press processes that affect lithography :

(i) Automated, mechanised planning and assembly systems
(ii) Demise of dichromated coatings - development of presensitised.
(iii) Scanner separated sets with computer generated register marks.
(iv) Development of improved U/V exposure units
(v) Use of Integrated light meters
(vi) Improved plate grains, use of anodizing
(vii) Developments with alternative plate-making methods including "film-less", laser scanned, electrophotographic systems.
(viii) Photo-composition, computer composition of film elements.
(ix) Automatic film and plate processors.

(x) Quality control guides and aids.

(xi) Oven stoving of presensitised plates.

Methods of creating accurately-imaged plates are mainly being influenced by the developments being made in the contemprory electronic, computer and micro processor industries.

It the aim of many to develop "filmless" or, for that matter, "plates" processes. In the latter, significant progress is being made in alternative printing processes that are emerging in specialised areas. Ink jet, electrophotographic and magnetic computer-driven printers are being developed, mainly in the information printing sector.

It is a priority of development to reduce both the "high cost" and "high skill" factors from the processes, by reducing labourintensive processes.

The development of "total" electronic imaging systems that are by passing the use of film are taking considerable time to develop and the number of installations will be governed by their high cost and the output of work. However, the electronic revolution continues.

The Opti-scan Camera, is a piece of contemporary technology in graphic arts, that is changing traditional practices and methods. The principle is a one- a micro- processor with a suitable software package of imposition knowledge is fed with the necessary basic information. The outer and inner imposition schemes print on a V.D.U. and on command, one at a time, in a pin-located, fixed position; the film holder in the dark room sections moves on XY axis as per program, to receive the exposed page.

Considerable Screen in Cooperation with Toppan Printing Co. Ltd., have launched a system intended to reduce film cost. The system is intended to be used in the book printing sector and utilises 35 mm film and projection platemaking. This again is a fully automated step and repeat machine which is programmed via a mag, card input. Moreover, specially-coated, projection speed lithographic plates are available from many suppliers.

Another piece of equipment that warrants attention and which is now available to the graphic arts industry is the Dale Graphic Equipment ruling and register table. It would be difficult to say which of the many capabilities is most impressive. The table appears quite standard until one notices the electrical, drives to the rail in the XY axis traverse. A micro processor (with built-in magnetic tape cassette recorder and memory facility) can be fed with a program. The head is capable of providing a variety of functions as it is adaptable to permit the most precise ruling, cutting and image positioning for step and repeat or process colour assembly described earlier.

Also mentioned earlier, but worth a mention under a section heading "development" is the Protocol table that has been linked in a micro processor driven capability, that performs the functions of automated ruling, masking, cutting image positioning. This wide ranging facility which is linked to the drilling registrator table should change the face of film planning departments in time.

Advanced automated assembly is an electronic facility available from more than one supplier. The planner sits at a console where he manipulates images via an alpha-numeric key board and an electronic cursor instead of the traditional scalpel, tape and linen tester. A typical programming unit with its double sided co-ordinate measuring table, paper tape punch and reader, line printer and VDT (Visual display terminal). The data generated from the microprocessor may be stored and used to drive the step and repeat style exposure unit.

Computer generated screens

It is not clear yet as to how these technologies might evolve. Initially, the development is seen provide "graphics" capability to ink-jet and electrophotographic computer generated images. However, this development could find its way into all manner of printing application.

The principle of the system is best described as follows. The computer is programmed with a "scanning algorithm"; the scanned continuous tone original image is transformed into a series of "on-off" decisions for the computer. The computer arranges the spatial frequency of the fixed-sized dots proportional to the density of the original. Certainly the possibility of long distance graphics transmission is worth consideration. Perhaps the day will come when a set of signals from a computer scan will be sent down a telephone line to an ink jet printer, where a randon screen positive could be imaged onto plastic base.

9

Camera

Photographs are often included in a printed product. They help printer to communicate graphically. When used properly photographs can clarify and add interest to a visual message.

If the photographs needed for reproduction are unavailable, they will have to be made to order. Photographs are made by exposing film in a camera, processing the film to make negatives and then converting the negatives into positive prints by contact or projection printing techniques.

Process Cameras

Cameras used to photograph line & continuous tone copy, are called process camera. Process cameras that are located in the dark room are known as dark room cameras.

There are two general categories of process cameras: 10 Horizontal & 2) Vertical cameras in both of these categories contain the same basic parts and operate in essentially the same way, but differ with regard to their construction.

The parts of a vertical process camera are arranged as a vertical axis.

The copyboard serves to hold the copy while it is being photographed. The copyboard usually consists of a flat, cushioned board with hinged glass cover. Copy that is placed between the board and cover in held in position by pressure, when the cover glass is closed. Some of copy boards are attached to a vacuum pump. Vacuum is used to hold the copy in positions.

The copy board as process camera is movable. The copy board can be moved toward or away from the lens. This movement allows us to produce negatives that are the same size larger, or smaller than the copy that is photographed.

The lights serve to illuminate the copy during exposure. Several types of lamps can be used for this purpose. These include incandescent, quartz iodine and pulsed Xenon lamps.

Metal reflectors located behind the lamps are used to concentrate the light on the copy board. These reflectors are generally movable.

The lens assembly consists of the lens, lens board, diaphragam and shutter. The lens is the most important part of the process camera. It is made up of several carefully shaped and polished pieces of glass that are held together in a tube or barrel. The barrel is attached to a lens board that serves to fasten the lens to the rest of the camera.

The lens is used to collect light rays that reflect from the copyboard and to focus them on the light- sensitive film. The distance from the optical center of the lens to the film, when the image first appears in sharp focus, represents the focal length of the lens.

Camera size : Process cameras are designated in size by the maximum sheet size of photographic material the film holder can accommodate, e.g. a 24" camera has a back big enough to hold a 20" x 24" sheet.

Camera optics : Camera manufacturers carefully match a process camera's optical requirements to the lens, or lenses they supply with it. Questions about lens resolution, copy reduction and enlargement ratios along with maximum film sizes, all contribute to the final lens selection. These factors, among others determines the camera's overall size.

Choice of camera type : The choice of a vertical or horizontal camera usually depends on the type & volume of copy to be handled, the space available and cost.

Vertical cameras : Vertical cameras are chapter and occupy less space but not versatile. These give limited copy size reduction ratios if the camera has only one lens. Vertical cameras are limited in size, also with about 24" the limit.

Horizontal cameras : These produce higher volume and better image resolution. They need more space and do not have as many size limitations.

State of the art cameras : Many variable have to be kept under control for decent results during exposure and processing. The present era of electronic microprocessors has been a tonic for the process camera. Current state-of-the-art cameras, when they have automated processing system, can bring these variables under control, high-quantity output is achieved at lower make-over rates, under the most difficult production condition of low volume and unskilled personnel.

Contact exposure facility : Image assembly facilities are involved in combining multiple images such as line work, graphics type, text and halftone into final layouts. The aim is to produce a finished, one piece film with all the image elements in place. Producing this involves photographic contact exposures. This transfers the image from previous film negatives or positives to a new sheet of film.

Image transfer using contact exposure methods reproduce a same size image. By using register plus multiple images can be positioned very accurately on the final film. Fidelity of detail can be maintained throughout these procedures. Equipment needed for contacting includes a vacuum contact frame, a contact exposure system and exposure control systems.

Vacuum contact frame : Vacuum contact frames are available in many sizes and from several different suppliers. Among their common features is a hinged frame that holds the cover glass . It is hinged along the back of the frame to allow it to be opened at the front so the operator can place and remove films.

Other feature is a rubber blanket with a rubber gasket around it. The blanket is the working surface on which film is positioned for exposure. There is a vacuum pump connected to the rubber blanket by tubing to get air out from between the blanket and glass after the frame has been closed.

Contact exposure system : Various light sources have been designed for contact exposure. Contact films and papers available for three different safelight environments and each one needs a different light source.

Exposure control systems : Just as in camera applications controlling exposures during contacting requires consistency from one exposure to another. There are several kinds of contact exposure systems. They have been beefed up by adding microprocessors and are available from a number of manufacturers.

Image Distortion : Image prepared for flexo printing often need image distortion because the distortion usually is related to direction of printing.

Several contact - exposure distortion machines are on market. Generally, image assembly procedure are carried out in conventional contact operations. The final films are contacted by using the distortion exposure system to produce the plate ready, image-distorted films. These systems can produce different amounts of distortion to a high degree of accuracy.

Machine processing equipment is available in assorted sizes & types of lith, or deep tank and rapid access processors. They all represent state-of-art technology of film transport, solution control, washing & drying.

Rapid - access processors generally are simpler, lower priced and easier to control than with processors. This partly accounts for the growing popularity of rapid-access photographic systems for image assembly. LIth system, although demanding more operator skills are the system of choice

Lith developers : Lith developers are packed in two parts and are designated "A" & "B"

Part "A" contains developing agents, part "B" contains a chemical accelerator, once the two are mixed, fairly brisk deterioration begins.

Another difference between lith developers and other types is that the sulfite preservative, commonly used to retard developer oxidation, seriously interferes with lith developers.

Therefore, the preservative content in lith developers must be reduced to about one-tenth the amount used in continuous tone developers.

Chemical mixing : There is a point to remember always, add the concentrate to the measured amount of water and never water to the concentrate. If a two-part concentrate is involved, such as Pith A &B, dilute each part separately before combining them to make the working solution.

After adding water to the concentrate, stir the diluted solution to produce a uniform dispersion of chemicals within the working solution.

Colour electronic prepress systems : Although these are expensive to install but they have image quality and production advantages with features for image capture, image manipulation, image assembly and creating special effects. The precision and ease with which image elements can be juggled into new design configurations is particularly adaptable to the packaging field.

Computer Aided Design (CAD) : New design -oriented systems are evolving with desk top publishing packages to bring electronic imaging into the creative end of packaging design. This system provide adequate image data handling to meet present sophisticated graphic arts needs.

Image -capture systems : Electronic color scanners have become the accepted norm for producing process color separations. They offer much flexibility in gaining good tone reproduction, grey balance and colour correction.

Image- assembly and image-manipulation systems : The systems handle complex image assembly with as screen monitoring. Simple keyboard commands can add masking, tints and borders including vignetted tints. In process colour, changes in tone reproduction and colour correction are evaluated on the colour monitor while in process, allowing visual control of their effects . Other editing features provide capabilities not available with conventional photographic methods.

There are three general types of copy images: Line copy, continuous-tone copy and combination copy

Line copy is all solid copy printed as the finished piece. Line copy is considered the easier kind to reproduce. All line copy should be solid black & white, have clear definition of detail, and contain no tonal values.

Type images are the easiest to recognize as line copy. Most type generation devices produce black type on a white substrate or reverse (white) characters surrounded by a solid background.

Rules or Borders : A second group of images with line-copy characteristics is, nonreading copy used as rules or borders. There are many ways to create these images, including type-generation devices, hand drawings, preprinted adhesives and transfer sheet.

Illustrative copy : This is the most complex group of images that can be used to create a line negative . To reproduce properly, the copy must maintain the general characteristics of line copy. Artist and photographic images can be produced as line copy. The illustrative copy can be an original photographic, piece of artwork or high- contract copy of the original.

Continuous -Tone copy : Continuous-tone copy must be screened, i.e. broken down into dots (halftoned) before reproduction.

Photograph and Illustration : The most common example of continuous tone copy is the black & white photograph. A photograph has a range of grey tones and will reproduce as a toned image only if made in halftone.

Combination copy, Continuous Tone & line : Combination copy includes any photographic or continuous-tone art images with type, rules or borders printing over or reversed out of toned areas.

Mechanical copy Assembly : Assembly of the mechanical requires bringing all the copy elements together in the right format a by using materials suitable for reproduction as either a line or halftone negative.

Image Assembly for reproduction: The layout of the image-assembly department the quality of the tools and equipment and the film assembler's skills, are equally important in producing high-quality work. Basically following equipments are required:-

I. Light table (stripping tables)
2. Lineup (layout) tables
3. Register punch
4. Work tables
5. Contact exposure frame
6. Tools

Stripping substrates : The basic film-assembly and masking materials are vinyl or paper masking sheets, clear film. photographic peelable or hand-cut peelable masking film.

Masking vinyl and paper have the same purpose to limit the actinic light (the action or property of radiant energy to cause a chemical change, as in photography) that exposes the emulsion on a film proof or plate. For most modern photosensitive plates, proofs and room light films, the activnic light is a strong exposure to UV radiation. Masking papers are generally available in yellow, orange or red.

Red is the best color for limiting UV, then orange yellow is least effective.

Masking sheets made of polyvinyl chloride are available in either red or orange. If possible avoid using vinyl masking material in a step-&- repeat machine because heat build up can cause softening of the vinyl mask.

Therefore, photography is essential to flexographic printing. Applied photography, in its reproduction state allows the conversion of pictures, line art graphic and type into an image that can be transferred to a flexographic printing plate.

The preparation of artwork for each design that is to be reproduced is a key element in the quality and cost of producing the finished piece. High quality (measuring high contrast) sharply defined type and line art, accurately positioned into a camera ready mechanical, are necessary to good reproduction.

Individual image elements in the form of line and halftone film are brought into a final assembly, to satisfy the design layout demands. This may involve additional photographic contact exposure, to combine the images onto one final film . Accurate image placement and the distortion free transfer of image elements are crucial.

10

Photographing Line & Continuous Tone Copy

Line & halftone negatives are needed to prepare several types of offset printing plates. These negatives are obtained by photographing line and continuous -tone copy

Process films:— It consists of an emulsion a transparent base, an antihalation backing and a protective coating.

The emulsion is the light- sensitivity layer, on which the image is recorded. Silver halide crystals (usually silver bromide) comprise the-sensitive component in this layer. The crystals are suspended in gelatine that is spread smoothly over a transparent base.

The transparent base can be made from several different types of plastics. Acetate or polyester, 0.003 inch to 0.008 inch thick, serves as the base for most process films. Although acetate base films are less costly, but polyster film are preferred for precision work.

The antihalation backing serves to keep light rays that penetrate the base from reflecting back and exposing the emulsion. The antihalation component, is a dye that absorbs light. By absorbing light, the dye prevents halation, which is a spreading or dubling of the recorded image.

The protective coating is an extra layer of gelatine that covers the emulsion. It served to protect the emulsion from scratches. It also helps to keep the film from curling.

There are three kinds of process films— blue sensitive, orthochromatic & panchromatic. Each is sensitive to a different portion of the visible light spectrum.

Blue—sensitive film react to blue, violet & UV light. These films are relatively slow & require a long exposure time. But this is ideal when making duplicates of other negatives by contact printing techniques.

Orthochromatic (ortho) film are sensitive to green & Yellows as well as blues, violets & ultra violet (UV) They are to handle in the darkroom under red safe light condition because ortho film cannot differentiate between these light-blue lines and the white base of the copy & moreover ortho films don't react to red light.

Panchromatic (pan) films are sensitive to all colours of light. They must be handled in total darkness. Generally pan films are used for color separation.

Steps to Expose a line negative

1. Prepare the process camera
2. Load the copyboard
3. Adjust the lens opening
4. Focus the camera
5. Check focus & Location of the image
6. Load the process film
7. Set the timer

   Exposure times for enlargements or reduction are computed by multiplying the basic exposure time by an appropriate exposure factor.
8. Expose the film

PHOTOGRAPHING COLORED COPY

A filter must be used to convert the color into either black or white. A filter is a thin sheet of colored glass or gelatine that allows light of its own color to pass but absorb other colors of light.

After selecting the appropriate filter, insert it into a filter holder, that is mounted on the camera's lens barrel. During exposure, light reflecting from the copy will have to pass through the filter before it strikes the film. Because all filters absorb some light, exposure times with filters must be longer than exposure times without filters.

Continuous tone copy images are made up of black areas, white areas & areas containing varying tones of grey. The blackest or darkest parts of the image are known as shadow areas. Grey in the image are referred to as middletones.

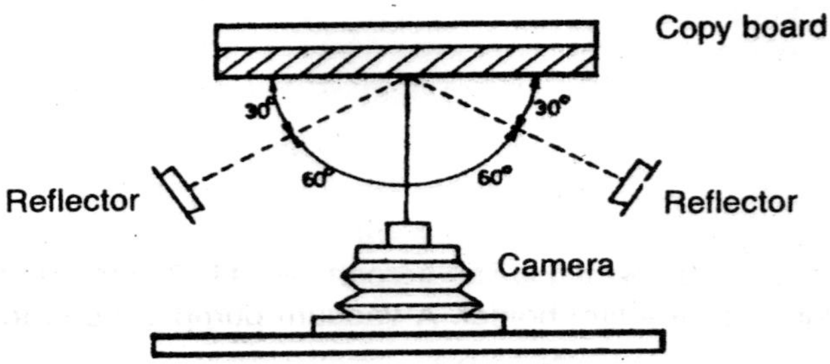
Copy board
30°
30°
60°
60°
Reflector
Reflector
Camera

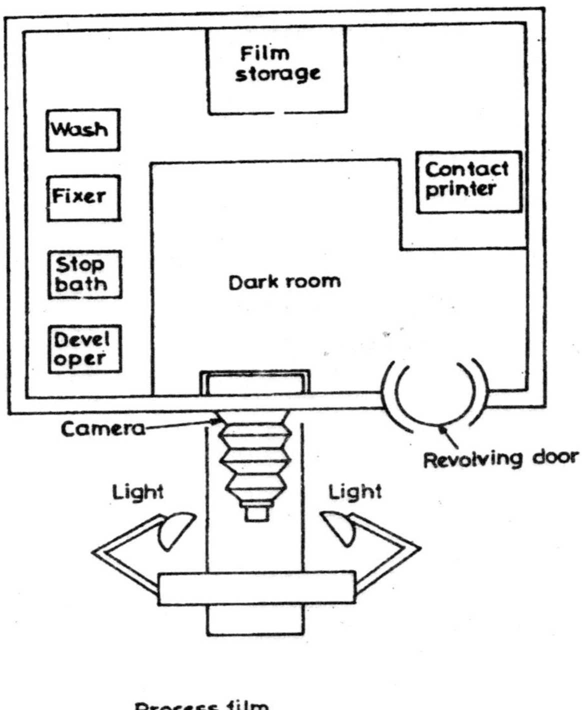
Film storage
Wash
Fixer
Stop bath
Devel oper
Contact printer
Dark room
Camera
Revolving door
Light
Light

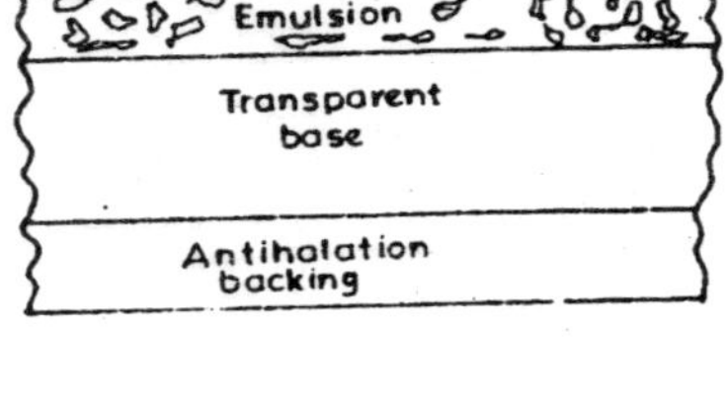
Process film
Protective coating
Emulsion
Transparent base
Antihalation backing

A printed reproduction of a continuous-tone image is called a halftone print or halftone. In order to make a halftone print, first prepare a halftone negative. This is done by exposing process film through a halftone screen, to the light reflecting from continuous tone copy. The screen creates the dot pattern on the negative, that is needed to make as offset plate.

A contact screen is placed in direct contact with process film when making a halftone negative. The screen actually covers the film, emulsion to emulsion, on the film holder. A vacuum pump is used to maintain the required contact during exposure.

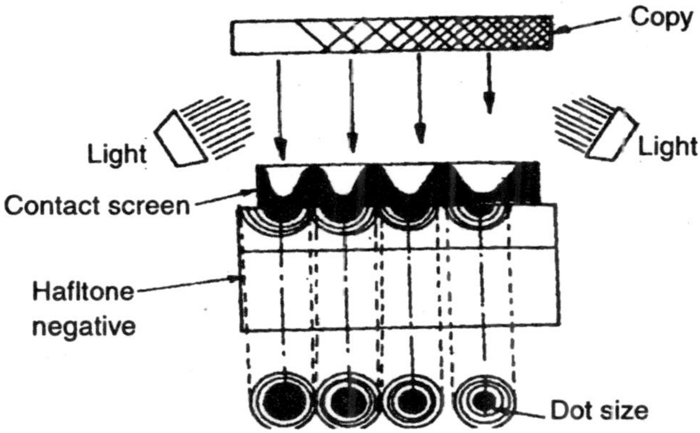

Figure : Contact Screen

As in the above figure, light reflecting from continuous-tone copy passes through the contact screen. It creates a latent image on the film. The latent image is in the form of a dot pattern. e.g. shadow areas reflect very little light. Shadow light can only penetrate the time transparent openings in the screen to form very small dots. The size & shape of each dot in the pattern is determined by the intensity of light reflected from a corresponding point on the copy.

Screen color : Contact screens come in two colors. Grey & Magenta (blue red)

Negative & positive screens : Negative screen are used when making halftone negatives from black & white continuous tone copy. Positive screens are used to make halftone positives by contact printing from continuous tone separation negatives.

Screen Rulings : The more dots per inch, the less noticeable the dot pattern in the printed halftone. Finely ruled screens create many more dots per inch than do screens that are more coarsely ruled. Screen rulings range from approx 65 to 200 lines per inch.

Dot shape : Dot shape must be considered when selecting a contact screen. Screen capable of producing square, elliptical or round dot patterns.

Round dot screens are most often used for halftones that are to be reproduced on web offset presses and high speed press.

Special effect screens are available for adding texture or an unusual pattern to the halftone reproduction.

Halftone negative : The image on an halftone negative is made up of black dots.

Half tone exposures : As many as three separate exposures may be needed to make a single Halftone negative. These include the main exposure, the flash exposure & the bump exposure.

The main exposure also called detail exposure, is made by placing process film in the path of the light that is reflected from the continous tone copy.

The flash exposure, serves to extend dot formation into the shadow areas on the film, is made directly through the halftone screen.

The bump exposure, also known as no-screen exposure, is used to improve detail and contrast in the high light areas of the negative. The time is generally less than 10% of the main exposure.

Steps to make a halftone negative

1. Prepare the process camera
2. Load the copyboard
3. Adjust the lens opening
4. Focus the camera
5. Check focus & location of the image
6. Load the process image
7. Make the pump exposure (if needed)
8. Position the contact screen
9. Make the main exposure
10. Make the flash exposure (if needed)

Filter Factors

Light source	Yellow K2 or 8	Deep Yellow G or 15	Blue 47B	Green B or 58
Pulsed Xenon	1.8	4.2	13.3	3.7
Quartz Codine	1.5	3.4	21.0	2.5
Carbon arc	3.3	8.7	16.8	5.0

STEPS TO PROCESS LINE & HALFTONE NEGATIVES

I. Prepare the chemicals : Three chemicals are needed to process line and halftone negatives—developer, stop bath & fixer. Each must be mixed according to the specifications. Equal quantity of developing agents and chemical accelerator first diluted with water to make working solutions. Once mixed, the developer will remain usable for about an hour.

The stop bath is a mild acetic acid solution. A stock solution can be made by adding 8 ounces of 28% acetic acid, to 1 gallon of water. One point should be noted always add acid to the water. Never add water to the acid and do not use concentrated acetic acid to prepare the stop bath.

Fixer is needed to remove unexposed and undeveloped silver halide crystals, from the developed film. Usually fixer is purchased in powder or liquid concentrate form and mixed according to the instructions.

A tray set up for processing line or halftone negatives, the trays that are used should be a little larger than the film that is processed. After cleaning the trays, place them in the developing ink and fill each with its proper chemical solution.

After filling the trays, turn on the water in the sink and set it to a temperature of 20° C (68°F). Let the water flow around the trays in order to raise or lower the temperature of the chemicals to 20° C. This is the temperature usually required for processing.

2. Prepare the darkroom : Line & halftone negatives are processed in the darkroom under safelight conditions.

3. Develop the negative : Raise one edge of the developer tray, until most of the developer flows to the lower side of the tray.

After noting the time, slide the negative, emulsion side up into the developer & quickly lower the tray. The developing solution must now be agitated throughout the development process.

There are three techniques used to determine when the film is properly developed.

(a) Inspection technique—allow the film to remain in the developer until black areas of the negative become completely opaque (about 2 to 2½ minutes). Then before the white areas of the negative begin to fill in, remove the film from the developer tray and transfer it to the stop bath. This method works well with line negatives. When developing a halftone negative, remove the film from the developer as soon as the dots appear to have reached their proper density. Then rinse the film in a water bath and inspect the dots more closely with a magnifying glass. If additional development is required, return the film to the developing tray. Otherwise place the film in the top bath tray.

(b) Using a developing aid technique - Negative made from normal copy should be developed until a solid appears on the grey scale.

(c) Time - and temperature technique consists of developing the process film for a determined length of time with the developer set to a particular temperature. This is the best & easiest technique to use, when developing a halftone negative, provided one use specific instruction in this matter.

4. Stop the action of the developer : After the image reaches its proper density in the developer, transfer the film to the stop bath tray. The stop bath solution within, serves to neutralize any developer that may remain as the film and thereby stops the development process. Let the film stay in the stop bath for a minimum of 30 seconds, while continuously agitate the tray.

5. Fix the film : Remove the film from the stop bath and place it into the tray containing fixer. The fixing solution serves to remove unexposed and undeveloped silver halide crystals from the film's emulsion. The fixer also hardens the emulsion so that it will be more resistant to scratches. Fixing times normally range from 5 to 10 minutes.

6. Wash the film : After fixing, the film is no longer sensitive to light. However it will begin to fog and discolor in a relatively short time if the chemicals that have been absorbed by the film's emulsion are not removed. To remove these chemicals, wash the film in a bath of running water for 10 to 30 minutes. The exact time will depend on the type of film used.

7. Dry the film : After washing the film may be treated with a wexing agent which will help to prevent spotting as the film dries. Now remove any excess water from the film's surface. A clean damp photo sponge or a film ster is generally used for this purpose. If a squeegee is used, place the film emulsion side down, on a clean glass plate, then run the squeegee the back of the film only.

Drying is accomplished either by hanging the film in a dust free area, or by placing it in a film drier which takes less than a minute to dry.

MACHINE PROCESSING OF LINES & HALFTONE NEGATIVES :

Film processing machines are sometimes used to develop and fix line and halftone negatives. The basic operation of a film processing machine is illustrated below. The most processing machines are also equipped with a control system that serves to mantain developer strength at a constant level.

To use this machine, the camera operator simply inserts the exposed film into the feed end of the machine. After a few minutes a completely processed & dried negative is ejected from the delivery end. During the period the film was developed, fixed, washed and dried all automatically.

ADVANTAGES OF MACHINE PROCESSING

a. Increased processing speed

b. Greater processing volume.

c. Ability to produce consistently good negatives

d. Time temperature, chemical strength & agitation are more accurately controlled in a machine than manual processes.

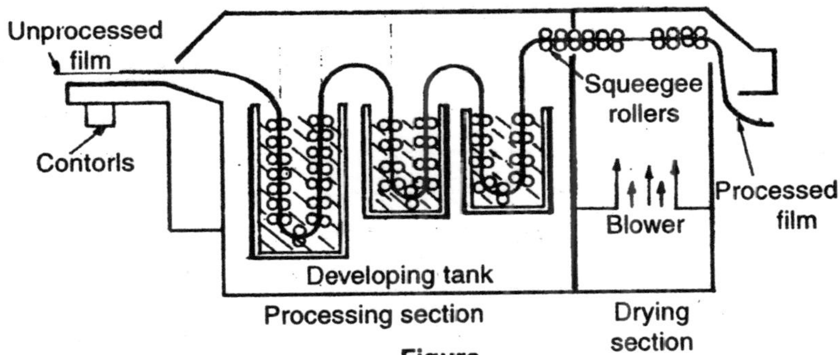

Figure

Contact printing

The contact printing process is illustrated below.

A light film negative or positive is in light contact, with an unexposed piece of film or paper. The light source serves to expose the unprotected areas of the film or paper and creating a latest positive or negative image on its light sensitive surface. Processing than turns this latent image into a visible one. The positive or negative image produce is the reverse of the negative or positive image from which it is made.

There are following reasons for making a contact print from a film negative or a film positive.

1. To produce a film positive from a film negative, it is required for certain types of offset plates.

2. To produce a paper positive from a film negative, it may be required for a camera copy.

3. To produce duplicate negatives or positives, it is used to create multiple images on single offset plate.
4. To produce screened paper positives from continuous tone negatives.
5. To change an image from right reading to wrong reading or vice versa.
6. To produce spreads & checks serve to improve registration when printing multicolor images.
7. To combine images from several sources.

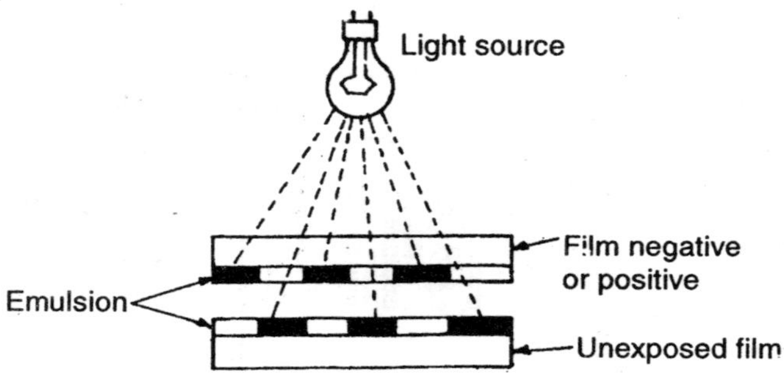

Fig. Contact Printing

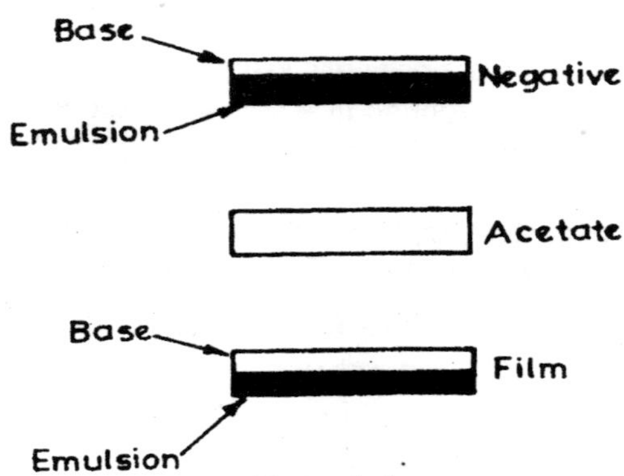

Materials

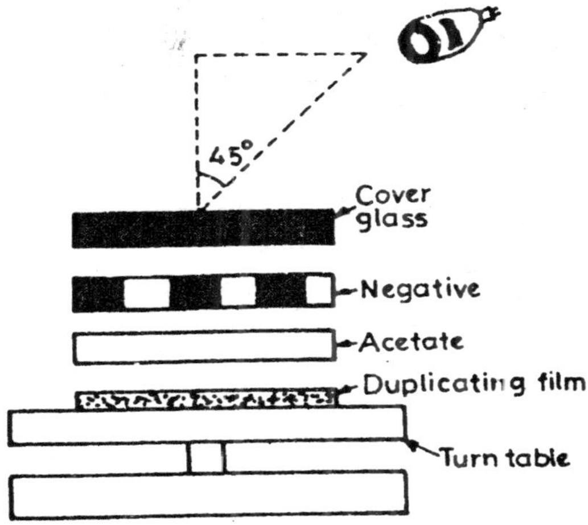

Expose and process

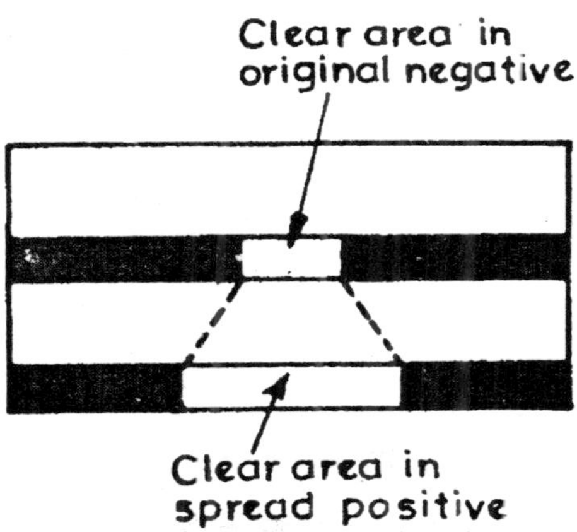

Result

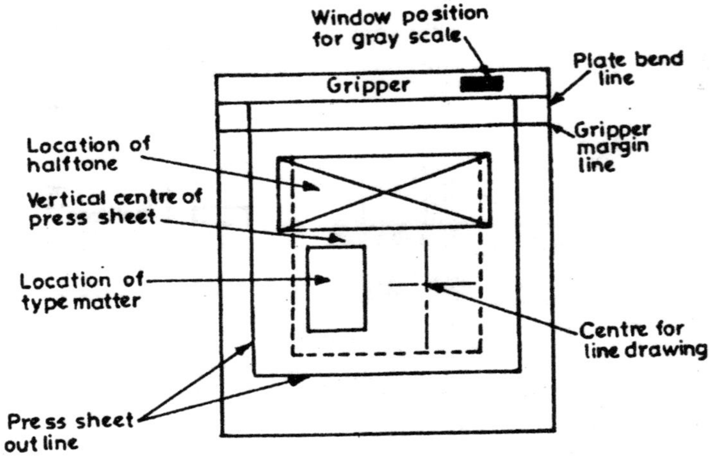

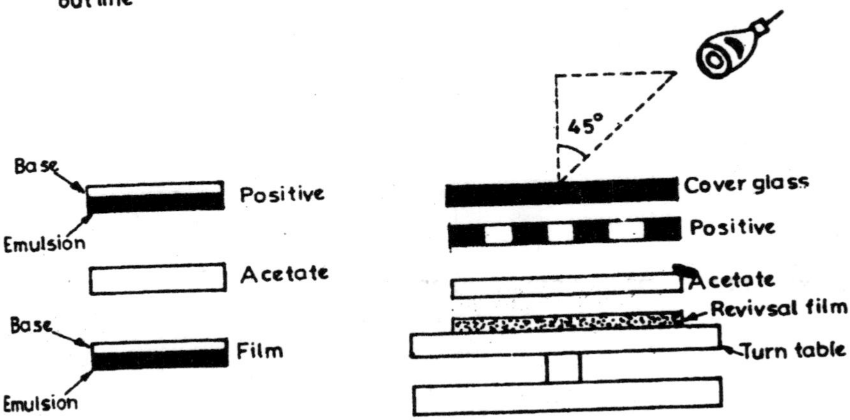

Materials

Expose and process

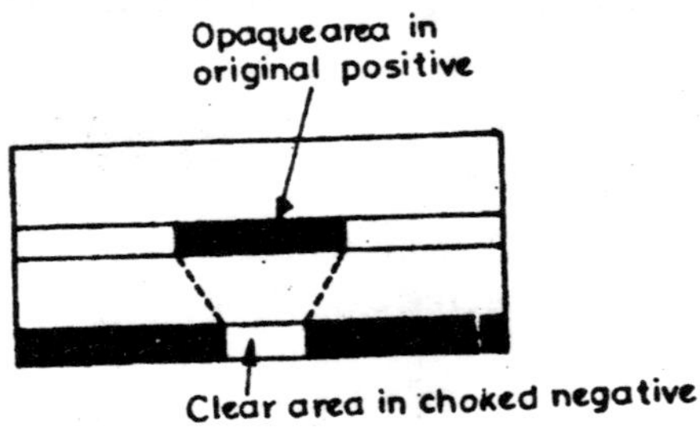

Result

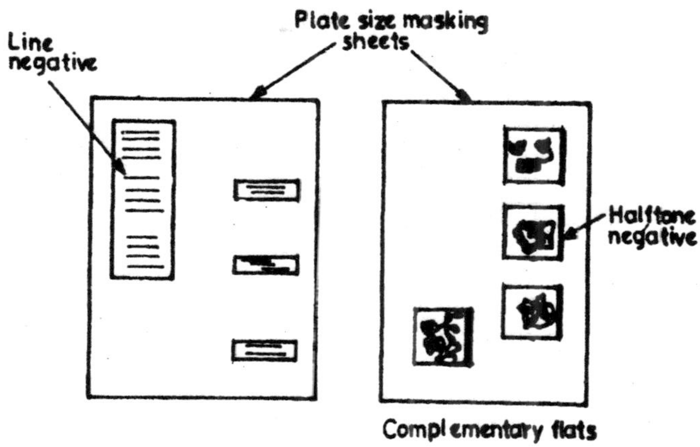

Contacting equipment

Contact printing is done in the darkroom under safelight. Two equipments are required

a. Light source

b. A device for holding the negative or positives and photographic film (paper). Point light sources are best for most types of contact printing. Because the light it emits is contained in a limited area and it should not offset other operations that are going on in the darkroom.

A vacuum or pressure-type contact printing frame can be used to hold the photographic materials together during exposure. A vacuum printing frame consists of a glass cover, a rubber cushion and a vacuum pump.

Stripping & proofing flat

Stripping is generally done on a light table or a line up table. The top surface of this table is made from frosted glass. A light located under this glass surface, helps the stripper to spot defects in the film.

A line up table is similar to light table but is also equipped with special devices for precisely measuring, ruling & scribing both vertical & horizontal lines.

Making sheets needed for stripping negative flats made from special translucent paper or plastic materials.

Etching & Opaquing

Etching removes density from the black areas of an image. Opaquing adds density to unwanted clear spot on the film.

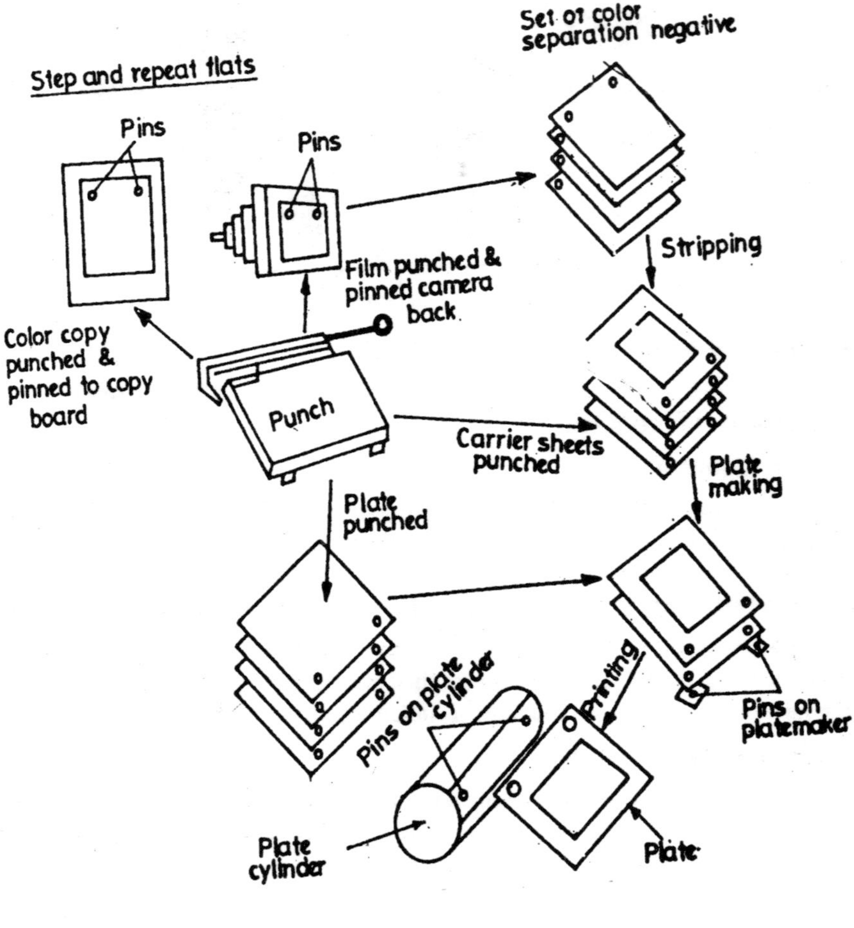
Step and repeat flats
Set of color separation negative
Pins
Pins
Stripping
Film punched & pinned camera back
Color copy punched & pinned to copy board
Punch
Carrier sheets punched
Plate making
Plate punched
Printing
Pins on plate cylinder
Pins on platemaker
Plate cylinder
Plate

Etching can be done chemically & mechanically. Chemicals are generally used when it is necessary to remove fog (grey tint) or unwanted specks of black emulsion.

Blackness can also be removed mechanically by scribing tools.

Opaquing may be done on either the base or emulsion side of the film-

Stripping positive flats

The steps for laying out and assembling a positive flat are somewhat different from those used to strip a negative flat. The main differences are:

a. Layout the flat on a piece of tracing paper

b. Cover the tracing paper layout with a clear plastic carrier sheet.

c. Using the tracing paper layout as a guide.

d. Use a clear plastic tape to attach the film positives to the carrier sheet below.

Stripping Halftones

a. Attach the halftone directly to the carrier sheet

b. Attach the halftone to a clear window

c. Prepare complementary flats.

Registration techniques - It refers to the proper placement on one image in relation to another image. Registration is an important factor whenever two or more flats are used to produce a single plate. Pin register systems have been developed to control registration accurately throughout all stages of the reproduction process. From copy preparation through printing (as shown in fig).

Step & repeat flats - may be used when it is necessary to reproduce the some image in several locations on a single printing plate. It consists of two parts 1, plate size masking sheet &2, a much smaller image carrier sheet.

11

Scanning

The purpose of scanning is to capture all the information from an original that will be required in the reproduction. The more details information about the colours, objects, texture tone and sharphen etc. that the reproduction contains gives more real, nice and original picture.

Information that is not there when the image is scanned cannot be replaced. If a digital image is lightened throughout the tonal range, highlight detail will be lost. When these details have gone they cannot be retrieved other than by returning to the mediated image. Image processing can add more information to any picture. But if an image has become below standard or degraded it is possible to use algorithms that reverse poor standard or the degradation. One should know about the process how the image has become of poor quality or degraded. Algorithms (Mathematical) solution of the type can restore and make it more acceptable. But one must remember that the new information added were not necessarily present in the original one.

Any transparency with middle tone areas should be scanned in such a way that maintains the separation between tones. By adding intermediate tones by interpolation will only make average existing tones but not actually adding any further detail.

One should not think that by capturing every minute detail of information in any image is sufficient. In every place there is a limit. Limit to how much details of information can be used in the reproduction. Otherwise files will be larger than necessary. It is always advisable to keep file sizes small as far as possible.

The advantage of using a scanner, in place of a process camera was that of colour correction was programmed into the circuit of the scanner. As a result retouching of films by masking and etching could be nearly avoided.

At initial stage scanners were analog rather than digital. The image was exposed on to film by the recording unit.

The development of digital scanner at first introduced by Crosfield in 1975, that made possible digital storage and processing of the image before bringing out to film. Since 1975 gradually scanner quality and productivity have increased. Preview stations have been added, modular and multi-station scanners have been developed to maximize output and work station for colour image assembly and a lot of planning have been incorporated. Simultaneously graphic arts scanners evolved into highly efficient and consistent workhorses for quality pre-press and now there is a need for small, simple scanning devices to record the image. With their resolution and colour capabilities flatbed scanners have become useful to the Desk Top Publishing market.

Scanners can be mainly divided into the following two types:-

(1) High-end Scanners that is designed for professional and commercial users. These systems are usually built as large drum scanners with internal programs.

(2) Desk Top Scanners that is clic fly used by designers and publishers. These systems include flatbed Scanners meant for reflection copy, transparency scanners for say 35mm and medium size, transparencies upto 125 x 100 mm or even a little more.

Desk Top drum scanner are very similar to the larger high-end scanner that accept both reflection copy and transparencies. Earlier desk top scanners were only for visualization scanning but now-a-days they are capable of handling more than 90% of the quality requirements of commercial colour job.

Scanning of an image involves analyzing the colour values of individual. It depends on the type of scanner used. It will then process the colour values before storing them in a file.

The basic elements of a scanner are as follows:-

(1) Light-sensitive capture elements

(2) An optical system to focus the image onto capture elements.

(3) A filter system to allow the red, green and blue components of the image to be recorded separately.

(4) A transport system to move the original in relation to the scanner.

(5) Controlling Software

(6) Processing software to carry out initial image processing before recording.

All graphic arts scanners use two systems either photomultiplier tube commonly named PMT or Charge-Couple device known as CCD. When photons strike the photocell in PMT or CCD, they give their energy as they disappear. As a result electrons to be emitted, turning the energy of the photons into electrical energy. The number of electrons that are emitted can he measured to determine how many photons struck the capture element. From this measurement the scanner generates a value for the intensity of light arriving from the point on the original analyzed.

Drum scanners are well equipped with one photo multiplier tube for each of the red, green and blue filter signals alongwith a separate tube for the unsharp making signal. The original transparency is mounted on the drum. A beam of light is transmitted from inside the drum, through the transparency and then onto the scanner optics. Now by rotating the drum at high speed and moving the analyze head slowly along the drum's axis and each pixel on the transparency can, in turn, be read. Splitting into red, green and blue beams, either by a prism, the light strikes the end of the photo multiplier tube. The photomultiplier tube consists of a photocell inside a vacuum tube, with a light sensitive cathode at one end and anode at the other. The photocathode emits electrons as it is struck by photons in the light beam. A services of electrodes emphasize the electron steam until it reaches the tube's anode. The smallest number of photons that can be detected, by indicating the darkest areas of the original, is limited by the amount of noise present in the system. The density that photomultiplier tubes (PMT) can detect varies from 3.5 to 4.0.

"Loop" film transport system.

The control panel. Automatic replenishment. Adjustable developing times and temperatures.

The racks easy to remove and clean. All guides can be removed without tools.

Figure: Film Processor

On a high-end scanner the analyze (input) mist is connected via the colour computer directly to the output unit. The colour computer calculates the correct dot values for each pixel.

On a digital scanner, scaling and resolution can be controlled by software. Most of the adjustments made for different image can also be software-controlled. In desk top systems, the scanner is controlled from a standard platform host computer.

Charge-coupled device (CCD) consists of an array of tiny electrodes packed closely together on a layer of silicon. When a photon strikes an electrode it causes an electron to be emited from the silicon layer.

It can detect the presence of a single photon. But interference between the signals produced by adjacent electrodes is higher than that of the PMT. The capacity of CCD to resolve detail in deep shadow areas in limited. Therefore the capacity of density range is between 3.0 and 3.5. Thus the PMT is roughly three times sensitive to shadow detail. High quality CCDs with elements of 10 microns and density as high as 4.0 have been also developed but are not economically viable but are rather expensive. When the imaging capability of CCD scanners rises to the level of PMTs. It proves expensive.

Graphics arts scanners have become capable of resolving from eight to twelve bits per colour. Earlier scanners were only able to resolve six bits of data for each colour

Bits available for each colour component	Grey level available for each colour component	limit of Sensitivity reflectance %	Density maximum
6	64	0.1	3.0
8	256	0.05	3.3
10	1024	0.025	3.6
12	4096	0.01	4.0

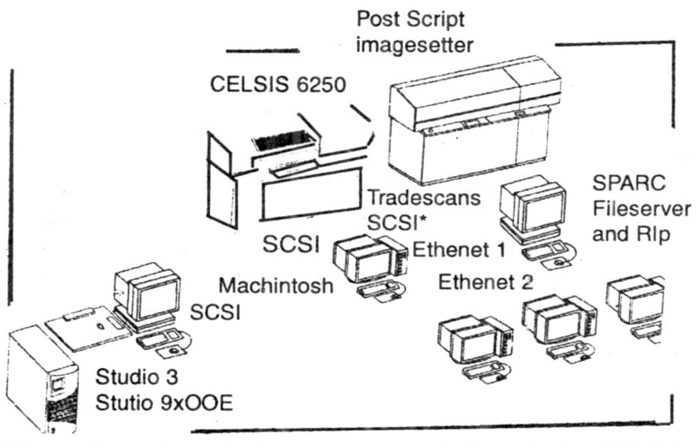

*In MagnaRip configurations, tradescans are carried outvia the SPARC station

In a CCD Scanner, the data is captured in parallel and the travel of the scanner head depends on the way the CCD elements are arranged. In a lineave array the scanner head makes a single pass over the original or three passes in the case of scanners that require a separate pass for each filter colour. While a two dimensional array captures the whole image in a single snap shot. Two dimensional arrays are used in a low resolution devices like camcorders, but are costly to fabricate in the higher resolutions required in the graphic arts. Systems that use two-dimensional CCD arrays include digital cameras and the linotype Hill Topaz and Kodak Photo CD Scanners.

The way in which the electron levels are translated into grey scale information by the ADC is flexible. The grey scale steps do not have to be the same size and the software can control tonal compression more flexibily than it can with a PMT Scanner, which produces a fixed logarithmic relationship between the light internsity and the output voltage.

Likely developments in CCDS include the use of mathematical techniques to filter noise and amplify the response in shadows, as well as continuing improvements in resolution and sensitivity.

Scanner Resolution

The maximum resolution of the scanner determines its ability to capture detail from the original, especially where the original is enlarged. Sampling resolution depends in a number of factors, including:

(1) The size of the CCD element or scanner beam.

(2) The mechanical precision of the scanner's moving parts.

(3) The focussing accuracy and depth of field of the optical system.

(4) The size of the stepping increments of the scanning head as it travels over the original.

Lower quality scanners do not have the mechanical precision to achieve high resolutions. Those that require three passes to capture all the colour information are most likely to lose register between colours. In a flatbed scanner, the resolution is controlled by software and un wanted information is simply thron away e.g. if a scanner has an addressable resolution of 1200 ppi and the user reflact 600 ppi for an image, the date from every pixel is discarded.

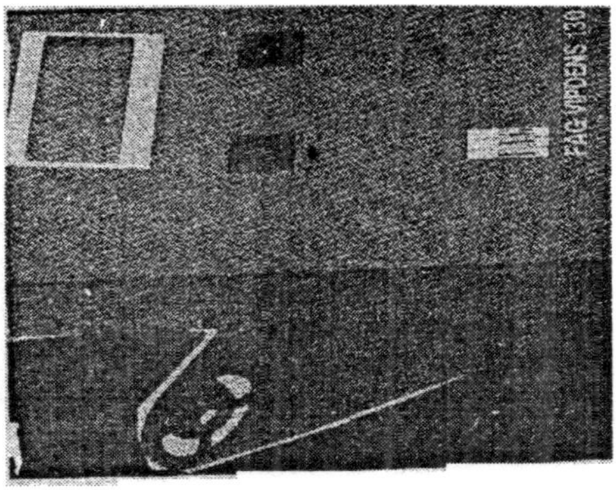

Figure Densitometer

In the drum, the analyze head moves along the drum while the drum is vitiating. Each revolution of the drum captures a like of data from the image. The resolution is determined by the speed of the analyze head in one direction and by the drum diameter and the way the data is digitized in the other. Input resolution in determined by the mechanical features of the scanner and is normally limited to a small number of possible resolutions.

The limit of a scanner's resolution, as determined by the scanner design, is often referred to as its optical resolution. If the scan resolution is not adequate e.g. if the image is being greatly enlarged or the scanner only supports low sampling resolutions, interpolation can be accomplished by the scanner software during scanning or by resembling in an image-editing program. Interpolation does not add any new information to an image and tends to make fine detail appear blurred.

Scanners for single colour line work have no need to capture grey scale information, but require high resolutions. They have small CCD elements with low sensitivity. Some line work scanners employ two or more linear arrays to achieve resolutions of 3000 ppi and more. Line work should be scanned at the highest resolution available to prevent the visibility of scan pixels oil the reproduction.

Scanner Choice - The decision to use a particular scanner will depend on number of factors :-

(1) Productivity

(2) Quality capability

(3) Customer's requirements

(4) Low capital cost for design

(5) Ease of use

(6) Support for creative design work

(7) The ability to work in open formats that can be transferred to the desk top without time penalties or the need for additional processing.

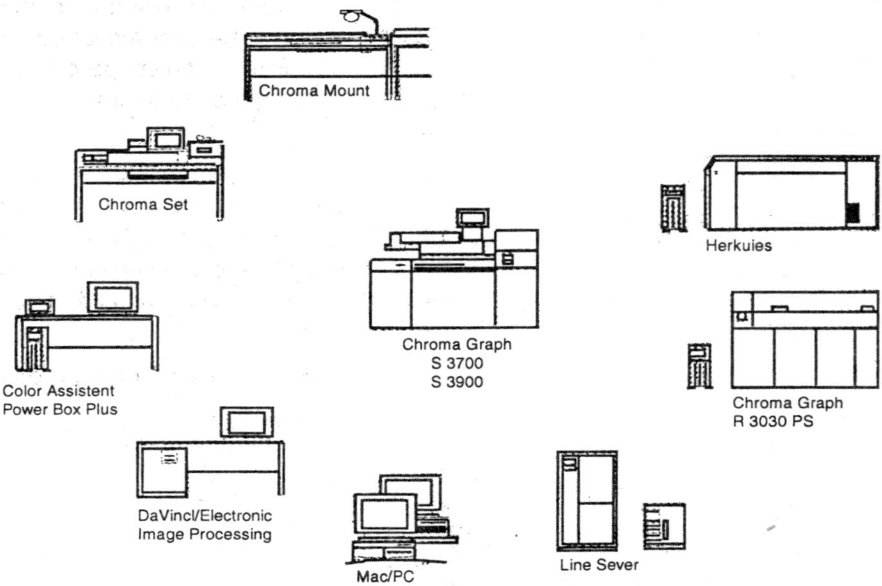

Followig Technical Capabilities are also to be considered:

(1) The scan area

(2) Maximum enlargement

(3) Density

(4) Bit depth available

(5) Sampling Resolution

(6) The scan controls including, colour cast removal and gradation control

(7) Productivity scan times. storage & buffering.

The Volume of Work- If scanning volumes are low, the productivity advantages of high-end scanners will be less relevant and can be traded against the convenience of being able to scan directly into the desk top environment without involving external suppliers

The level of quality needed—It is important to be realistic about the output quality that can be expected from different scanners and to match this carefully to the quality requirements of the work being produced.

Broadly, high-end scanners are appropriate to trade shops, service bureaus, and printers. Designers and publishers, as well as some printers and service bureaus, will employ mid-range desk top systems, while low-end scanners will be mainly used for visualization purposes and for work where reproductions quality is not important.

The following are advantages of Drum Scanners :-

(1) Low noise

(2) High resolution.

(3) High Security

(4) High Density

(5) A High productivity for volume scanning

(6) Produce colour-separated files ready for output

(7) Accept large-format originals.

The following are advantages of Flatbed Scanners :-

(1) Ease of use

(2) Compact construction

(3) Good Price/Performance

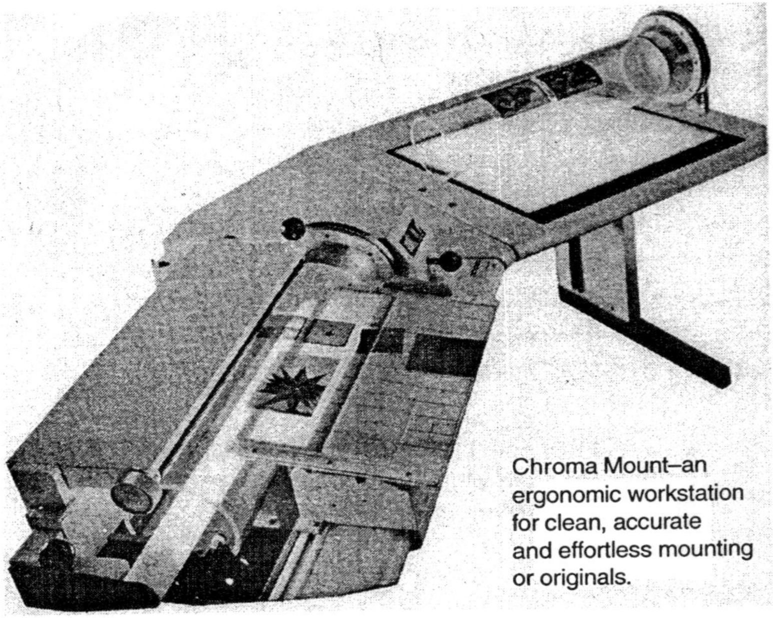

Chroma Mount–an ergonomic workstation for clean, accurate and effortless mounting or originals.

ECRM manufactures a full line of equipment for printers, typesetters, and trade shops. Including Autokon® camera/scanners, PelBox® recorders, and ScriptSetter® imagesetters. And each is designed to deliver the maximum performance for a substantially smaller investment.

(4) Faster Scanning of Single images

(5) Solid-State electronics that do not need replaced or recalibratred

(6) Accept rigid originals.

The productivity advantages of the drum scanner lies in its ability to scan multiple originals simultaneously and output colour-corrected images directly to disk or film without the need for further processing.

For some users, these productivity advantages are irrelevant, as the number of scans they produce does not justify the capital investment.

The time involved in dealing with outside suppliers and in transporting files and originals, can outweigh any savings made during production. The control over image reproduction that in-house scanning can bring may be a far more important consideration.

Internal Colour Models—Scanners handle data internally in ways that are most efficient for the scanner hardware and software to process. The output from both CCDs and PMTs is initially an anlog voltage that is digitized and then converted into a common format. On high-end systems, these formats are often propertary and cannot be accessed directly by other applications.

If this is the case, some form of translation will have to take place to convert the data into an open file format in order to link to high-end scanner with the desktop. The internal colour models of scanners are based on the RGB, CMYK or CIE Colour spaces.

All scanners initially capture the colour information that is output from the CCDs or photomultipliers as RGB values. The RGB values are device-dependent— They relate to the way in which the individual scanner transforms light intensity into output voltages and codes them as RGB values—if they are to be printed they must, at some point, be transformed into the device space of the printing system. Most desk top Scanner do not carry out this transformation, but simply record the RGB values in a file.

High-end drum scanners usually carry out colour separation as the image is scanned, translating the output voltages of the photomultipliers directly into CYMK dot values. The CYMK dot values are stored in the devices space of the printing system, giving more efficient image handling but requiring that the target printing system is known at the time of scanning and the image is intended for output to one printing system of a single size and resolution.

If any of these output parameters change, the image will have to be edited or rescanned with the potential loss of image data and productivity.

Scanner Adjustments—Scanning resolution needs to be carefully to make sure that all wanted detail is captured without making files larger than necessary. The accepted rule in the graphic arts is that there should be two scan lines for every line of the halftone screen in the printed reproduction. For a reproduction at the same size as the original, this means that the sampling frequency should be twice the halftone screen frequency.

For example, a picture that is to be enlarged by a factor of 1.2 and printed at 200 lines per inch, should be scanned at 2 x 200 x1.2 or 480 ppi.

Sampling frequency (in ppi)=2 x screen frequency

(in ppi) scaling factor.

This formula has been found to give optimum reproductions from all kinds of originals. Increasing the resolution gives absolutely no increase in image quality where normal halftone screens are used. In fact it may lead to over sampling effects and a loss of quality.

But when the requirement to make the scan resolution twice the screen ruling imposes demands that the system cannot meet e.g. when very fine screens are used or when a high magnification in required. The "twice screen-frequency" principle is not a rigid rule but simply a guide to achieving the optimum reproductions. Reducing the sampling resolutions causes a gradual loss of detail and sharpness, although at 1.2 x screen frequency, the loss of equality is still quite marginal.

The use of stochastic screening in place of the conventional halftone screen allows the scan resolution to be reduced without losing any information from the original. File sizes can be reduced by approximately half without loss of quality in comparison with conventional screens.Stochastic screening technology can also make use any extra information that is captured and the "twice screen-frequency" principle does not apply.

Where images are kept for archiving instead of immediate use, they should be scanned at the highest resolution that is likely to be required and then resampled down to the appropriate size as necessary. Assuming a limit of 10x enlargement and a screen frequency of 200 lpi, the maximum scan resolution likely to need for the majority of work is 4000 ppi.

As an image is scanned, it can bring to the size at which it will be printed. If the exact proof is not known at this time, cropping can be completed in image-editing or page-layout applications. Cropping the sampled image and discarding the unwanted image areas reduces the image file size.

Sharpness

The senstivity of the visual mechanism to edge definition makes sharpness a priority of all images, except those that are intentionally soft or defocused as shown in the figure below.

Electronic scanners achieve the same effect with unsharp masking (USM). On a high-end system, a slightly out-of-focus signal is captured by a separate photomultiplier and added to the sharp signal.

The result is increased contrast at the boundaries between different parts of the image. Virtually all sampled images gain from sharpening and it can bring out the detail in an image better than increasing the resolution as shown in the figure below.

Distance across image

On images captured by CCD, unsharp masking is not carried out optically but by computation. A digital file that has already been scanned has USM applied by calculating the difference between adjacent group of pixels and increasing the contrast between them at the boundary.

Unsharp masking comes into play when there is a significant difference between two adjacent tones, but it ignores minor difference that are caused by small variations of tone within an object. Where all tonal values need to be given increased separation, a sharpening filter should be applied instead.

Again excessive use of USM can cause unwanted effects, such as random noise and the exaggeration of blemishes in flesh tones. Unsharp masking or sharpening should ideally be applied after any scaling that is necessary.

The tone controls of a scanner have two main functions as following:-

(1) They are used to set the location of the endpoints (The two extremes of highlight and shadow) of the original and control the tonal compression

(2) They allow the user to adjust the distribution of the intermediate tonesin the original through the tonal.gradation.

Both gradation and end points can be altered later in an image-editing package but correct adjustment during scanning will capture more wanted detail from the originaL

White point adjustment—If the image is being captured in RGB or independent colour space, the white point is set to the lightest tone in the original. If there is no white in the original, the white point is set from a white patch on a photographic grey scale.

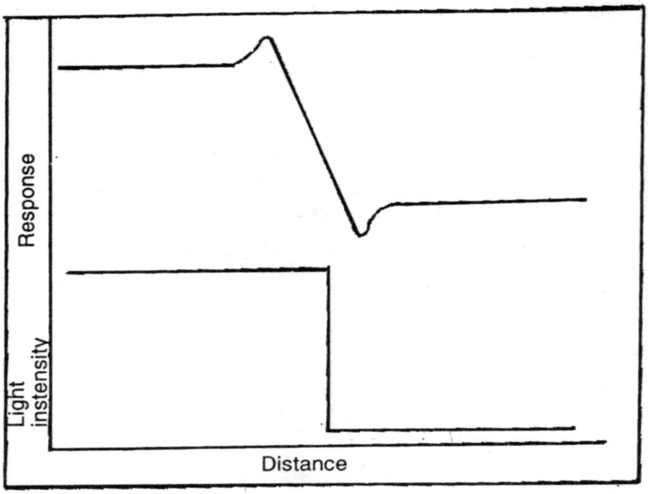
Response
Light instensity
Distance

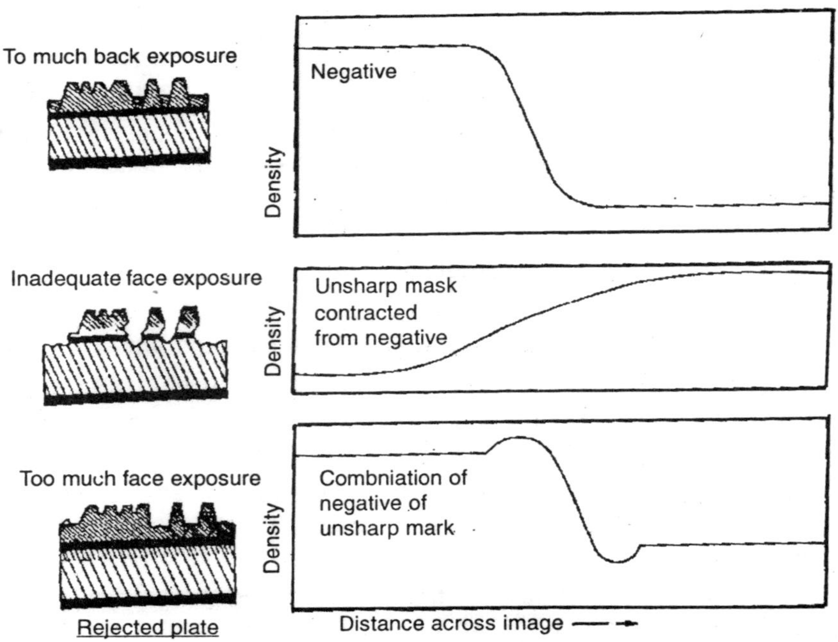
To much back exposure
Negative
Density
Inadequate face exposure
Unsharp mask contracted from negative
Density
Too much face exposure
Combniation of negative of unsharp mark
Density
Rejected plate
Distance across image

If the image is being scanned into CYMK, the white point adjustment is slightly more complicated, as the objective is to make the diffuse white point on the original correspond exactly to the smallest dot that can be printed.

If the original has a tone that can be used as the diffuse white point, set the smallest printing dot at this point e.g. 5% cyan, 4% magenta, 4% yellow.

If there is no white point, use complementary colours. Set the smallest dot for cyan in a red part of the image, for magenta in a green and for yellow in a blue. Alternatively, choose the lightest tone in the image and set the CYMK values by reference to the closest colour in a printed tint chart.

An accurate white point setting is the most important scanning adjustment, and it ensures that there is good contrast in the reproduction, with highlight detail, spicular whites where necessary and good tonal rendering in high lights and midtones.

Setting the white point to a tone on the original with a density that is to high will cause the reproduction to look light and washed out, with a loss of highlight detail. Setting it very low, in a white that should print as a spicular, will cause the reproduction to be dark and lacking contrast.

Black point adjustment—Adjusting the black point controls the way that the density range of the original is compressed into the density range that will be printed. Correct black point adjustment ensures that the scanner is capturing only wanted detail and not wasting available grey levels by using them to record tones that will not be printed.

If the image is being scanned in RGB or independent colour space, set the black point to the darkest tone in the original. If there is no black in the original, set the black point from a photographic grey scale.

If the image is being scanned in CYMK, set the darkest tone in the original to the largest printable dot, e.g., 95%. If there is no black tone in the original, use a tint chart to set CYMK values for the darkest point available.

One should avoid using the black patch on the photographic grey scale or the black border of the transparency, if there is no black in the original, as it will have the effect of setting the largest dot to a density higher than is needed. Thus compressing the shadows more than is necessary.

Saturated colours are also affected by the black point setting. If it is set too high, saturation is reduced and colours look washed out. Again if the black point is set too low, the colour contrast of the reproduction may seem excessive.

The very darkest tone in the reproduction should in fact reproduce on a sheet fed press on coated paper, as something like 90% cyan, 80% magenta, 80% yellow, and 90% black. These values will be calculated by the scanner depending on the way that values are set for black generation, grey balance, undercolour removal and grey component replacement.

If the contrast of the original is too low, adjust the end points by setting the white point to a higher original density, and the black point to a lower original density.

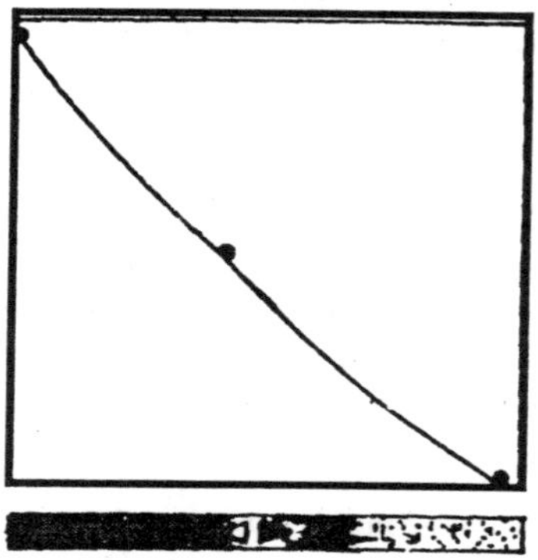

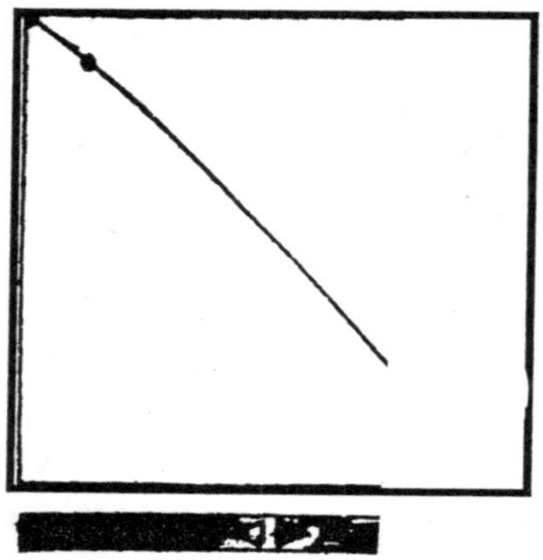

Gradation— If necessary the gradation of the scan can be adjusted after the end points have been set. The gradation adjustment can be used to control the setting of the midtones and the distribution of tones on the reproduction. It can expand the number of grey levels in a specific region of the tonal range to emphasize the interest areas of the original or to compensate for originals with unsatisfactory contrast or exposure. As it is shown in the two gradation curves.

The number of grey levels, the amount of details that are captured can be increased in a part of the tonal range by making the gradation curve steeper in that area. That part of the density range of the original is then

The curve on this left will increase contrast, whereas the curve on the right will make an image less yellow (more blue) distributed over a wider range of grey levels in the resulting file.

The most efficient way of working with gradation curves is to store a set of basic curves as shows in the above figures. Adjustments can be made to a basic curve to suit the particular requirements of the original if necessary.

Colour Adjustments—The majority of transparencies can be scanned without the need for colour correction. On a high-end scanner, correction can take place during scanning. But on desktop scanners colour correction is normally applied after the image has been captured. If the original has an overall colour cast, it should , if possible, be handled during scanning. If a colour cast has been identified on an original, the alternatives are to ignore it, reduce it or remove it altogether.

If the intentions of the designer are not known when the image is scanned, the cast should probably be removed if it gives an objectionable appearance to the image or has the effect of reducing the colour contrast. Otherwise, it may be safer to leave it for adjustment in an-image-editing application.

Colour cast is removed by manipulating the red, blue and green output of the scanner, either by controls on the scanner itself or in the controlling software. Preview scans are useful in identifying colour casts and a software densitometer can evaluate colour components much more accurately than simple visual assessment, since the eye will tend to view any near-white as a neutral white.

Preparation — Before scanning, originals should be assessed and any special requirements noted. They should be cleaned if necessary, as removing dust or marks at this stage will be more productive than pixel editing in an-image-editing application.

Transparencies should be held firmly on the drum or transparency holder so that there is no danger of the distance to the scanner optics varying. The depth of field of scanner lenses, is quite small, and focus can be lost if the original is not held flat.

In most scanners, the original is held against a smooth plastic surface. This helps the original remain flat.

Usually, a film is scanned with the emulsion up but if it has been duplicated, it must be flopped during scanning. The emulsion side looks duller than the other side and when viewed from the emulsion side the film manufacturer's coding appears reversed. The image should read correctly from the side opposite to the emulsion (known as the film side, the non emulsion side or the reading side unless it has been flopped during duplicating.

	Problems with originals	Possible Solution
(1)	Original artwork is too rigid to curve around drum	use flatbed scanner use gallery camera Make transparency from the artwork
(2)	Original is too large for the scanner	Make a small transparency format
3)	The scaling ratio is too large for the scanner	Make an intermediate sized transparency
(4)	Original has been damaged	Restore by pixel-editing in image application.

Scanner Workflow :- A colour scanner is one of the most expensive pieces of pre-press equipment. But most users will want to maximize their investment in it. In many instances, the economic case for employing in-house scanning instead of using external suppliers is not convincing. Since scanning is a very competitive market and prices are not high, there are few tips on managing the workflow most efficiently—

(1) How to handle originals are made in advance or scanning to save unnecessary rescues.

(2) Use batch scanning where possible

(3) Use intelligent software for adjusting tones and removing colour casts in routine images.

(4) Provide training and retraining to scanner operators.

12

Offset Plate Making

The most common printing process in commercial purpose now-a-days is offset lithography, which is capable of producing high quality work, at a nominal cost from posters to postage stamps and catalogue to cartons.

This chapter elaborates the steps to be taken to progress work thorough the various stages of sheets fed offset printing, from preparation and making of the litho plate, to proofing, printing and finishing.

Plate & Platemaking :

Aluminium is today the most common base for litho plates due to price, stability and surface properties. Plate manufacturers will supply their plates already grained, and dized and coated.

Almost all lithographic printing today, is done from plates made from paper, plastic or metal materials. A plate may be made from a single material e.g. aluminium or by combining different materials e.g. plastic-impregnated paper or aluminium laminated to a plastic base. Multimetal plates are also available and are used when extremely long run printing are required. Such plates are made up of two (bimetal) or three (trimetal) layers of different metals.

Lithographic plates vary in thickness from about 0.005 inch to 0.012 inch, for duplicator size plates to as much as 0.030 inch for plates, used on large offset process. Plate thickness too must be carefully controlled during the manufacturing process, to ensure that each plate will print with a uniform pressure across its entire surface.

Graining :

The surface of an offset plate may be rough (grained), or smooth (grainless). Graining is done during manufacturing process by either etching (chemical graining) or abrading (mechanical graining) the plate material. Most metal plates are provided with a grained surface mainly because it serves to make such plates more water-receptive and it improves their printing characteristic.

Anodizing :

A raw aluminium plate exposed to the atmosphere, will automatically react with the oxygen in the air, to form a film of aluminium anodize on its surface, so that it protects from other atmospheric or chemical attack. Anodizing strengthens the surface of the plate and gives it added resistance to atmospheric or chemical attack. Since it is porous, its water retaining properties are enhanced.

Coating :

Early lithographies made their own light sensitive coatings and applied them to the plate by hand or by means of a mechanical wheather, which spread the coating evenly over the surface. To-day the majority of plates are supplied pre-coated by the manufacturer as pre-sensitized plates (PSP). The first PSPs introduced were, based on light sensitive dizonium (diazo) compounds, which were found to have greater uniformity, longer shelf life and consistent quality, including the ability to hold a good halftone lot, Diazo coatings have continued to improve, but a number of other light sensitive coatings are now available including on photopolymer, imagined by contact printing through either a negative or positive flat. Plates used with negative flats, are coated with a negative working emulsion. Negative working emulsions are designed to produce a positive plate image when the plate is exposed through a negative flat. The image is formed by the action of light. Exposed parts of the plate become image areas after the negative-working plate is processed. And those exposed through positive plates are coated with a positive-working emulsion. Positive working emulsion are designed to produce a positive plate image when the plate is exposed through a positive flat. The image is formed by the absence of light. Exposed pares of the plate become non-image areas and unexposed parts is processed. Additive and subtractive processing is available for both negative and positive working plates.

Prepare a presensitized plate :

A vacuum frame is needed to hold the flat and plate in tight contact, throughout the exposure process. A source of actinic light is also needed. Actinic light sources limit ultraviolet light.

To prepare the plate maker, clean both sides of the vacuum frame over glass. Use a good quality glass cleaner and paper towels for this purpose. Presensitized plates should be kept in their protective packaging and removed from wrapper only in light. Yellow safe light are preferred.

Place the plate, emulsion side up, on the rubber blanket of the vacuum frame. Then position flat, emulsion side down, on top of the plate. Film emulsions on the flat must contact the plate's emulsion during exposure.

Now align the top and right edges of the flat with the corresponding edges of the plate. Once aligned, position a plate maker's gray scale in the window that was cut for this purpose during the stripping process. Then close the vacuum frame and turn on the vacuum pump once the air is evacuated the flat and plate will be held tightly together. If a flip-top plate maker is used, it is must to turn the plate towards the light source before proceedings.

Set the timer for the required exposure time and expose the plate.

Remove the exposed plate from the vacuum frame and place it, emulsion side up, in a plate-developing sink or any clean, flat surface. The plate is now ready for processing.

Desensitize :

Pour a small quantity of desensitizer (sometimes called developer or process gum) into the surface of the plate. Then spread the desensitizer evenly over the entire plate surface use clean sponge a cotton pad, or a special applicator for this purpose.

The desensitizer serves to remove the emulsion from the non-image areas of the plate.

Lacquer :

Before a desensitizer has a chance to dry, pour a small quantity of lacquer (plate developer) into the plate. Now with a clean sponge and cotton pad spread the lacquer over the entire plate surface. Use a firm, circular motion as you apply the lacquer. Continue doing until a strong and uniform image appears. If the lacquer starts to dry before a strong image is achieved, add a small amount of water and continue rubbing.

Wash :

Remove excess developer from the pate's surface with a sponge dampened with water. Press lightly to avoid scratching the image.

Develop :

Pour a liberal quantity of subtractive developer on an applicator pad. and on the plate. Now use the pad to spread over the entire surface of the plate. After giving the developer a chance to soak into the emulsion (about 10 seconds), re-rub the plate. Rub with a tight circular motion as you apply moderate to firm pressure on the abdicator. Continue rabbing until all the lacquer is removed from the non-image areas of the plate.

Squeeze :

Use a clean rubber squeeze to remove excess developer and lacquer from the plate's surface. Clean the applicator pad and squeeze.

Wash :

Remove excess developer from the plate with top water and squeeze the plate dry.

Finish :

Spread a small amount of finishing solution (suletractive gum) over the surface of the plate with a clean sponge or litho wipe. Then buff the plate's surface dry, with a fresh litho wipe.

Print or gum the plate :

The presensitized plate is now ready for printing. If it will not be printed immediately, coat its surface with a gum arabic solution (14 Baume'). Gum arabic serves to protect the plate from scratches and prevents it from oxidizing, while in storage. Therefore, plates may be exposed either to a negative film image or a positive film image, depending on the preferred method of working.

Negatives :

Negatives are the normal first stage product from original copy via a camera and it would seem obvious to use these to produce the plate. It is also easier to see minor spots, scratches and other bolemishes in the non-image areas on a negative, and to correct these by spotting out with an opaque medium.

Positives :

Positives contrary are preferred by most as they are easier to see, specially for superimposition of colour half tone images but they may require a secondary exposure from the negatives and more detailed attention to the non-image areas to correct minor blemishes.

The nature of the coating and the method of working will depend largely on whether the plate is to be exposed to negatives or positives.

Inking-in :

If the plate is not to be used immediately, the image area should be protected with a purpose-made inking-in solution, which is applied to the plate by a wiping action and which will adhere to the image but not to the non-image area.

Automatic plate processors

Automatic plate processors may be used for batch production or high volume work. The various steps of development, washing out, desensitize action etc. are carried out, in a fraction of the time taken to process plates manually. Automatic processing will also introduce a standard uniformity to processing as temperatures, times, chemicals and other variables may be more accurately controlled than with manual processing. Not all processors can cope with both positive and negative working plates, as they require different chemistry. Maintenance of the equipment, monitoring and replenishment of the chemicals, as well as the initial investment, must all be taken in to account when considering the purchase of an automatic processor.

To determine correct plate making exposures, will depend on the type of plate and light source being used, as well as the distance between the two. Exposure requirements for your equipment for each particular kind of the plate may be determined by making a series of test exposures, processing these tests and then comparing the results to a standard.

Step 1 : Make a Series of Test Exposure

Cut a 1-inch wide strip from type of the plate you wish to test. Then cut the strip into three equal parts. Now make a series of test exposures by locating each plate part through a plate maker's grey scale for the time indicated. Expose part 1 for 30 seconds.

part 2 for 60 seconds, and part 3 for 90 seconds.

Step *2 : Process the Test Strips*

Process the test strips according to the directions supplied by the plate's manufacturer.

Step 3 : *Compare the results to a Standard*

Exposure requirements for a particular plate may be determined by reviewing the specifications as supplied by the plate's manufacturer. A step number on the grey scale serves to indicate correct exposures e.g.., develop to a solid step 6. Most negative working additive plates require solid step 6 development. Negative working subtractive plates are normally developed to a solid step 6 or step 7. Positive-working plates generally require an open step 3 or open step 4 (steps above these numbers become solid).

To determine correct exposure, locate the test strip that developed to a step that matches or come, closest to matching the required development step (the standard). If the step on the test matches the standard, then the time used to produce that strip represents the correct exposure time for plate making.

If the development step on the test strip is lower than the standard, then additional exposure is required. The new exposure is calculated by multiplying the time needed to produce the test result by an appropriate exposure factor, that will bring development upto standard. Exposure factors used to increase grey scale readings are listed below

Platemaking Exposure Factors

TO INCREASE GREY SCALE READING BY:	1 STEP	2 *STEP*	3 *STEP*	4 *STEP*
MULTIPLY ORIGINAL EXPOSURE BY	1.4	2.0	2.8	4.0

Example : *If the* test strip that was exposed for 30 seconds developed to asolid step 5 and the standard call for solid step 6, then one additional step is required therefore.

Original exposure	30 sec
X Exposure factor	X 1.4
Exposure that will	42 sec
Produce a solid step 6	

If the development step on the test strip is higher than the standard, then a shorter exposure is required. Again exposure factors are used to calculate the new exposure time.

Example : If the test strip that was exposed for 90 seconds, developed to a solid step 8 and the standard calls for solid step 6, then exposure must be reduced by two steps. Therefore :

Original exposure time	90 secs.
X Exposure factor	X 0.5
Exposure that will	45 sec.
Produce a solid step 6	

PLATE MAKING EXPOSURE FACTORS

TO DECREASE GREY SCALE READING BY :	1 Step	2 Step	3 Step	4 Step
MULTIPLY ORIGINAL EXPOSURE BY :	0.72	0.5	0.36	0.25

In offset lithographic plates, the image areas grease receptive (oleophillic) and water-repellent (hydrophobic). They attråct oil-base ins and repel water. Non image areas are water-receptive (hydrophillic) and grease-repellent (oleophobic) Non image areas attract water and repel oil-based inks.

A variety of techniques are utilized in order to form the image and on areas that are required for printing with a lithographic plate.

COMMON TYPES OF OFFSET PLATES

SURFACE PLATES	**DEEP+ETCH PLATES**
Direct image	Single metal
Presensitized	Bi-metal
Wipe-on	Tri-metal
Diffusion-transfer	
Electrostatic	
Photo-direct	

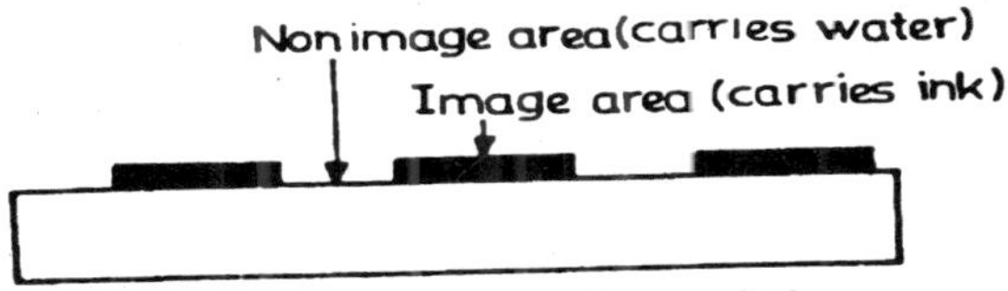

Cross section of a surface plate

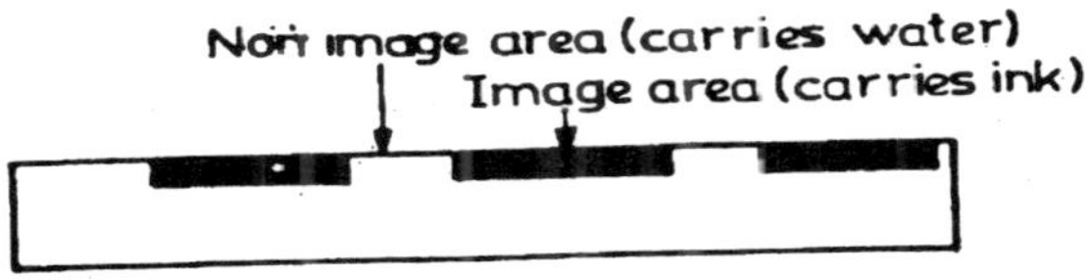

Cross Section of a deep–etch plate

Plates in the surface category forms images that are raised slightly above their non-image areas. The amount of image raised is so small, that it cannot be distinguished by feeling the plate's surface.

Deep-etch plates have image areas that are slightly lower than their non image areas. Because of their structure, deep-etch plates have a greater ink-carrying capacity **than** do plates in the surface category. Some deep-etch plates are capable of hundreds of thousands, if not several million, impressions, before image break down occurs. Brilliant halftone reproduction results from the use of deep-etch plates. There are two basic types of deep-etch plates: single metal and multimetal, single metal plates are normally made from aluminium. Multimetal plates are made from either two layers or three layers of different metals.

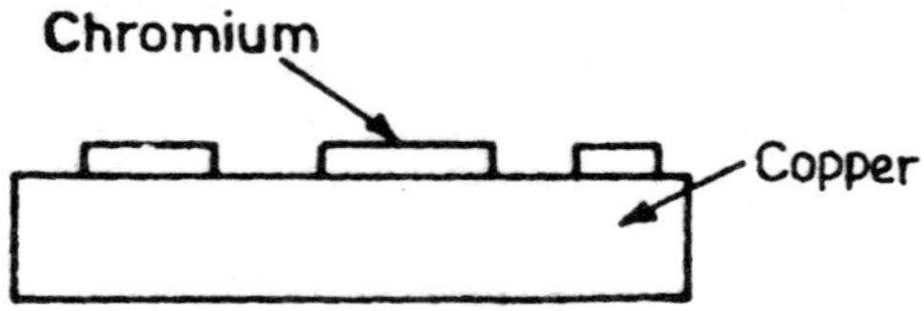

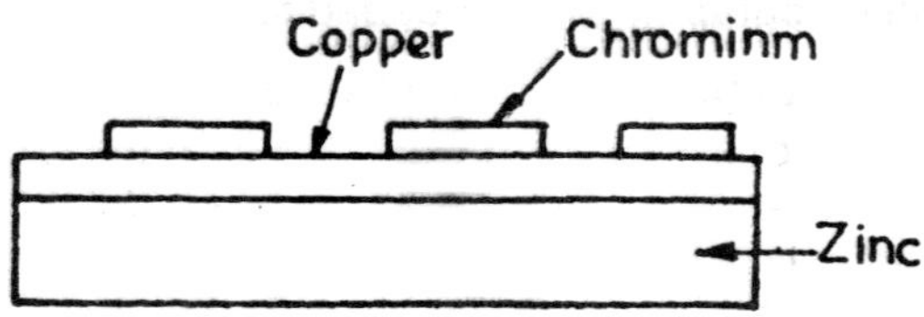

Multimetal Plates :- A different approach to producing a lithographic image is based on the fact that different metals have varying degrees of receptiveness to water and to printing ink. For instance etched chromium is water receptive and ink repellant; stainless steel has similar properties. Copper and brass on the other hand are ink receptive.

A typical multimetal plate might comprise, a stainless steel base with a layer of copper formed by electro-deposition and a further layer of chromium electro deposited. A light-sensitive coating on top of the chromium, completes the plate in pre-sensitized form. Exposure to ultra-violet light, in a positive working situation, softens the coating, which may then be washed away to leave the chromium, non-image areas exposed, while the image areas are still protected. The chromium is then removed by a suitable etchant, to reveal the copper which will form the image. The coating on the remaining areas is then dissolved and removed to leave the plate ready for use.

There are a number of variations with various combinations of metals, but the main advantages of multi-metal plates are :

1. The durability of the metal surfaces ensures long life of the plates, upto a million or more.
2. The uncomplicated nature of the surfaces, makes for cleaner printing with fewer problems.

Waterless plates :- Another departure from the traditional presensitized plate pattern is a lamination, which includes a layer of siliconized rubber banded to a photosensitive material on an aluminium base. The siliconized rubber will form the non-image areas and the photosensitive material will be developed, to form the image area. Positive and negative revisions are manufactured and the processing steps upto development, are similar to those for other plates.

At the development stage the silicone rubber layer is removed from the image areas and the photosensitive layer is reinforced, to form the ink receptive areas. The unique properties of the silicone layer causes it to reject ink, even without the presence of water; thereof this type of plate can be run on lithographic press needed for a damping system.

Thermal Hardening :- Thermal hardening (baking) is the technique of heating the processed plate in an over to harden the image area. This will increase the image life possibly upto three times. It is important to have a totally clean plate before baking, as the image becomes so permanent that any minor flaws will be difficult. Plates should therefore be proofed and checked carefully, so that any minor deletions or corrections may be under taken before committing the plate to the oven.

But again not all plates are suitable for baking and there is no standard for time and temperature in the oven. Test plates should be tried with temperatures ranging from not less than 180 C to not more than 250 C, and time between 5 minutes to 15 minutes.

TYPES OF OFFSET PLATES

Plates may be straight serrated, punched or sloted. The style selected will depend on the clamping mechanism on the press that is used to hold the plate.

The image area of the plate are grease receptive (oleophitlic) and water -repellent (hydrophobic). They attract oil base inks and repel water. Non-image areas of the plate are water receptive (hydrophilic)and grease-repellent (oleohobic). Non image areas attract water and repel oil-base inks.

A variety of techniques is utilized in order to form the image and non image areas that are required for printing with offset plate.

Plate in the surface category all form images are raised slightly by above their non image areas. The amount the image is raised is so small that it cannot be distinguished by feeling the plate's surface.

Deep-etch plates have image areas that are slightly lower than their non image areas. Because of their structure, deep etch plates have a greater ink—carrying capacity than do plates in the surface category. Some deep-etch plates are capable of hundreds & thousands if not several million, impressions before image breakdown occurs. More brilliant halftone reproductions result from the use of deep-etch plates.

There are two basic types of deep-etch plates

(1) single metal

(2) multimetal

Direct-image plates

Direct key plates may be used from paper, plastic, plastic-impregnated paper or aluminium foil laminated to a paper base. Their printing surfaces are specially treated to accept a grease image, that is applied mechanically. This treatment also serves to make non image areas more receptive to water. Press runs with a direct-image plate may range from 50 to 5000 impression, depending on the quality of the plates used.

Presensitized plates (PS Plates)

P.S plates are purchased already coated with light-sensitive diazo or similar emulsion. One or both sides of each plates may be coated in the manner. Most PS plates are designed to imaged by contact printing through either a negative or positive flat. Plates used with negative flats are coated with a negative working emulsion. Those exposed through positive flats, are coated with a positive working emulsion. Additive & substractive processing is available for both negative & positive working plates.

To prepare a PS.plate

A vacuum frame is required to hold the flat and plate in tight contact throughout the exposure process. A sources of actinic light is also required. One should know that actinit light sources emit ultraviolet light. UV light can damage eyes. Do not look directly at the light that is used to burn (expose) P.S. plates, clear both sides of the vacuum frame cover glass.

Presensitized plates should be kept is their protective packaging and removed from wrapper only in subdued light, yellow safelights are preferred.

Place the plate, emulsion side up, and the rubber blanket of the vacuum frame. Then position flat, emulsion side down on top of the plate. Film emulsion on the flat must contact the plate's emulsion during exposure.

Now align the top and right edges of the flat with the corresponding edges of the plate. Then close the vacuum frame and turn on the vacuum pump. Once the air is evacuated the flat and plate will be held tightly together. If a flip top plate maker is used, be sure to turn the plate toward, the light source before processing.

Set the timer for the required exposure time & expose the plate. Remove the exposed plate from the vacuum frame and place it, emulsion side up, in a plate developing, ink. The plate is now ready for processing.

1. Desensitize : Pour a small quantity of desensitizer, also known as developer or process gum on to the surface of the plate. Then spread the desensitizer evenly over the entire plate surface. Use a clean sponge, a cotton pad or a special applicator for this purpose.

The desensitizer serves to remove the emulsion from the non image areas of the plate.

Lacquer- Before the desensitizer dry, pour a small quantity of lacquer (plate developer) on to the plate. Now with a clean sponge or cotton pad spread the lacquer over the entire plate surface. Use circular motion while applying lacquer. Continue rubbing until a clear and uniform image appear. If the lacquer starts to dry before a clear image appears add a small amount of water and continue rubbing.

Wash - Remove excess developer from the plate's surface, with a sponge dampened with water press lightly to avoid scratching the image

1. Developer - Pour a liberal quantity of subtractive developer as an applicator pad and on the plate. After giving the developer time to soak into the emulsion say for 10 seconds re-rub the plate. Rub with a right circular motion. Continue rubbing until the laquer is removed from the non-image areas of the plate.
2. Squeegee - Use a clean rubber squeeze to remove excess developer and lacquer from the plate's surface.
3. Clean the applicator pad & squeegee.
4. Wash - Remove excess developer from the plate with tap water and squeegee the plate dry.
5. Finish - Spread a small amount of finishing solution (substracrtive gum) over the surface of the plate with a clean sponge or litho wipe. Then buff the plate's surface dry with a fresk litho wipe or a piece of cheesecloth.

Print or gum the plate- The P.S. plate is now ready for printing. If it will not be printed immediately, coat the surface with a gum arabic solutio (14 Baume'). Gum arabic serves to protect the plate from scratches and prevents it from oxidizing while in storage.

To prepare wipe-on plates

Wipe-on plates are grained aluminium plates, that are not light-sensitive when purchased. Each plate must be coated with emulsion just prior to use.

1. The emulsion is usually prepared under yellow or red safelights, by mixing a dry diazo powder with a special liquid carrier.

2. Under appropriate safelight conditions, pour enough emulsion on to the center of the plate to coat its entire surface. Now use a fine cellulose sponge to spread the emulsion across the plate. A circular motion and firm pressure is recommended after spreading, wring out the excess emulsion from the sponge. Then with the flat side of the sponge, wipe the plate's surface from one edge to the other until coating is smooth. When it is smooth, fan dry the plate. After the plate is dry it can be exposed.

Processing

Procedures are essentially the same as those mentioned for presensitized plates.

Diffusion-Transfer Plates

Diffusion-transfer plates are produced directly from mechanical rather than from process negatives or flats. Plate images are formed in much the same way as are diffusion- transfer prints.

1. Expose a paper negative

A special photographic paper is used to record the images that appear on the mechanical (copy). This paper is coated with negative-working, light-sensitive emulsion and should be handled under subdued light only.

Place the paper negative material, emulsion side up, on the glass exposure surface. Position the mechanical, image side down on top of the

To prepare distfusion transfer plate.

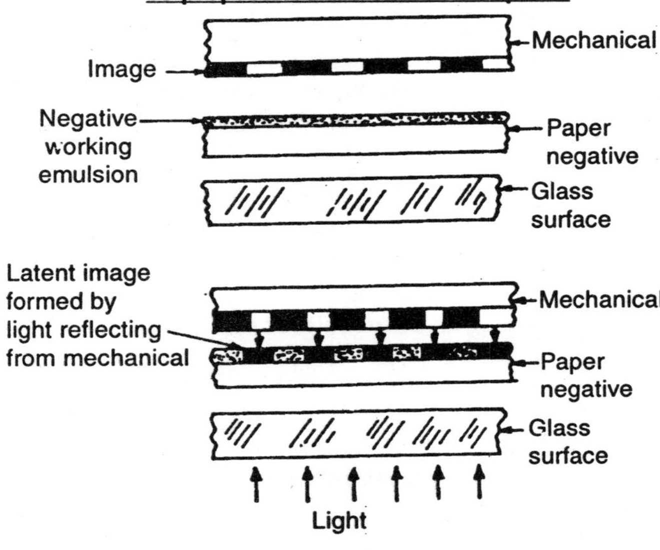

paper negative (as shown in the above figure). After aligning the edges of both the mechanical and the negative, close the lid on the platemaker. When the lid is closed, a vacuum pump will automatically turn on. Vacuum serves to maintain tight contact between the mechanical and the negative throughout the exposure process. Once proper vacuum is achieved, the exposure lamp will be automatically activated. Exposure results when light passing through the back of the paper negative, reflects from the white areas of the mechanical and strikes the emulsion of the negative as in the figure.

2. **Transfer the unexposed silver from the negative to the plate—**

Position the metal printing plate, with the grained side up, at the entrance to the processing section of the platemaker. Then plate the paper negative, emulsion side down , on the plate. After aligning their edges, feed the plate and negative into the developing bath as shown below.

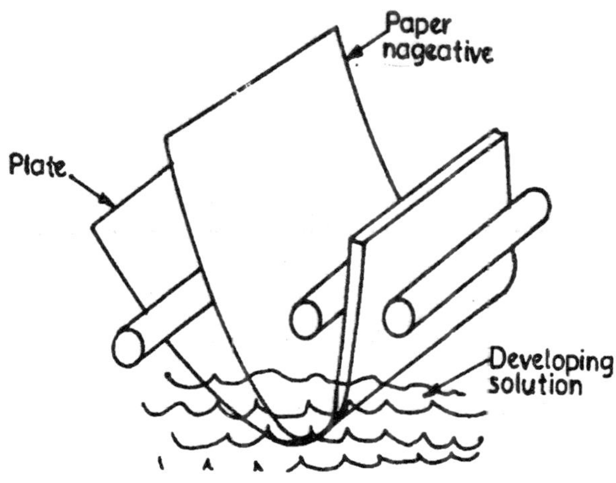

As they enter the bath, the plate and negative are separated slightly. Separation is needed so that the developer can contact the emulsion side of the paper negative. The developer serves a three fold purpose. It develops the latent image on the paper negative. It removes or dissolves the unexposed and undeveloped silver halide crystals from the same paper negative. It separates the silver from the unexposed halide crystals, that have plate, pass through the developing bath as in the figure.

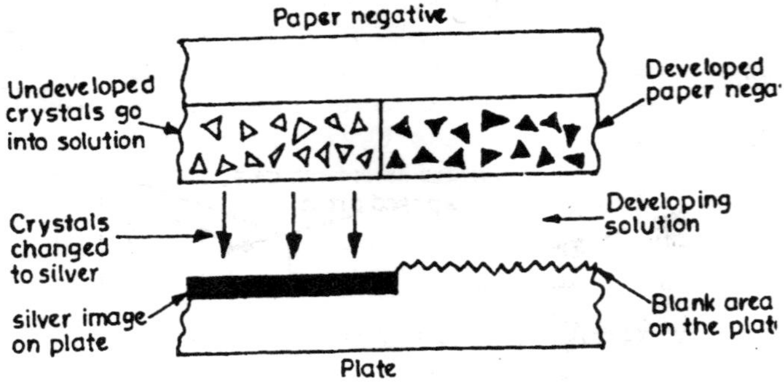

The image formed, as the plate is made up silver that was liberated from the nonimage areas of the paper negative. Image transfer take place as the plate and negative are pressed tightly together between rollers as they emerge from the processing bath.

3. Process the plate

After about a minute strip the negative from the plate. Then desensitize and lacquer the plate. The plate is now ready for the o ffset press. If the plate will not be printed immediately, coat it with a gum arabic solution.

Electrostatic plates

Most electrostatic plates are designed for short run work, say less 5000 impressions. They are produced directly from a mechanical using an electrostatic platemaker. The plates are generally made from a zinc oxide-coated paper material, that is supplied in roll or sheet form. The process is as follows :

1. Charge the plate - The zinc -oxide coated surface of the paper plate is given a negative electrostatic charge.

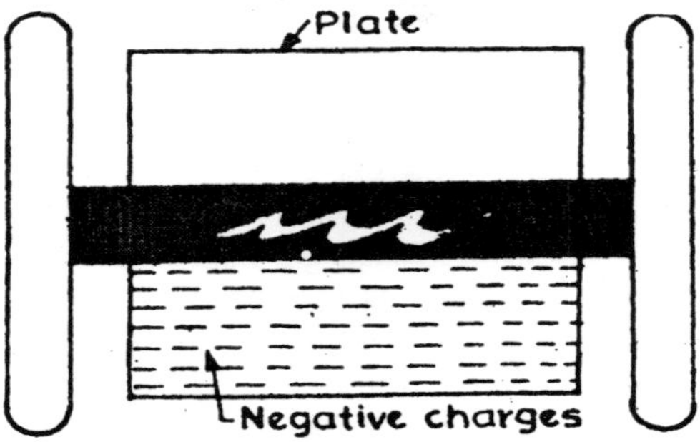

Figure: Charge the plate

2 Expose the plate - The charged plate is exposed through a system of lenses to the light reflecting from the white areas of the mechanical where light-strikes the plate, it removes negative charges. The plate is now left with a largest negatively charged image.

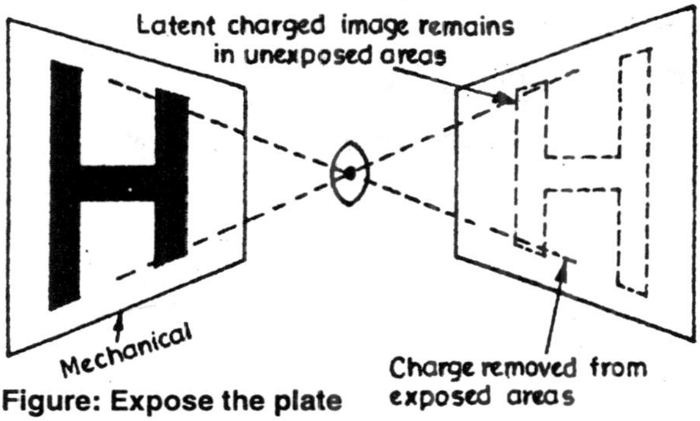

Figure: Expose the plate

3. Develop the image - Either a liquid or powdered toner may by used to develop the latest image on the plate. The toner is given a positive charge. Because the toner is charged positively. It is attracted to the negatively charged image on the plate-causing the image to become clearly visible.

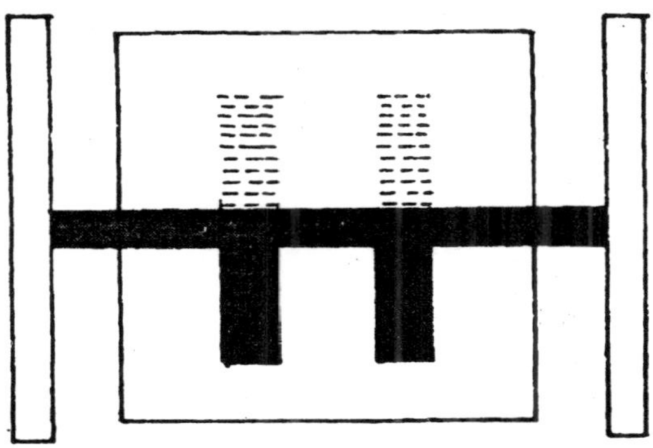

Figure: Develop the image

4. Fuse the image - Radiant heat is used to fuse the toner permanently to the plate.

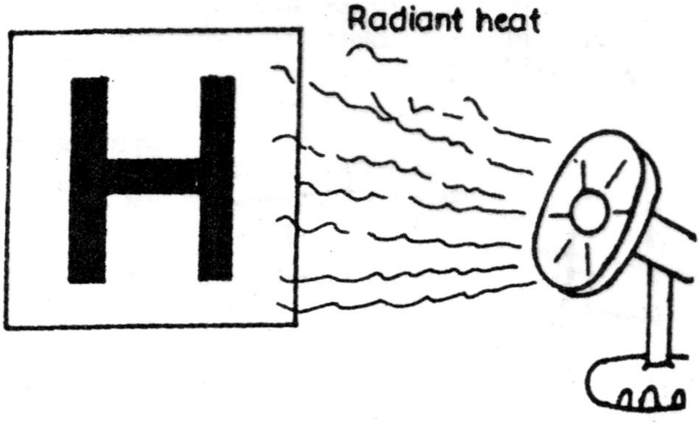

Figure: Fuse the image

5. Process the plate - Processing requires the use of a special converter solution, that serves to remove the zinc oxide coating from the non-image areas of the plate. Processing may be accomplished by hand or by machine.

Photo-Direct Plates

Photo-direct plates are exposed and processed in a camera type plate maker. This plate maker is ideal for quick print, in -plant, & commercial purpose. It can produce upto three plates per minute. It offers reduction and enlargement from 45% to 150% and will accept originals upto 24" x 38".

Exposure is accomplished by projecting light, that is reflected from the original (mechanical) onto the prepensitized surface of the photo-direct plate. Because the projection printing technique is used, the iamge that is recorded as the plate may be larger, smaller or the same size as the image on the mechanical.

After exposure, the plates is cut to length and automatically transferred to the processing section of the platemaker. Here an activator solution develops the plate & makes the image areas in receptive and the non-image areas water receptive. A second solution which is combination stop bath, fixer and stabilizer, serves to ready the plate for printing or for storage.

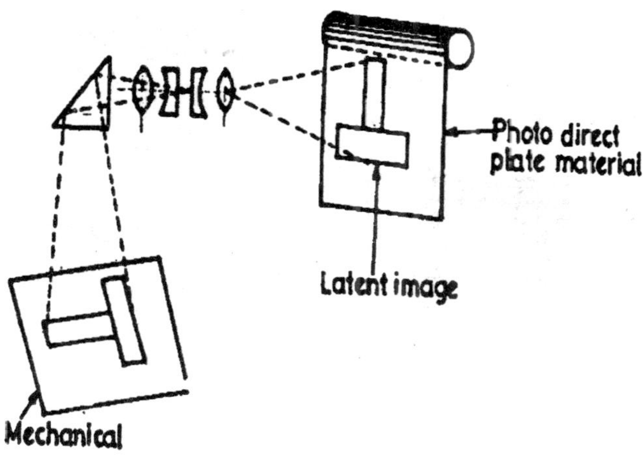

Figure: Exposing a photo direct plate

Step & Repeat platemaking

This technique are used when it is necessary to reproduce the same image is several location on a single printing plate. This may be accomplished manually by exposing the plate several times through complementary step and repeat flat, or automatically by using step & repeat machine.

Platemaking procedures

Almost all offset printing today is done from plates made from paper, plastic or metal materials. A plate may be made from a single material e.g. aluminium or by combining different materials. Such plates are made up of two or three layers of different metals.

Offset plates vary in thickness from about 0.005 inch to 0.012 inch, for duplicator size plates, to as much as 0.030 inch for plates used on large offset presses. Plate thickness must be controlled carefully during manufacturing process, to ensure that each plate will print with a uniform pressure across its entire surface.

The surfâce of an offset plate may be rough (grained) or smooth (grainless).

Graining is accomplished during the manufacturing process, by either etching (chemical graining) or abrading (mechanical graning) the plate material. Graning serve to make plates more water-receptive and improves their printing characteristics. Paper & plastic plates are generally grainless.

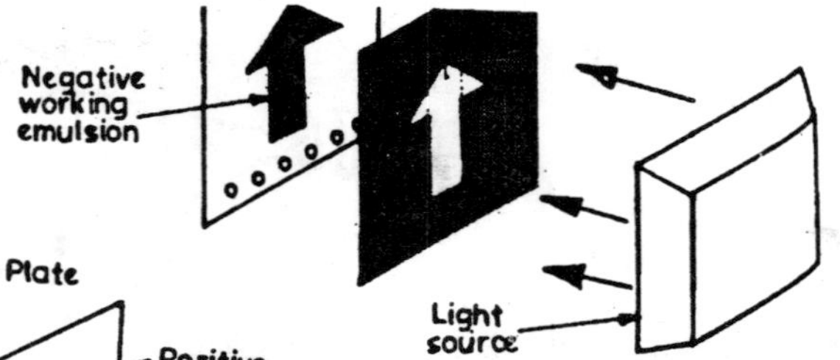
Negative
working
emulsion
Light
source

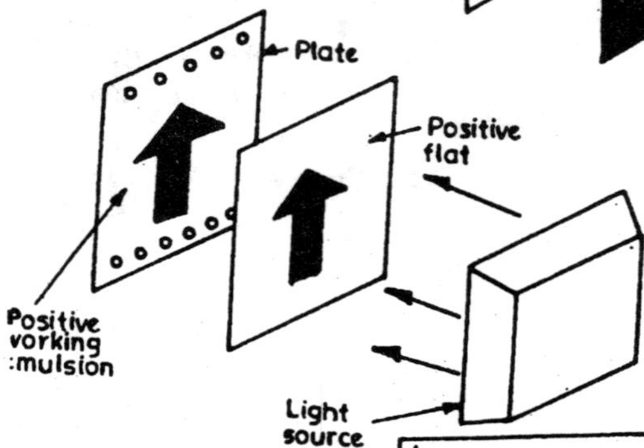
Plate
Positive
flat
Positive
vorking
:mulsion
Light
source

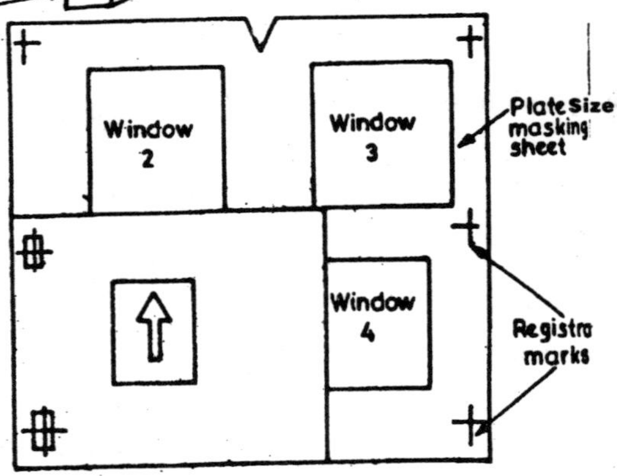
Window
2
Window
3
Plate size
masking
sheet
Window
4
Registra
marks

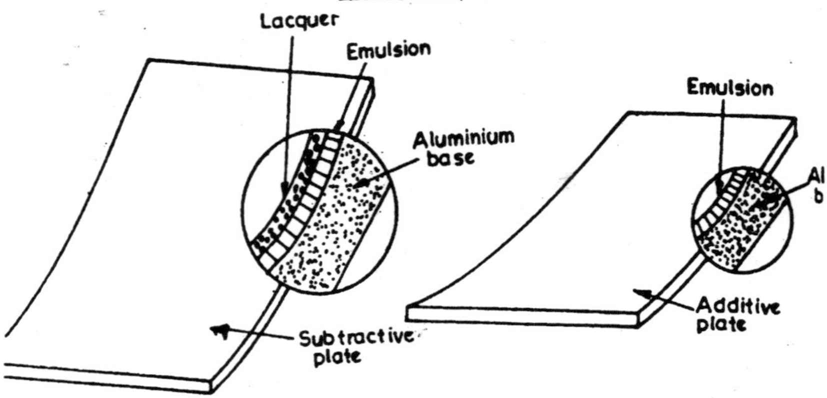
Lacquer
Emulsion
Aluminium
base
Subtractive
plate
Emulsion
Additive
plate

Most offset plates are coated with a light sensitive diazo emulsion. Both negative-working & positive - working emulsions are used.

Negative -working emulsions are designed to produce a positive plate image when the plate is exposed through a negative flat. The image is formed by the action of light. Exposed parts of the plate become image areas and unexposed part become non-image areas after the negative working plate is processed.

Positive - Working emulsions are designed to produce a positive plate image, when the plate is exposed through a positive flat. The image is formed by the absence of light. Exposed parts of the plate become non-image areas and unexposed parts become image areas, after the positive working plate is processed.

Lacquer is used to intensify & improve the printing characteristics of the image areas of an offset plate. This is true for both negative & positive—working plates. Lacquering is accomplished in two ways 1, by adding lacquer to the image area or 2, by removing lacquer from the non-image areas of the plate.

Plates that have lacuqer added to their image areas during processing are called additive plates and that have lacquer removed from their non image areas during processing are called subtractive plates.

13

Presses

Letter Press

Printing from a raised surface is termed relief printing. It is called letterpress printing. In this process ink is applied to the raised surface of a printing plate. When paper is pressed against this surface, the ink transfers to the paper. It is to note that the image on the plate is "wrong-reading" or a mirror image of what is to be printed. Printing reverses this image to form a right-reading image on the paper.

Hand -set type is assembled in a composing stick to create words and sentences. Each piece of type consists of a raised character as a wood or metal body. All the characters needed to form words and sentences are stored in special shallow drawers called job cases.

Machine-set type is generated by typesetting machines that store matrices (molds) of letters, numbers and punctuation marks. A machine operator depresses the keys that send select matrices to a casting area where they are filed with molten metal. After the metal solidifies, a type slug containing the desired words and sentences is delivered to the front of the machine. Matrices are then automatically returned to their storage area and may be reused over and over again.

Photo engravings are relief plates that can print drawings and photographic as well as words and sentences. The plate are prepared using both photographic and etching techniques. A photographic negative containing the desired copy is placed in tight contact with a metal or plastic plate that has been coated with light sensitive emulsion. The plate is exposed to light through the negative and then processed. Processing removes the emulsion from the unexposed area of the plate only. The exposed emulsion remains on the plate and serves as a resist during etching. The unprotected areas, however, are lowered or by the etching solution.

Three types of presses are used to print from relief plates—platen, cylinder and rotary.

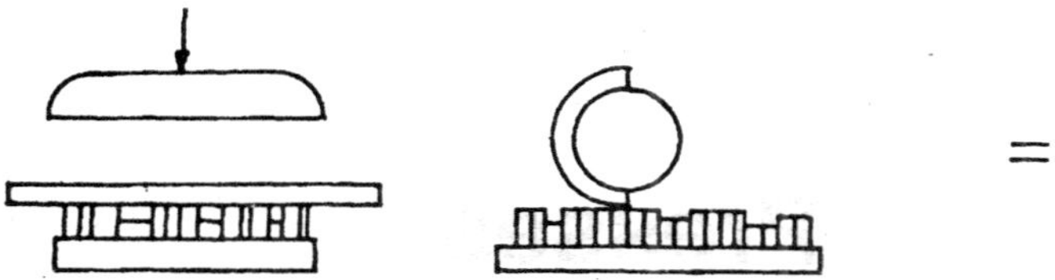

The plate and the bed are flat in flat bed cylinder presses | The bed is flat but the platen is replaced by an impression cylinder | The plate cylinder takes the place of the flat bed

Letterpress was the first printing method. The relief printing surface is inked with a paste ink and pressed onto the paper.

In 1455 in Mainz, Germany, Johann Gutenberg produced a two-volume Bible, the first book ever printed with movable metal type. The irony of fate was this venture practically bankrupted him but the printing process grew step by step.

Until the 1950s, letterpress was the only printing process for printing. In 1950s offset printing got started and eventually became the major printing process. In 1980s the share of letterpress market declined and web-offset replaced it at most newspapers and magazines. Small jobs could be done as high speed offset duplicators or electrostatic copiers.

The print station includes an ink fountain and a steel fountain roller turning in contact with the thick paste ink. The ink is picked up by a roller that conducts the ink to a series of oscillating steel rollers, with rubber rollers in between the ink is thinned out and transferred to rubber form rollers which then ink the type or plates. The image is pressed into the substrate against cylinders covered with rubber or a make ready packing paper, while under pressure as shown in the figure (Letterpress printing station)

The sharp image for which letter press is known is slightly below the surface of the substrate. For fine line screen printing, a smooth substrate is necessary, the smoother the substrate, the greater the detail.

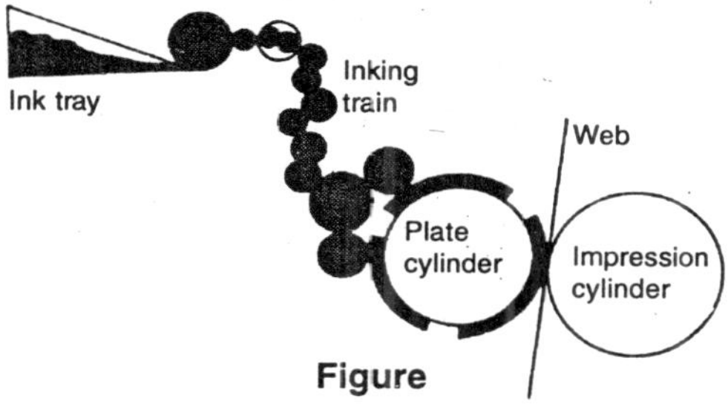

Figure

The Offset Process and it's Principle, Advantages and Limitations

Printing process which transfer ink directly from a plate's image to the paper or other substrates require a resilient material to back up the paper. In the case of letterpress machines the impression cylinder which carries the sheet or web to the printing form is covered with packing of paper and cloth or rubber blanket. Such type of resilience is essential to compensate for in accuracies in the caliper and in the surface of the printing plates and substrate materials and also in the press construction itself. With offset, the impression cylinder packing is eliminated and the necessary resilience provided by means of an intermediate surface between printing plate and sheet material. Thus, the ink and the image is transferred twice—first to the intermediate surface then to the material receiving the image. Therefore, it can be said that ink is offset' from plate to paper.

The intermediate surface commonly takes the form of a rubberised cloth with a comparatively thick layer of rubber (generally a synthetic polymer) on its surface, mounted on a metal cylinder or roller. This is so arranged that it alternately contracts the plate and the sheet material. The arrangement may be flatbed or rotary. A consequence of the double transfer is that, the image laterally reversed twice during printing. Thus the image of the plate must be right reading and not wrong reading as required for direct printing. Such reversal has some advantages in plate-making, but more of any importance in press work.

Offset printing was first adopted by the metal decorating industry, for applying lithographic images to tin-plate. There are several reasons why it has developed. It's first advantage was by comparison with direct lithographic printing and it did not compete seriously with other printing processes for many years. Early litho plates did not have very durable images, particularly half-tone images. Offset extended the life of such plates and was instrumental in enabling lithography to reap the full benefit of the photography to printing. The modern lithographic plate provides very long runs.

It was also observed, that offset was capable of printing fine line and tone work on to comparatively rough or hard surfaces. A direct process such as letterpress using hard metal type or blocks, requires a smooth surface on the sheet material if very fine lines or halftone screens are to be reproduced without loss of quality. Thus for high quality halftones (say 120 or 133 screen) the usual specification is for a coated paper such as brush coated art. If lower-grade paper is necessary, the screen must be much coarser e.g. 65 for newsprint. Coarse screens do not reproduce a photograph's tonal values without loss of much original details, for this reason web offset has been adopted for some newspapers. The offset process permits the use of screen rulings of 100 and even 120 as newsprint.

Offset's ability to print effectively on uncoated papers led to lithography's use of cartridge and machine finish stocks. Because these are cheaper than coated papers, the common impression of a cheap process prevailed, together with the idea that offset and low quality are synonymous. Today, of course, there is widespread evidence that offset can print most types of work at a quality to equal and sometimes to surpass other printing processes. Each process has its advantages and limitations.

Other offset advantages relate to the lithographic aspect of the process e.g.., the use of cheap, thin, flexible metal plates. It has already been stated that the principle of offset was to shift the resilient material from behind the paper or cord stock to between it and form or plate. The resilience thereby conforms more effectively with the stock surfaces irregularities, generally with less pressure than required to the direct print. But the offset process need not be confined to lithography, and is in fact applied to letterpress increasingly, particularly since the advent of thin flexible plates for letterpress (indirect letterpress or letterset).

But since such plates are more expensive than equivalent lithoplates, the advantage remains.

Lithograpgy Offset

In 1798, Alois Senefelder wrote on a flat stone with a grease pencil. He dampened the limestone surface with water and inked the writing with greasy ink. He pressed paper against the stone and the inked image transferred to the paper. The image printed in reverse

Actually what happened was the water welled the stone's non image areas but was repelled by the greasy image areas. Then the grease ink was repelled by the water and was only attracted to the image areas.

Later Senefelder wrote on paper with greasy ink and then pressed the image to the dry stone surface. In doing so, the image reversed itself when transferred to the stone. He walled the stone and inked the surface's reversed image. When he pressed paper to the stone , the image it produce was the first readable, direct stone lithographic print ever. Thus Senefelder had discovered the basic principle of lithography.

Lithography prints from a flat surface with the image areas being neither raised, as in letterpress and flexography, nor lowered below the surface as in gravure printing. For many years a special Bavarian limestone was used for the image carrying plate from which the process got its name taken from the Greek words "litho" (stone) and Graphein (to write)

Based on the principle that oil & water repel each other, grease pencils were used to draw images or to do hand lettering on paper. The image then

SWIFT OFFSETS

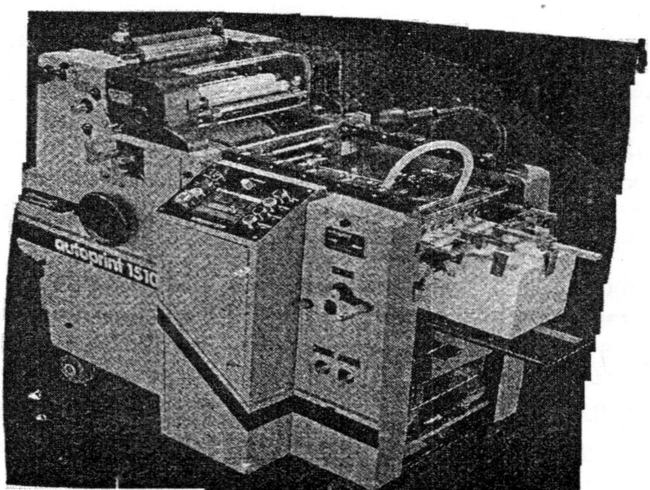
autoprint 1510

MR 237
MR 237

Printoram - 164

was transferred to the stone. Sometimes, images were drawn in reverse directly as the stone with greasy ink. Water & greasy ink were then alternately applied to the entire surface of the stone. As in Senefelder's experiments, water was quickly attracted to the non-printing areas while the ink was drawn to the image areas only. The paper to be printed was than heavily pressed against in surface.

To day, stone lithography is very rare and is only being done by a small group of professional artists to produce limited edition prints. Most modern lithographs are made from thin aluminium plates. At one time zinc was widely used. The plate images could be rubbed off with abrasives, dried, recoated with photo emulsion and reused.

Thin-gauge aluminium plates are relatively inexpensive and are not reclaimed. In some cases, plastic-coated paper printing plates are covered with a photo emulsion and used for short-run jobs. Offset printers buy pre sensitized aluminium plates. That they expose through negatives, under vacuum contact, using bright white light. After exposure the latent image is developed with a greasy developer and dried. On press, the aluminium plate is dampened with a water fountain solution and inked by rubber form rollers. Faithful to the process, only the ink is attracted to the image, since the water repels it from the non-printing areas.

The image is positive-reading on the plate surface. Both the inked image and non printing areas are on the same plate. The plates are attached to the plate cylinder by clamping the plate's leading & trailing edges. Because of the gap between the clamps, continuous design patterns are impossible to print with the process.

As the plate cylinder turns , it's dampened with a water-welled roller & immediately inked. The plate cylinder then comes in contact with a rubber blanked cylinder. The positive printing plate image is transferred or "offset" to the blanket surface in reverse. The blanket in turn transfers the image to the substrate against an impression cylinder in positive, readable copy. Offset presses can be either sheet-or web -fed.

Offset presses first appeared in the 1960s. Out of a need for higher press speeds most publication work is being done with web-offset.

Presses (offset) or duplicators

Duplicators & presses are used to print offset plates. Normally duplicators are utilized for small printing jobs no large than 11" x 17" Presses are generally identified by the maximum sheet size, that the machine can handle. 'These range from approx 10" x 15" upto 54" x 77". Mini offset are manufacturing by MOI, **Swift Auto Print** & others.

The smallest size presses are used for general commercial work, where as the large size machines are capable of printing books, magazines & newspapers.

Press & Duplicator operating System:—

A wide variety of machines may be used but all contain the same five basic operating systems 1, the printing system 2. The inking system 3, The dampening system 4, The feeding system and the registering system - 5. The delivery system.

Perfector & multicolour presses

Perfector presses are capable of printing both sides of a sheet or web of paper in one pass through the machine. Perfector press is equipped with two plate & two blanket cylinders. 'The blanket cylinders, serve as impression cylinders for each other.

Multicolour presses are designed to print two or more colors of ink, as one side of a sheet or web is one pass through the machine. Both in-line and common-impression cylinder printing system arrangements are used.

The common - impression cylinder arrangement for multicolor printing serves to back up the two blanket cvlinders. Perfector -type multicolor presses are also available.

14

Offset Press

Most sheet fed printing now-a-days is by offset lithography. Offset lithographic printing press will comprise the following three elements :

(1) The sheet feeder, which holds the bulk paper stock and issues sheets, one at a time to the printing units.

(2) The printing units, which will comprise plate, blanket and impression cylinders, plan inking and damping arrangements.

(3) Sheet transport and delivery.

Sheet Feeder :- An automatic pile raising mechanism, based on a sensing device which detects the position of the top sheet in the pile and raises the paper table as sheets are consumed.

A quick change over system as the stock is depleted., on most presses it is possible to truck in a press pile to replace the expiring stack without stopping the press.

Sheet Separator :- The second requirement is a sheet separator. This is usually a set of pneumatic suckers and ancillory equipment, designed to separate the top sheet in the pile and forward it to the press. The most common technique, is to blow air through the top few layers of the stack, in order to give an initial separation. The top sheet is then isolated from the stack and pneumatic suckers will pick up the top sheet and forward it to the press.

There are two basic types of sheet separation in common use. First the front separator in which gripper edge. This is a simple and efficient system, for smaller sizes of sheets are separated and forwarded from the front gripper edge. This is a simple and efficient system for smaller sizes of sheets can operate almost ten thousand sheets per hour. However, this method does have a disadvantage when it comes to larger sheet sizes, as the suckers cannot pick up a second sheet until the first sheet has passed.

Then the back separation technique is employed, in which the sheets are separated at the rear of the pile, the advantage here is that the second can be picked up immediately after the first has cleared the back edge. Sheets can therefore be forwarded to the press in a continuous overtaping shingle. This type of sheet feeder which may be capable of feeding even the largest sheets and up to 15,000 per hour. Having separated and forwarded the individual sheets, it is now essential to ensure that they are presented accurately to the printing unit.

PRINTING UNITS : There are three main cylinders in an offset Lithographic printing unit :-

1. Plate Cylinder
2. Blanket Cylinder
3. Impression cylinder

An inking and damping systems will complete the unit.

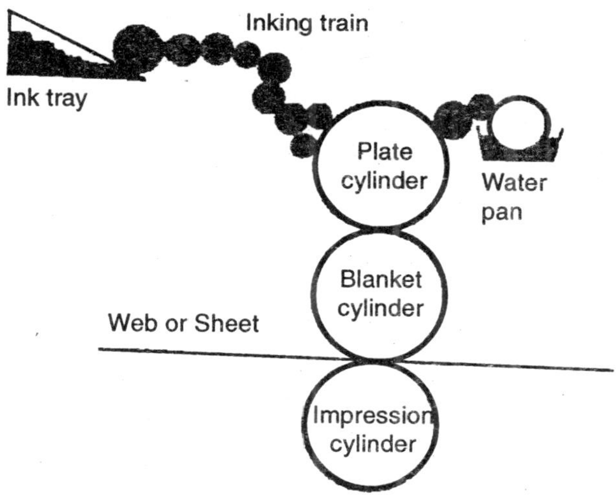

Figure: Offset printing station

Plate Cylinder :- The plate cylinders must have position for quickly and accurately locating the printing plate. This is by means of front and rear clamps which grip the ends of the plates and tension it around the cylinder. Arrangement for fine adjustment is usually necessary in order to accurately register and print with another and a pin locationing arrangement may he provided for instant location of the plate, where this has been provided for at the platemaking stage.

The stress on precision also applies to the caliper of the plate. Most presses have bearer rings at the end of the cylinders and these indicate the pitch circle of the gears, which drive the cylinders.

Inspect and Measure the Plate :- The plate caliper should be ascertained and required under packing prepared. In those machine-rooms where only are size and caliper of plate is used it is often assumed that measuring the plate thickness is a super fluous operation but mistakes have been done by the plate supplies and plate making departments. Sometimes same question does arise to plate size but it is very time wasting then to get a plate partly through the clamping procedure and then find that it is fractionally either two short or too long for the clamping system.

The types of packing material suitable for use as under packing are limited. A supply of packings of different calipers should be available, especially where close register work is regular printing or where plates vary in caliper. As for example, for a normal packing and blanket packing 0.005 (0.0125) is specified then a range consisting of 0.010 (0.25mm); 0.008 (0.2mm); 0.007 (0.175cm); 0.005 (0.125mm); 0.003 (0.075mm) and 0.002 (0.05mm) will be required, Again packing of 0.001 (0.025) is impracticable and the use of two sheets together is not always satisfactory. Therefore the above mention range can be considered only as a compromise. If a system of packing storage is adopted and the packing sheets cut to correct size and if possible they are of different colour for each caliper-then much time can be saved. And as we all know in management time means money.

The packing material itself can be manilla paper, oiled manilla paper or a form of plastic sheeting. The materials are expensive, particularly the

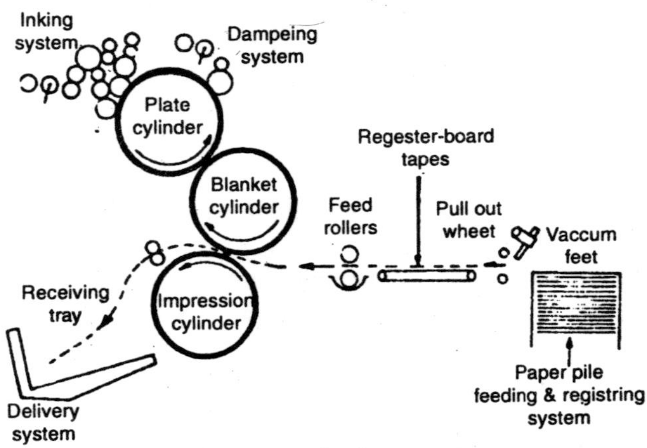

Figure: Offset printing system

plastics, but provided care is taken in handling they are re-usable. The cut size should be long enough in the round cylinder direction (side to side of the machine) The packing should be cut to slightly less than the plate dimension in this direction say about 10mm, so that the sheet does not extend the plate-edges, where it will be affected by the damping rollers and so that it is sufficiently inside the edges to minimise attack by the solution and yet not reduce pressure of the damping rollers at these points. Few machine are fitted with any form of mechanism on the plate cylinders for gripping the plate under packing. Thus there is always the risk of packing slippage or creep packing which falls sort of the trailing edges of the cylinder is always liable to creep forward, resulting in bunching at the leading edge with consequent damage to plate and print quality.

Markout "pitch" lines'check registration system :- In order to align the plate image, accurately to the plate cylinder axis and consequently to the grippers edge of the stock to be printed, various system are in use but these will differ from press and sometimes, a system has been fitted to conform with the overall plate registration method adopted by the particular organisation punch, hole registration systems are an example. Holes, usually two, are drilled in a convenient position on the leading edge plate-clam in to which fit accurately long pins with flag shaped knobs on their ends to aid indentification and to act as warnings, corresponding holes aligned to the plate image will have been punched in the plate gripper edge during the platemaking stage. In the system no making of scribe lines on the plate is required.

In other cases it will be necessary to mark lines on the plate side edges, parallel with the leading edge at distance corresponding to measurements given in the press manufacturers specifications. The time should be parallel to the image, not necessarily to the plate leading edge and will be in position to time up with corresponding marks on the plate cylinder. A center mark should be put on the plate to line up with the plate clamp center mark.

Clean the back of the plate and the plate cylinder-Before cleaning the back of the plate it is suggested that the old gum is washed of the face of the plate, fresh gum applied and wiped down to a thin film and dried. The image than can washed out with a shphaltum or similar wash out solution. This is helpful if the plate is out of store and the ink on the image hard and dry. In such cases washing out is much more easily carried out with the plate cylinder. The plate should then be turned over on to a sheet of clean paper on the bench to facilitate cleaning. If the plate making and proving departments have done their job properly, little cleaning should be needed. But there may be small bits of hard gum or ink adhering to the metal and these should be removed or wiped over with a mixture of water and in solvent.

On all presses with adjustment for the plate on the cylinder, a system of alignment and of applying tension to the plate should be worked out. An ideal arrangement would be to use foroque wrenches for tightening the lamp both and tensioning screws/ the most common fault relates to over-tensioning which can cause plate splitting. It is often found that out of a series of tension screws on one clamp, one will be over tensioned and the rest too slack, almost faulty register and a real risk of the plate breaking loose during printing.

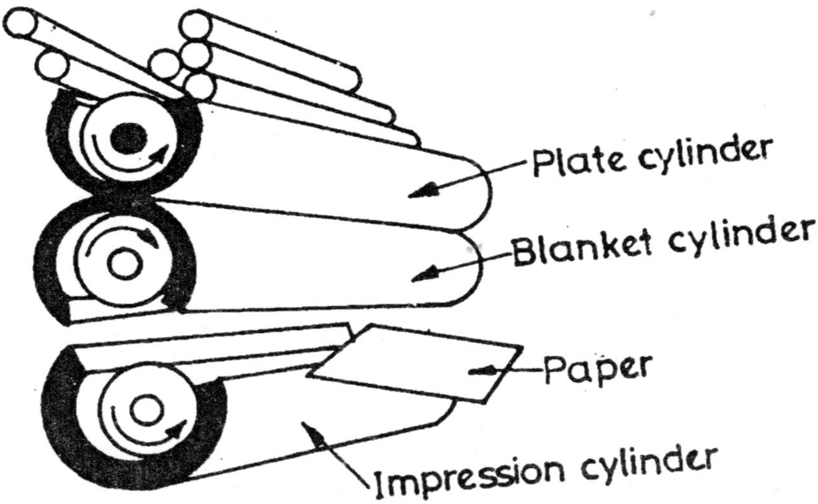

Figure: Three cylinder printing system

Blanket Cylinders— Offset are manufactured with considerable care and at first sight appear to be an exclusive item. With proper care in use, the cost can be more than regained in the useful life of the blanket over a long period.

The blanket is constructed together with ruldoer, and the carcases and it is thus nows, is spread with an overall layer of to a depth of approx. 0.025". Finished blankets have an overall caliper of between 0.065" and 0.075"; depending on the

70 deg. shore = soft

75 deg. shore=medium

80 deg. shore=hard

Many printers engaged in general printing use blankets in the medium range of hardness for all purpose quite successfully but it should be borne in mind that other are available if difficulties are encountered.

During the run of most jobs, the blankets will need to be cleaned periodically, and this should be carried out in a methodical manner. As there almost certainly will be an accumulation of flutt from the paper on the blanket of coating from such stock. This must be removed by first washing the blanket of coating from such stock. This must be removed by first washing the blanket with a swab soacked in water, and than following with a thorough

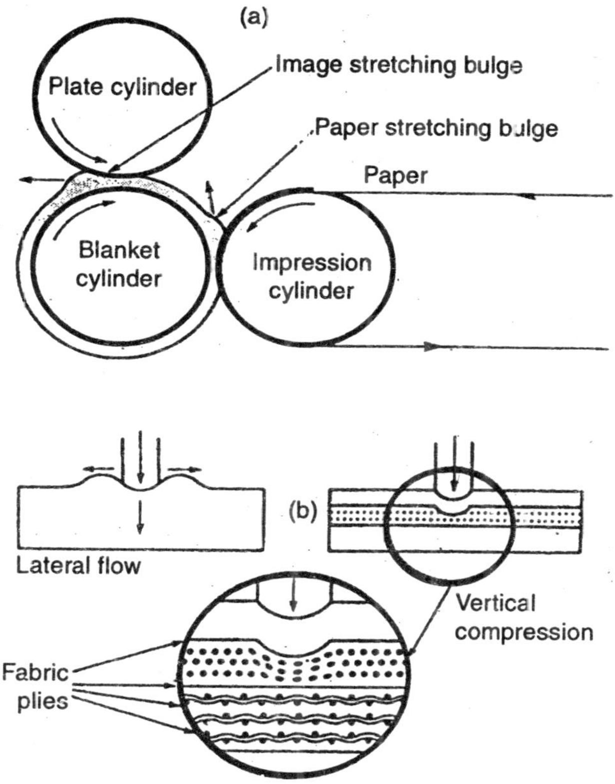

Figure : Litho blankets. (a) Conventional rubber blankets will distort under printing pressure. (b) Compressible blankets are design to give under pressure rather than distort Courtesy : Dunlop

washing with a swab soaked in a good quality blanket wash, of which there number in the market. The procedure should be to wash an area and immediately dry it with another awab. This prevents the blanket wash from evaporating too quickly and leaving behind residue of ink in a thin film. This thin film can quickly develop to form a glaze over the whole surface of the blanket, which results in a poorly printed sheet, as the image is formed on a glazed surface instead of the rubber and is not picked up and transferred properly. The periodic cleaning of the blanket with methyl ethyl ketone (M.E.K.) is a help in removing glaze, which is also caused by the natural oxidation of the rubber.

Impression Cylinder— The impression cylinder carries the sheets through the nip, where the ink is finally transferred under the pressure to the printing stock. This is the moment of truth where the smallest dots, the finest lines, the lightest tinits and the heaviest solids are laid upon the paper. The slightest variation in pressures, packings, inking, clamping will be recorded at this juncture.

A set of grippers will take the sheet as it is presented to the impression cylinder and will release the sheet to the transport system after printing. The timing of the opening and closing of these grippers is critical in order to ensure accurate register and fault-free sheet transport. The grippers must hold the sheet firmly, but without damaging it and it should be noted that the section of the sheet held by the grippers can not be printed upon a factor which needs to be under stood when designing and planning work for sheet-fed printing.

Most machines use the Three-cylinder printing systems. The plate is wrapped around the plate cylinder. Clamps on the plate squarely and firmly in place. As the plate cylinder turns, the surface of the plate contacts a rubber blanket that surrounds the blanket cylinders. Ink is than transferred from the image areas of the plate to the rubber blanket. The image formed on the blankets is reversed or wrong reading. This inked image is then offset on to the paper as the paper passes between the blankets and impression cylinders. The image that is printed on the paper is the same as the image on the plate. Both are right reading images. The impression cylinders serves to apply the necessary pressure against the paper and blanket to ensure cylinder image transfer-like devices on the impression cylinder called grippers are used to pull the paper thought the printing system. After image transfer is accomplished, the grippers automatically release and transfer the printed sheet to the delivery system.

The Inking System— The inking system serves to coat the image areas of the plate with ink. There are two basic techniques employed for the purpose: conventional and combined.

In the conventional technique, ink is carried from an ink fountain to the plate by a series of roller. The fountain roller transfers ink from the ink fountain to a ductor roller. The ductor roller moves back and forth between the fountain roller and a train of ink distribution rollers. The vibrating and oscillating rollers serve to coat the form rollers uniformly with ink. The form rollers are the ones that actually transfer ink to the image areas of the plate.

The combined technique also makes use of a series of rollers to transfer ink from the fountain to the plate. Some of these rollers, however, also carry water to the plate. The water rides on top of the ink and transfers to the non-image areas.

The Damping System— The method of applying moisture to the lithographic plate has changed very little since the rotary press became accepted. Two rollers with cotton material as covering contact the plate when printing. These two rollers receives their supply of moisture from a feed, or supply rollers which alternatively contacts the fountain roller, which rotates in a reservoir and the distributer. In most cases the amount of turn of the fountain roller is adjustable, and is the means of regulating the supply of moisture to the plate. There are other designs whereby the amount of dwell of the feed roller on the fountain roller is a system of regulation.

The forgoing description of damping is known as the "conventional" method of damping, and is often criticised as being the main cause of most

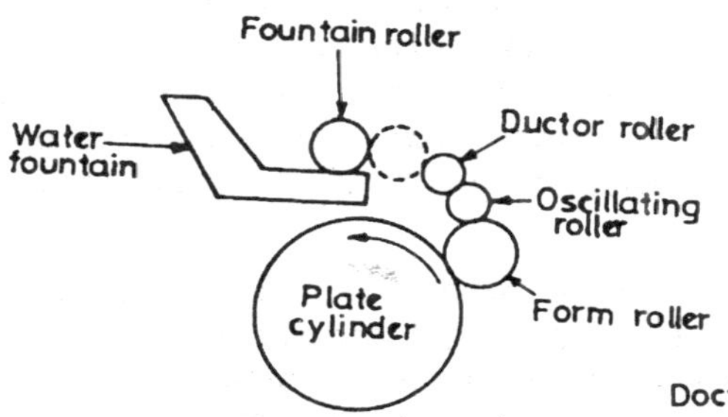

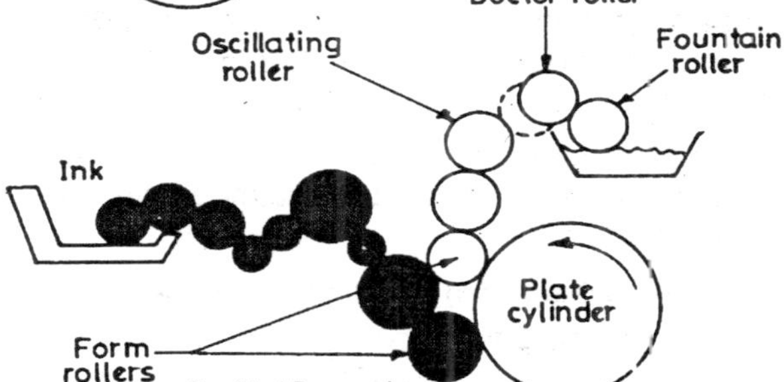

Combind Dampening

of the trouble in litho printing. The basic requirement of good printing by litho is to maintain the correct balance between ink and water, and this is not always easy. The cotton covering of the dampers soak up and retain certain amount of moisture, so that when, through experience the operator wishes to reduce the amount of moisture to the plate, the change will not be immediate. The covering, acting as a reservolour, will retain a certain amount of moisture so that a time lapse occurs before the dampers give up the moisture at it holds and commences to function with the lesser amount for which the adjustment was made.

The problem has been the concern of lithographic over the years, and, almost all the changes of damper covering materials and methods of control have been made with, and this is the fact that the change in the type of litho plates are aimed at using less and less moisture as well as giving greater printing life. The correct balance of ink and water is of the utmost importance to the lithographic and the ability to quickly achieve a balance and to maintain the balance can only improve the print quality.

No covering :

Recently a new type of damper has appeared, requiring no covering at all. The stock is covered with a grained surface. These seem to be meeting with some success, particularly, if isopropyl alcohol is used in the water of damping, as this helps to reduce the surface tension allowing the moisture to spread in a thinner film. The Roland company have introduced a damping system of this type which they call "Roland Auto-Damp system". This has damper only, which is an oversize, uncovered, rubber roller, the length of the damper has been increased and has the ends built up to form a ring, which actually runs on the plate bearer rings on the plate cylinder. This gives a positive drive in the cylinder gap. The water feed roller has a conventional cover. They advice using 20% the isopropyl alcohol to 80% water.

The manual operation of the dampers on the off the plate has been motorised, and when a press check occurs for any reason, the dampers lift automatically. Upon re-starting they are also automatically applied. There are adjustments provided to allow for a given number of revolution before printing re-starts a 'linkage' roller is incorporated so that after removing the feed roller the damping roller is washed up with the rest of the machine.

Some advantages have been found with refrigerating the system to cut down alcohol evaporation. It is claimed that this system offers the following advantages-

1. Less spoilage from colour variation, particularly when the checks.
2. Very little maintenance or damper cleaning required.
3. Press can easily revert to conventional damping at any time.

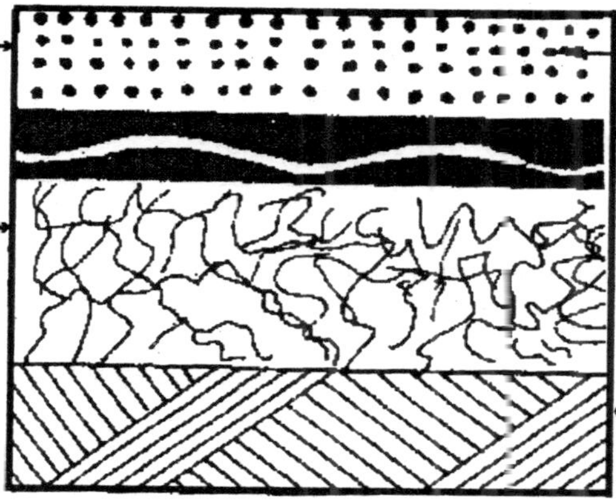

Figure: Compressible blanket

Brush damping — With this system the feed of water from the fountain roller is done by a brush roller instead of the feed damper. As the fountain roller revolves the brush roller sprays fine droplets of water on to the distributing roller. An advantage of this system is that no dirt or paper lint can work back into the fountain. The amount of feed is controlled by varying the speed of the fountain roller.

Cotton covers — Originally these were sewn on to the damper stock by hand. Later 'sleeves' of suitable materials became available-these were more accurate and gave a better damper. The range of materials, that have been used as sleeves varies considerably, and those that retain less water are the most popular.

Dampers of this type have to be cleansed at regular inter vals to remove ink and paper fluff etc, collected during printing.

Paper damper — These appeared sometime ago, and required a specially soft stock. High wet strength the paper approximately wide, is soacked in water and then wound in a spiral around the damper with edges overlapping by to cover the full length of the damper. The advantages of this type of covering is that the paper does not hold a great quality of moisture, but merely conveys it to the plate, and consequently responds more readly to changes made in the moisture feed settings of the damping rollers to both the distributing roller and the plate are more easily maintained, and do not require the frequent adjustment as do dampers covered with thicker material. Cleaning of these dampers is not necessary as the covering can be stripped and replaced with new materials in a matter of minutes, unlike the dampers with thick coverings which must be scrubbed clean readily.

Stockkinette sleeves — These coverings have tended to super sede the paper damper, which, in the opinion of some tend to disintegrate at moments. They consist of sleeves of thin cotton material and are pulled on to the rubber stock. Like the paper dampers, these also tend to convey rather than hold the moisture, but are longer lasting and to not dry out as rapidly as paper dampers during machine stoppages.

3M paper sleeves — This type of covering again requires a special stock. The sleeve should be slid on to a dry damper stock and when wetted shrinks on to the roller making a tightfit. The sleeve does not become loose again ever when allowed to dry out, and the ends of the sleeves do not required any fixing. As with paper dampers and stockinette sleeves, these covers tend to convey rather than hold moisture.

One clear difference between offset printing and other printing processes is that mater must be continuously fed to the image-free areas of the plate in the same plane as the printing areas of the plate.

From the earliest days of lithography many attempts have been made to improve control of this peculiarity of offset printing.

Two ways of achieving this aim are particularly obvious

1. Many types of dampening unit have been developed in an effort to apply the water to the plate in the most uniform and thinnest possible film.
2. It is being attempted to make the damping solution itself more suitable for offset by various chemico-physically effective additives. Acids, salts or mixtures of the two, or substances such as gum arbic (colloids) and alcohol are mixed with the damping water.

Conventional damping units have been used for many years on many rears on various makes of offset presses. In accordance with their construction they are also called ductor damping units. They consist of a fountain roller, a ductor, a vibrator and one or two damping form rollers. The ductor transfers the water received from the fountain roller to the vibrator. Each stroke delivers as much water to the vibrator and damping from rollers as is required by the printing plate for one or more revolutions. The ductor and damping form rollers are fibris covered (molletion, sponge cloth etc.) for transport and storage of the water. During printing these coverings absorb an increasing amount of ink and must therefore be washed by hand or with the damping roller washing machine from time to time. Fibers work loose during this process and later re-appear on the printing products as hickeys.

The continuous stressing causes the damping roller coverings to loose so much material that they can no longer store sufficient water. In addition, the coarser layers of fabric in the covering leave marks in the printing image. The damping rollers must then be re-covered.

Many devices have been invented for better adoption of the water feed to the requirements of the printing plates e.g., water striper, squeeze rollers, fountain roller blowers etc.

On many machine the water level is always kept constant by circulating pumps and overflow pages, so that the fountain roller always immerses to the same depth in the damping solutions. Sometimes a filter is also incorporated in the water circuit to remove mechanical impurities from the damping water.

IN the "Plant-o-damp"method a parchment-type material is wrapped spirally round the damping rollers.

With 3-M Damping the fabric covering are replaced by coverings made from a plastic fibre material, which are shrunk on the damping from rollers.

Dampening Solutions

Dampening solutions serve two important functions 1, to wet the non-image areas of the plate so that these areas will repel ink & 2, to keep these non- image areas from becoming oleophillic (grease receptive) due to the chemical action of the ink on the plate.

The dampening (fountain) solution consists of water and a small amount of acid. A quantity of gum arabic may also be added to the solution, in order to help maintain the hydrophilic quality of the non-image areas. Some press dampening systems operate best, if a small quantity of alcohol is added to the fountain solution. The alcohol allows for a finer film of water to coal the plate , which results in a faster drying printed sheet.

Fountain solutions are generally purchased in concentrated form & then mixed with water according to the instructions. When mixing the fountain solution, add the recommended amount of concentrate to the total volume of water that is required. Then test the PH of the fountain solution.

The PH refer to the degree of acidity or alkalinity of a solution. The PH scale runs from 0 to 14. A neutral solution has a reading of 7. The lower the ph number under 7, the more acid the solution. The higher the number over 7, more alkaline the solution.

Specific PH requirement will depend on the type of plate that is used to print the job. The general recommended fountain solution PH reading for most plates is between 4.0 and 5.5.

Alcohol Damping is the simplest method of printing without coverings on the damping form rollers. Both damping form rollers are removed from a conventional damping unit and replaced by one uncovered "alcohol roller" (shore hardness 20°). As the outer ends of this rollers are 0.3mm. thicker for small and medium formal machines. The necessary friction for uniformal drive is also achieved even when passing the cylinder gap one large format machines the alcohol roller is maintained at the machine speed by a geared drive and friction clutch.

If an uncovered roller is used for printing ink soon accumulates on the alcohol roller. To enable the damping water to surmount this barrier, the surface tension of the water is reduced by adding about 10% is propyl alcohol.

When alcohol damping is used, the ink water balance is achieved more quickly. Consequently, the amount of water is reduced. The damping film or the plate is thinner as a result of the reduction of water surface tension by alcohol, hence the inks are printed in more brilliant colours. Cleaning and maintenance of the fabric coverings are eliminated. Fibres from the damping coverings can no longer deposit on the plate of rubber blanket.

If the storage effect of the uncovered rollers is inadequate for critical printing images, e.g. solids with negative type-it may be advantageous to install a fabric covered roller in the damping unit, it is adjusted so that it comes in to contact with the vibrator, but not with the plate.

In the course of time evaporation may cause reduction of the alcohol concentration of the damping solution can quickly be determined with an areometer. Damping water is sucked into the pipette of the areometer, and the percentage alcohol content can be read directly off the calibrated as required.

The Autodamp unit is a further development of the alcohol damping system. The printer compensates for any damping solution that has evaporated during machine stoppages by adjusting an automatic system in a way that a few strokes will pre-and postmoisten the plate.

Some typical representatives of many other types of damping unit are described below—

Brush damping unit— The required quality of damping water is sprayed onto the chromium roller by means of the brush roller and baffle.

The Hydro-colour system—The damping solution mixed with "Rota-Fluid" is transferred directly to inking unit by a separate roller. Good results have been achieved on small duplicators. However, it has not been widely adopted for longer-format offset machines.

Condensation damping unit —In practice it is used on lithographic proof presses. The fixed plate base and thus the ambient air are cooled to dew point. The water in the air condenses on the plate surface and serve as damping film.

The Dahlgran unit-has been widely used in the U.S.A. and Europe in recent years. It consists of the water fountain roller with chromium surface and metering roller. The damping solution is applied to the plate by the first ink from roller. At least 25-30% alcohol must be added to sure that the water can reach the plate via the ink from roller which is continuously supplied with fresh ink by the ink vibrator. In this system the ink from roller fulfils a double function: it operates practically as an inking roller and partially as a damping roller. This fact combined with the high alcohol contents may cause ghosting.

Matic damping unit with this modern damping unit the damping solution is applied-separately from the ink. The unit consist of the water fountain roller which is driven by a separate motor. The roller squeezes off the excess damping solution entrainment by the fountain roller in the water fountain until only a thin uniform film remains, which is transferred to the vibrator (stainless steel). From here the damping solution passes via an uncovered damping from roller to the plate. The damping form roller and vibrator are driven by gear-wheels at the machines speed. The ink accumulating on the damping roller is distributed uniformly by the distributor roller, hence possible ghosting is prevented.

Because of its constructional features the matic damping unit fully utilizes the advantages of alcohol added. With a cooling system the percentage of alcohol may be substantially lower.

The damping solution is fed continuously to the plate, i.e. without ductor rhythm.

On account of the small storage capacity and simultaneous feeding the ink water balance is achieved very quickly.

For the same reason the matic damping unit also reacts almost immediately to any desired change in the water supply.

As fabric coverings are not used all the associated disadvantages, e.g. fibre formation separate washing and replacing covers, are eliminated.

At the end of a shift or when changing to otherr inks, the matic damping unit can be washed at the same time with the inking rollers.

Feeding & Registering System

The feeding & registering system is used to transfer blank sheets of paper from a paper pile to a predetermined position on the press. From this position the papers is fed into the printing system. Air & vacuum feeding technique used on most duplicators & presses. An air blast serves to lift the top sheets from the paper pile. Sucker then picks up the uppermost sheet by mean of vacuum and transfer it to a set of pull out rolls. The pull-out rolls then roll the sheet onto the register table.

On the register table the paper first passes through a multiple-sheet detector. If more than one sheet is detected, the deflector plate will open and the paper directed into the base of the press. A properly set multiple sheet detector serves to ensure that only one sheet of paper will be printed at a time. It also prevents jam-ups from occurring on the press. Single sheets that pass under the detector are conveyed into a preset position before being fed into the printing system.

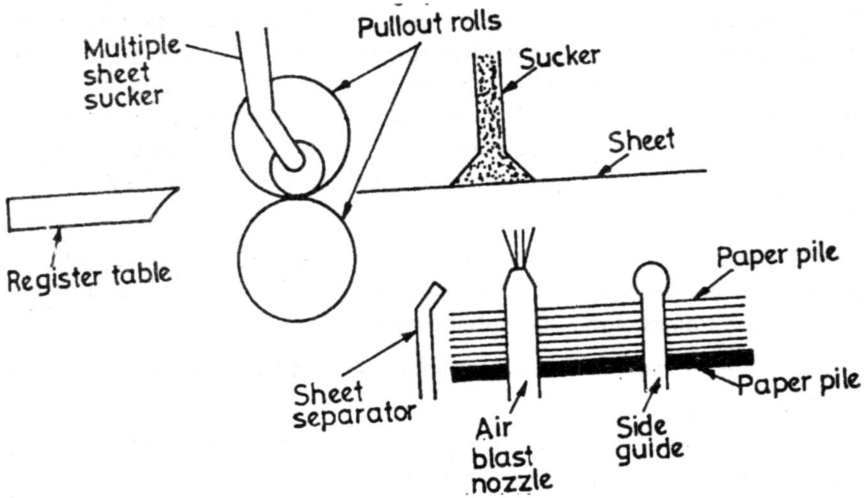

Figure : Air Vacuum feeding

Sheet, Stream and Web-Fed Presses

Presses & duplicators may be sheet-stream or web-fed. Sheets of paper are fed into a sheet-fed press or duplicator, one sheet at a time. Because the sucker must lift each sheet of paper by its leading (gripper) edge, a second sheet cannot be transferred until the tail edge of the first sheet clear the sucker.

Stream-fed machines on sheets. As soon as the first sheet begins its forward motion the tail of the next sheet is exposed, lifted and then forwarded to the registration table. Stream -fed presses are much faster than those that are sheet-fed.

Web-fed presses print on rolls rather than as sheets of paper. The leading edge of each roll (web) of paper is actually threaded through the machine. Tension on the web must be carefully controlled throughout the printing process. After printing into signatures or rewound on a roll, depending on the design of the particular press.

The delivery system

After image transfer, the sheet is transported from the printing system to a tray or table.

Today, the chain delievery is common. Gripper bars are used to transport the printed sheet from the impression cylinder to the delivery table (receding slacker). These grippers bars are suspended between two delivery— gripper chains that are located at either end of the delivery cylinder. The delivery table is called a receding attacker because it automatically lowers as the height of the paper pile increases.

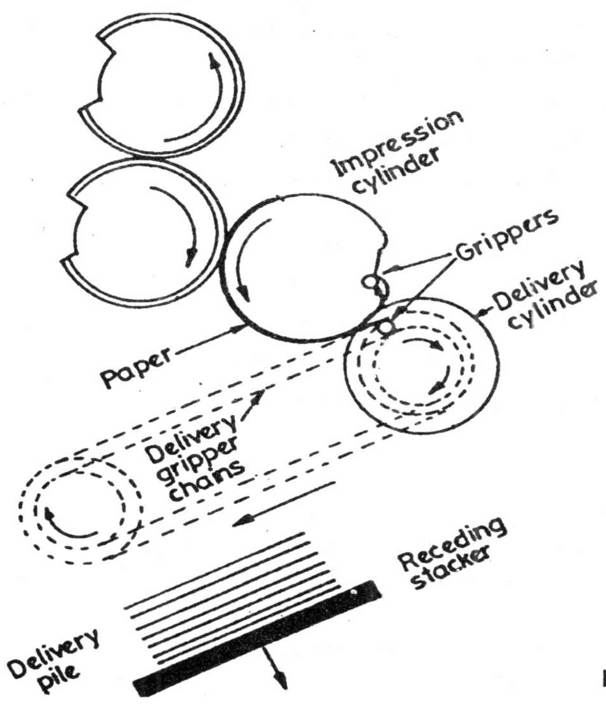

Figure: Chain Delivery

Pre- make ready check list

1. Oil the press. Lubricate properly to function the mechanism efficiently
2. Clean the press of dirt, dried ink, grease and excess oil.
3. Check pressure setting of all rollers & cylinders
4. Check the blanket
5. Check the inking rollers for glaze, if glaze has developed remove it with deglazing compound, otherwise glaze rollers will not transfer a uniform film of ink.
6. Check dampener roller covers & change them if required.
7. Check the PH of the fountain solution.
8. Check the image and non-image areas of the plate to be printed.

Make ready check list

1. Add ink to the ink fountain and adjust the ink feed.
2. Fill the water fountain bottle with dampening solution and place the bottle over the fountain trough.
3. Load the feed table with stock and ensure paper pile sets square.
4. Mount the plate on the plate cylinder.
5. Adjust the plate or move the paper pile on the feed table, to register the image with the sheet.
6. Raise the top of the paper pile to its feeding position.
7. Adjust the air blast to riffle in top few sheets.
8. Position the suckers & adjust the sucker vacuum .
9. Set the multiple sheet detector mechanism.
10. Set the jogger, side guide, hold down straps & conveyor belts on the register table.
11. Set the joggers & reeding stacker.
12. Etch the plate.
13. Turn on the press and lower the dampening rollers to wet the non image areas of the plate. Then lower the ink form rollers to bring ink to the image.
14. Run several sheets through the press to check these sheets and make adjustments if required to improve ink, water feed, cylinder pressures and image registration.

15. Run the job.
16. Preserve the plate and clean up the press.

Clean up check list

1. Remove printed sheets from the delivery tray and place them in a rack to dry.
2. Remove unprinted sheets from the feed table.
3. Remove the fountain solution bottle.
4. Drain the fountain solution tray & wipe dry.
5. Remove excess ink from the ink fountain.
6. Clean the blanket wash.
7. Remove the ink ductor roller and clean with cloth moistened with blanket wash.
8. Wipe the ink fountain roller clean, with a cloth moistured with blanket wash.
9. Clean all dampening rollers.
10. Clean the blanket with blanket wash.
11. Clean the impression & plate cylinders with blanket wash
12. Replace the fountain solution bottle, ink fountain & rollers & ensure that the rollers are returned to their correct positions.

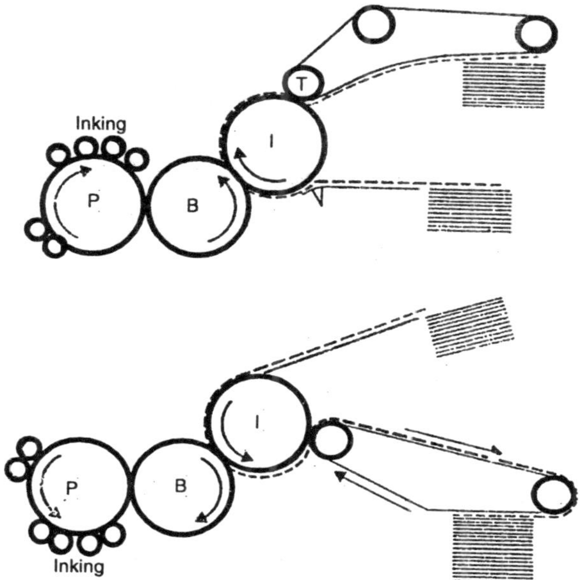

Figure : Alternative cylinder arrangements for single colour sheet fed offset machines

Anti-set-off sprays, ink duct agitators, fountain solution recirculating pumps and fountain trough rollers all have small independent electric motors needing their own switch gear, but they will not all be found on smaller, less expensive machines.

Electrical and electronic devices are widely used for "trippint" systems, i.e. devices which automatically operate the printing cylinder pressure, inking roller application, etc, some are coupled with manually-operated controls, able sometimes to over-ride the automatic devices. Electrical over-ride systems are also used. The term "over-ride" describes the

(iii) feeder setting for sheet size and type of stock (including layboard mechanism)

(iv) delivery joggers and gripper timings

(v) register controls.

Whenever it is necessary to "trip" an offset machine-either because of faulty sheet feeding or immediately prior to stopping-the printing pressures must be released. To effect this, the printing cylinders are moved out of contact by a very small amount, usually by means of eccentric bearings fitted to the blanket cylinder journals. This cylinder is moved away from the plate and impression cylinders because it is normally most convenient but variations apply to particular functions, e.g. sheet-fed perfecting presses. However, the main purpose of such movement is to avoid a "miss" on the impression cylinder surface. The release of pressure also reduces the risk of damage to printing plates and blankets, whilst facilitating machine and make-ready adjustment. Other movements must take place simultaneously if full protection is to be afforded and wastage of sheets reduced to a minimum. An obvious trip is the inking roller lift, and coupled with this must be an ink feed and a fountain solution feed trip. A trip to the impression counter is also essential to ensure a reasonably accurate count, although so far no means has been devised to enable the counting mechanism to differentiate between good and bad sheets. There should be a simple method to stop and start the counter (and perhaps some form of reminder).

The damping roller lift is almost always manually controlled, i.e. does not operate with the automatic trips: this is partly because of the inefficiency of even the best conventional damping mechanism. It is impossible for such systems to apply an effective and correct moisture film to a dry printing plate within one cylinder revolution; hence, unless some system of delayed trip timings can be devised, i.e. applying the damping rollers for three or four revolutions prior to the inking rollers, it is far simpler to leave the plate-damping rollers directly under the operator's control. The usual method is for these rollers to remain in contact with the plate while the press continues

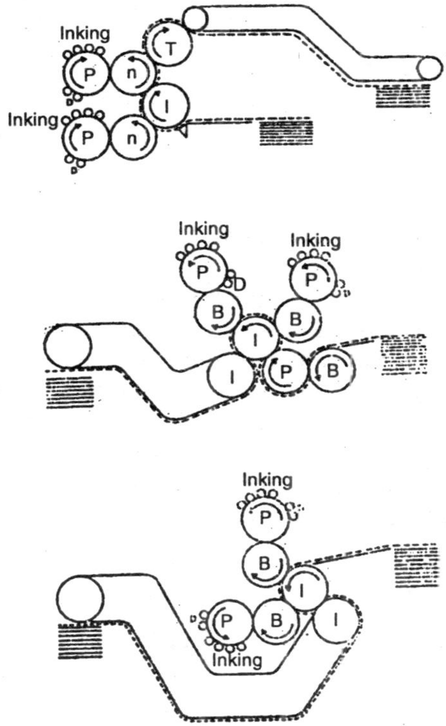

Figure : Three examples of common impression cylinder two-colour sheet-fed offset machines

idling in the trip position during the short delays that occur in feeding, due to faulty sheets etc.

Although this unfortunately results in smudging and thickening of a heavily -inked plate, this can be overcome if the inking rollers are applied to the plate before the cylinders go into contact. Many presses are not designed for this to happen with the automatic trips, and it is left to the operator to "drop" the rollers by the hand-lever mechanism, a control which will be discussed later. New forms of damping systems tend to reduce this problem.

The foregoing trip features are the minimum for a simple offset press, and as attachments are added to increase productivity so further trips become necessary.

An obvious requirement is the automatic feeder, and it is usual for this to cut out with the machine trips; this prevents a pile-up of sheets on the feed-board, with the danger of serious damage if they are drawn into the machine. Another safeguard is a trip to the feed grippers, a system fitted to large machines with an over-ride to permit feeder testing

Purpose— e.g. during make-ready, missed sheet, etc-this is not desirable, but for printing on heavy card and similar material can be a valuable safeguard. If the device can be controlled at the operator's discreation, it is of course an asset.

Some variations exist in inking roller hand leavers. The hand lever may over-ride the automatic operation. If used to roll up the plate before taking impressions it may be found that the automatic trip does not lift the rollers. To overcome this difficulty, some makers provide an additional hand lever or plunger; otherwise the roller drop lever must be re-set. Ideally the rollers should rise and fall with the automatic action and still allow for necessary hand application. Moreover, it should be possible to lock the rollers in the off position. Roller drop timing has already been mentioned, but another aspect is whether or not the automatic drop allows for a complete roll-up of the plate. In other words, roller application, i.e. rise and fall, should as far as possible take place in the cylinder gap.

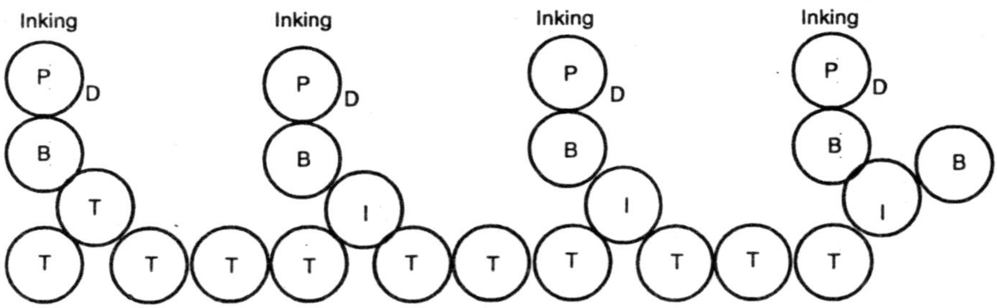

Figure : Four colour unit type sheetfed offset machine. Transfer cylinder equal in diameter to printing cylinder.

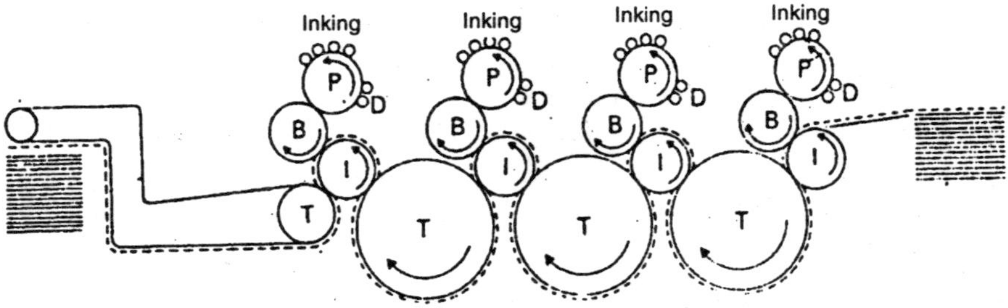

Figure : Four colour unit type sheetfed offset machine. Transfer cylinder three times printing cylinder diameters.

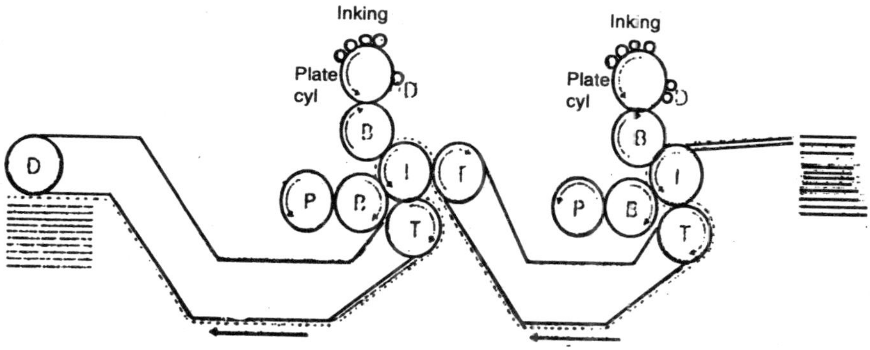

Figure : Four colour sheetfed offset machine, two common impression two-color units coupled with chain gripper transfer system.

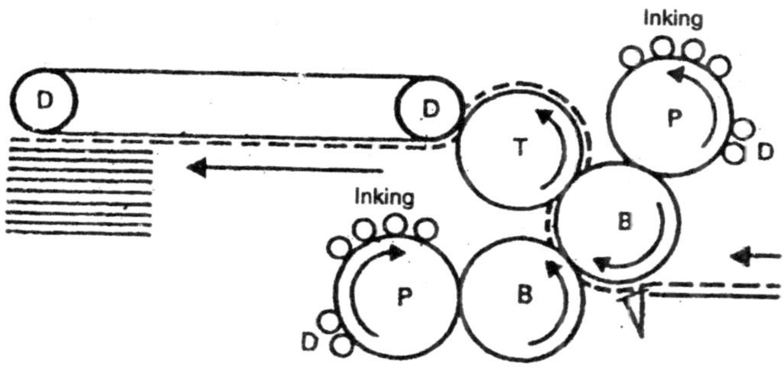

Figure : Blanket to blanket sheetfed perfecting offset machine

Another manual control concerns the printing pressures, but few machines now built provide a hand lever for this purpose. The whole automatic trip action brings the press into printing position, when running, by feeding sheets down the feed-board, provided the first sheet to the lays is correctly registered. Thus by using the feeder starting handle the whole printing operation is actuated. Sometimes a foot pedal or small hand lever must also be moved to put the machine on impression. Whatever system is used, it should be possible to run sheets through to delivery without bringing the cylinders into contact, if the operator wants to test the feeder settings on a new type or size of stock.

Whether maintenance, including lubrication, is the operator's direct responsibility will depend on the machine-room organisation . In larger printhsops and on large presses, a maintenance system involving separate staff is likely. However, the operator is responsible for production, and if his press does not function correctly it is he who suffers. Certain procedures can only be undertaken by the operator, e.g. setting of damping rollers, inking rollers, register devices etc. Other functions involving cleaning of inking systems, automatic feeders and other parts are sometimes delegated to assistants. Mechanical and electrical maintenance is usually the province of trained technical staff. Some printing companies do not employ full-time maintenance engineers, even though possessing a large amount of equipment, but rely upon machinery manufacturer's service departments. They may either pay for a regular service of planned preventive maintenance, or merely call in engineers as required.

Whatever the maintenance system, lubrication is vital. The operator must be able to recognise the symptoms of a lack of oil or grease, such as oil leaks, overheating, noisy running and sluggish movement in reciprocating parts. On most presses there are built in systems for lubricating such essential parts as the main drive shafts, cylinder bearings and inking roller drive mechanism. Many machines use plain bearings for these parts with large areas of metal-to-metal contact. Direct contact must be prevented by using an oil film between the two parts of the bearing and this film must be maintained. Pumps working from oil reservoirs circulate the oil through pipelines to all main parts, after which it is filtered as it is returned to the reservoir (sump) . Old-type presses do not have a sump, and the oil is deposited on the floor or drip-tray . Some small presses still use this system; others do not have an oil pump but rely on hand-oiling which must be undertaken before starting-up and during running time. Many small presses and most older models use cylinder drive gears bolted to the cylinder body; these are exposed to all contamination surrounding the printing operation. Lubrication of such gearing is haphazard and cleaning extremely difficult.

Improved machine designs use ball and roller bearings (instead of plain bearings) for heavy duty parts, as well as for lighter bearings such as inking-roller ends. These bearings are packed with grease, and efficient oil-seals prevent loss so that topping-up is required only at intervals of, say, six months. Cylinder gears and other gearing systems are outside the machine's main-frame, enclosed with oil-tight guards containing oil bath, circulating pump and oil filter. Oil change is required only at fairly lengthy intervals, but by using pre-packed ball-bearing with oil seals it is possible virtually to eliminate daily oiling. Lubrication can then be scheduled weekly, monthly and six-monthly, perhaps using coloured indication on the machine

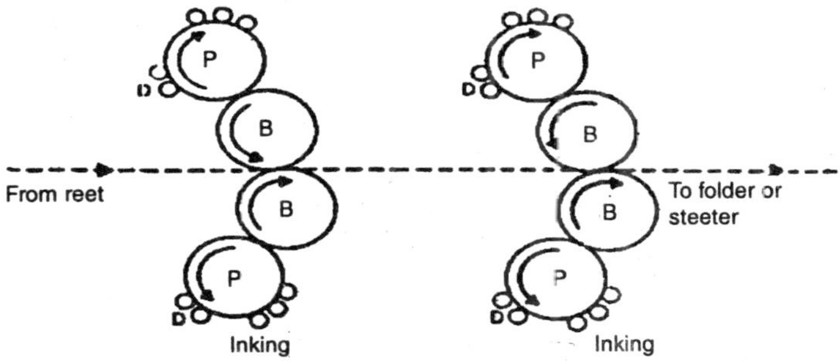

Figure : Horigental web lead two unit blanket to blanket perfecting web-offset machine, printing two colours each side of the web.

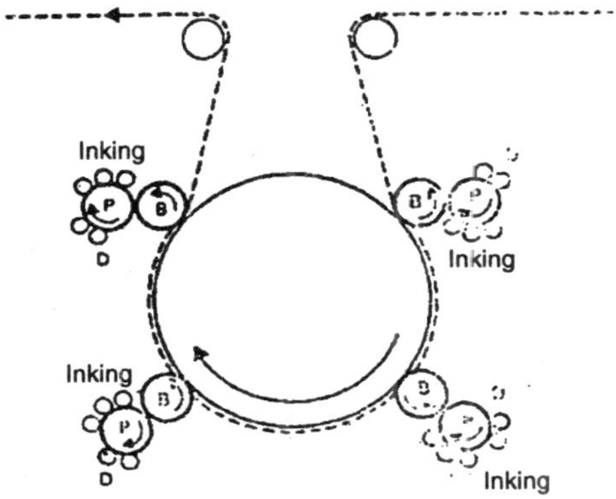

Figure : Four colour satelite (common impression) web-offset machine printing four colours on one side of the web.

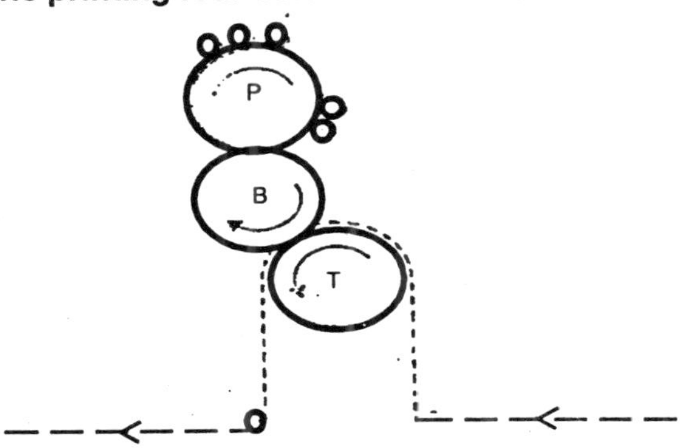

Figure : "Open" type web-offset machine

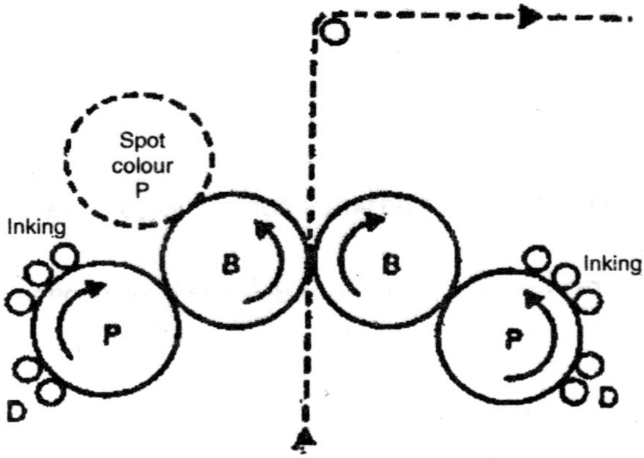

Figure : Arch type blanket to blanket perfecting web-offset machine with facility for spot colour.

related to an oil chart and different types of grease guns and oil guns so that the correct type of lubricant is used at the correct points. The main problem with long interval systems is to ensure that the 4 schedule is maintained.

Any hand-oiling required should always be carried out before starting up after an overnight or longer stoppage. It is the start-up period which is most critical for heavy- duty plain bearings, the residual oil having drained out during standing time, oil pipelines and unions develop leaks and should be kept under constant surveillance. Damping roller bearings require frequent attention; the plain variety should be wiped clean and re-oiled after the roller-covers are cleaned. If ball -bearings are fitted, these need to be removed or protected during damping-cover cleaning. Very small parts such as those on automatic feeder separators and feedboard transport, should be cleaned and oiled periodically, depending on the amount of paper dust and/or spray powder accumulated.

Similar considerations apply to the gripper systems and parts of delivery mechanisms. Care must be exercised in lubricating these, in case the stock receives oil streaks as it passed through the press. Air pumps may require lubrication, although many are of the "oil-less" type. Oil used must be of the correct type, and heat-resistant in most cases.

In addition to lubrication and necessary attention to the machine's lithographic elements, a number of items require periodic maintenance, mostly connected with cleanliness. Maintenance tables published in journals, text books, instruction manuals etc. lay down the frequency of cleaning and inspection. Of course, the frequency depends upon the volume and type of work printed, and particularly the sort of stock. Long and regular runs of very dusty stoc necessitate cleaning at very frequent intervals, not only of all machine's parts but of its surroundings too. This is a saftey as well as an efficiency precaution. Oil, dust, spray powder and French chalk accumulating on footboards and floors commonly cause accidents and spoilt work.

By acquiring an understanding of the principles of offset litho machines, the printer can apply his knowledge to any type of press. On taking over a particular machine, he can soon "find his way around" by studying the controls and adjustments. Also necessary is a grasp of the particular specifications of the machine, normally published by its manufacturer.

The essential specifications are :

Maximum sheet size.

Minimum sheet size.

Maximum stock caliper/weight.

Minimum stock caliper/weight.

Maximum printing area.

Gripper margin, maximum/minimum.

Plate size.

Plate thickness

Plate packing.

Distance from plate edge to first line of print.

Types of plate handled by clamping system.

Offset blanket size.

Underlay blanket size (if fitted).

Blanket pacing.

Floor space.

Height.

Weight.

Power requirement.

Maximum no. of units

The foregoing may be regarded as a press's "vital statistics", and should be available to estimating, planning, and platemaking departments, as well as to those responsible for ordering supplies (plates, blankets, spare parts etc). Other useful data for supply purposes includes the number of inking and damping rollers, their covered length and diameter, and their overall length.

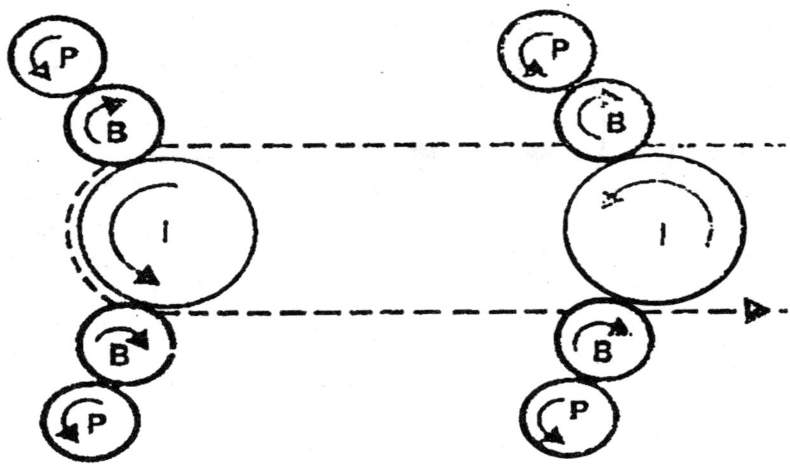

Figure : Double cylinder web-offset four colour machine

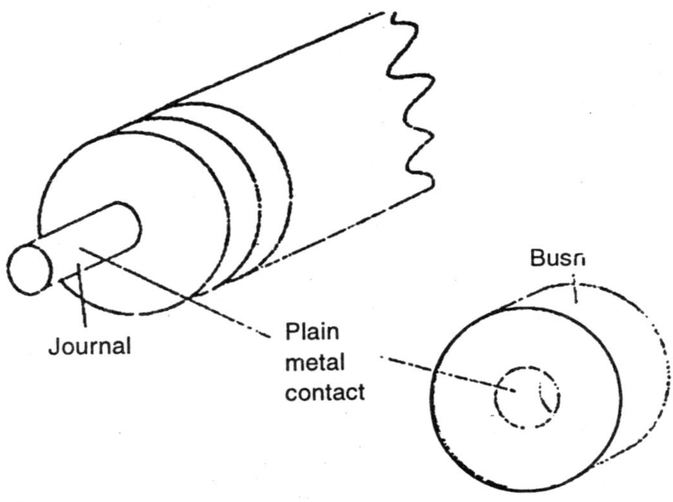

Figure : Features of a plain metel bearing

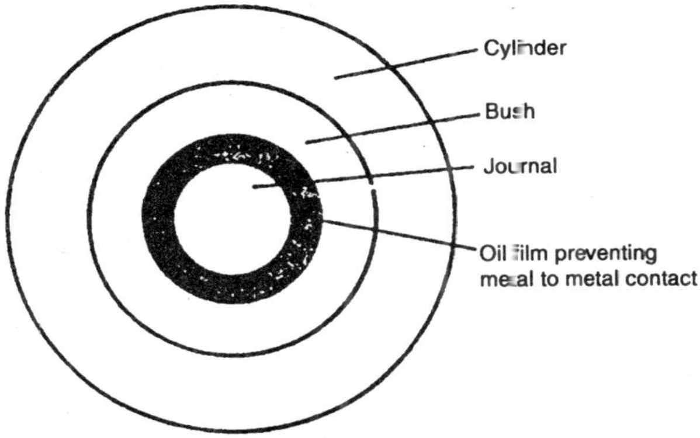

Figure : Lubrication of a plain metel bearing

Two-colour and multi-colour sheet-fed presses,cylinder arrangements.

OFFSET machine design and construction becomes rather more complicated for two-colour, multi-colour and perfecting presses. The principle of two "sections" still hold good viz .(i) printing units arranged to print on to (ii) an impression cylinder or cylinders which are part of the sheet-handling section carrying the material to be printed. Two colour and multi-colour machines may be classified thus:

1. Common impression; i.e. one impression cylinder for two or more printing units;
2. Unit construction; i.e. each colour unit or desk is a complete offset press in itself, with a means of coupling to permit sheet passage between units.

Most older two-colour machines were of classification 1, a method still widely used by machine construction engineers. It is not used to any great extent for sheet-fed multi-colour models (i.e. those with three or more printing units), but is frequently adopted for web offset presses on process colour work. A high percentage of two-colour and multi-colour machines are unit-type, but several designs use combinations of each class, i.e. two or more two-colour presses, each with common impression cylinders, are coupled by means of transfer system , usually of the chain gripper type.

Each class has advantages and limitations; some are easily defined, but others become evident only when handling specific classes of work. A major advantage for common impression is that the sheet is held by one set of grippers during passage through all printing units. This reduces the

Numbering unit & Perforating Attachment (Optional)
PRINTOMAX

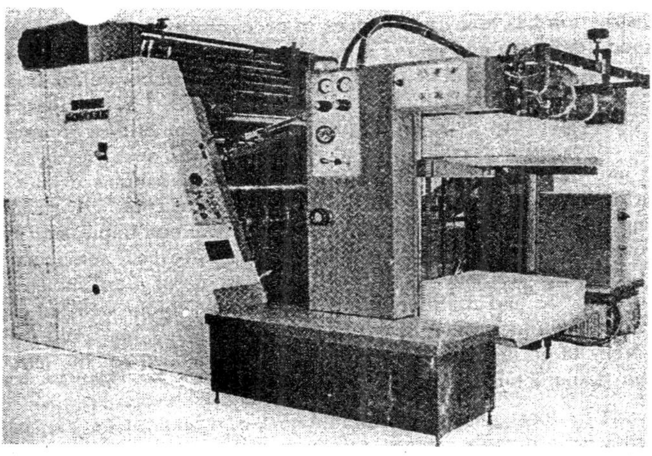

risk of mechanical variation affecting register, but does not totally eliminate the risk of slippage in the grippers. Variation in tension on the sheet as it is pulled from each blanket may affect the gripper "bite", normally this should be uniform once the press starts to run with correctly adjusted ink. The machine cycle duration between each colour on a common impression machine will be shorter than on the unit type. This may be advantageous where the stock is dimensionally unstable, but can aggravate superimposition problems in wet-on-wet printing on low absorbency paper.

Unit construction has many advantages in accessibility and ease of control. Using identical cylinder, damping and inking arrangments for each unit also eases the task of achieving ink/water balance, and helps plate installation. But complicated sheet transfer systems increase the risk of marking the wet ink on the sheets as they pass between the units. On the credit side, however, the surplus ink which sometimes accumulates on the side and back edges of impression cylinders (outside the sheet area) is not carried forward form one unit to the next, because the transfer elements do not contact the impression surfaces. Another attribute of the unit-type press is the greater latitude when making packing adjustments for altering print length. Common impression systems with the bearer contact method of cylinder setting are in this respect somewhat restricted.

Web offset presses are made in varied designs, some of them shown in bigs they include unit construction and common impression. A four-colour web-fed machine with common impression cylinder design is more easily constructed than a four -colour sheet-fed machine. Very large diameter cylinders would be needed for the latter, although such have been manufactured.

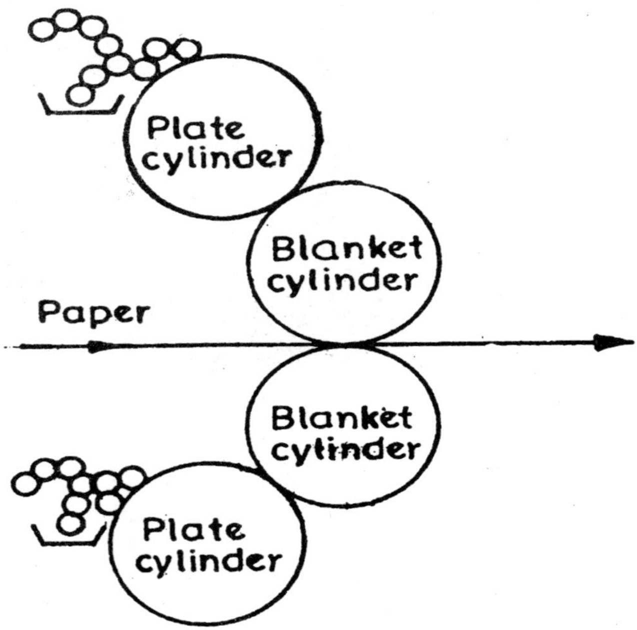

Perfector & Multicolour press

The section for handling the pile or stack of sheets consists of a board or framework supported on two metalbars, which are designed to book on to lugs fixed to chains acting as a hoist. The chains are moved by turning the sprockets over which they travel, either by hand or electric motor. The sprockets are also turned through a mechanical linkage (the automatic pile-raising device) fitted to the separator mechanism. Vertical bars, adjustable for spacing to the size of sheet, act as guides to facilitate loading of the sheets and to maintain the pile in correct position. A device for moving the whole pile sideways by small amounts is sometimes included. Operating methods for these devices differ; some are actuated by a hand-wheel which, working through a screw arrangement, moves the guide-bars and load simultaneously, while others require that the guides be moved separately prior to adjustment of the load position.

Loading systems for large, fast -running offset presses are frequently fitted with "automatic" or semi-automatic changing systems; these enable second and subsequent stacks of sheets to be brought into use without interrupting printing. They all require the preparation of loads either in the feeder loading area or outside the press, on a loading gauge system which

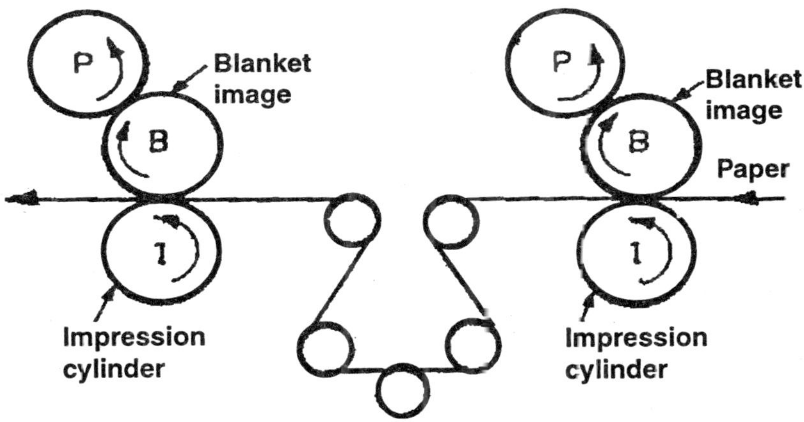

In-line multicolour printing system

is a duplicate of that within the feeder framework. Precise details of the procedure for change-over vary, but in most cases a secondary hoist system, or fitting a series of rods through grooves which are then attached to the secondary hoist.

This operation takes place when the load in current use has been run to the last 500-1000 sheets. The latter are then temporarily supported and raised by the secondary system, while the main board and bars are removed and the chains attached to the next load (this having been wheeled in or prepared as the previous load was raised). The main (primary) hoist is then removed, the secondary hoist being switched out of operation and the main hoist resuming its normal role.

Separator mechanisms of the friction type are used chiefly for small offset-duplicators. Large versions are still found in printing finishing , mainly for folding machines; these work on a principles different from the small offset types and need not be discussed here, small friction feeders copy the action of a hand feeder's thumb and forefinger when separating the sheets (or a cashier counting notes). The friction feeder's forwarding action is crude by comparison with the hand feeder's lifting and "floating", but is equally effective so long as long it maintains control over sheet. Shows one commonly -used type. Friction feeders working on this principle do not permit very accurate register and are unsuitable for numerous stocks. Moreover, some stocks can be marked or damaged by the friction bars action.

Pneumatic separation methods use a blast and suction principle. Compressed air blown through small nozzles or slits is directed towards the leading edge of stack's top sheets. These sheets are therefore "fluttered", and as this happens suction tubes or pads (commonly fitted with rubber washers or flanges and known as suckers) positioned above the top sheet at the point of disturbance pull the sheet upwards and keep it firmly in position. The suckers then lift and move forward, placing the sheet's leading edge between a pair of wheels and a roller carrying the transport tapes.

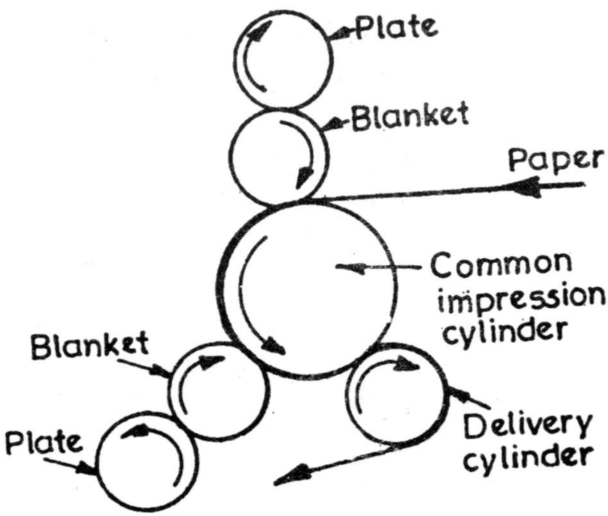

Common impression multicolour printing system

For a back separation movement the principle is similar, but for all types except the smallest a separate set of suckers is used for the forwarding motion. In addition, the airblast nozzles are usually adjustable for direction, the aim being (a) to disturb the sheets as for front separation, and (b) to provide a cushion of air below the lifted sheet and assist its forward passage (as well as ensuring that the front edge is lifted sufficiently to clear the top of the loading gauges). Back separation will recommence the operating cycle on the next sheet before the previous sheet has cleared the top of the stack, so producing the underlapping necessary for stream feeding.

For all air systems it is important for air in sufficient volume to be available and with enough pressure for maximum requirements. Very fine

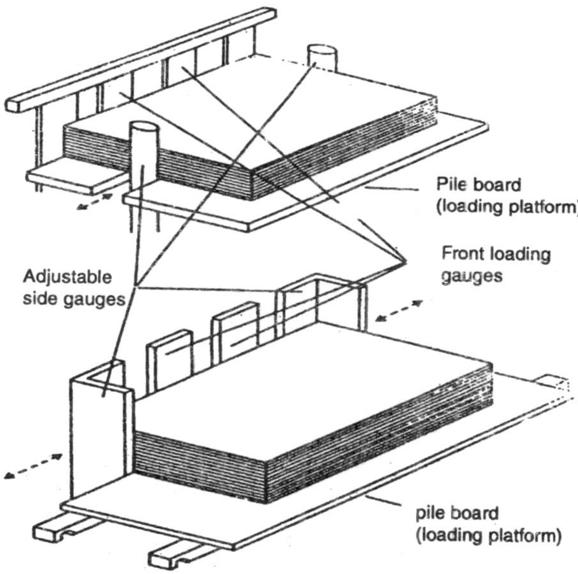

Figure : Pile loading gauges (Examples)

adjustment is necessary, usually via relief valves with screw adjustment. Graduated scales to the adjustments and/or dial guages are very useful, but not fitted to many small machines.

Most systems use small rotary compressors or pumps which need little attention apart sometimes from cleaning of filters and lubrication. The compression are driven by separate electric motor, although in a few machines the pump drive is taken from the main drive. On certain older feeders a piston/cylinder method gave suction on one stroke and blower pressure on the other, the reciprocating action timed to suit the feeder action.

Timing is necessary for the air supply or vacuum requirements from a rotary compressor, in order to apply or cut-off in the correct sequence. For example, there are two stages for suction and two for blast. This is accomplished by in-built valve arrangements which should not be interfered with by the printer (apart from routine maintenance). In more complex feeders the suction and blast system is fitted with individual adjustment, a necessary feature where a light blow may be required for separation but a stronger blast for forwarding. Similarly, suction for separation must be varied to suit the stock; e.g. an open, absorbent - surfaced stock tends to permit the suction to pull through more than one thickness, thus picking up two or more sheets together.

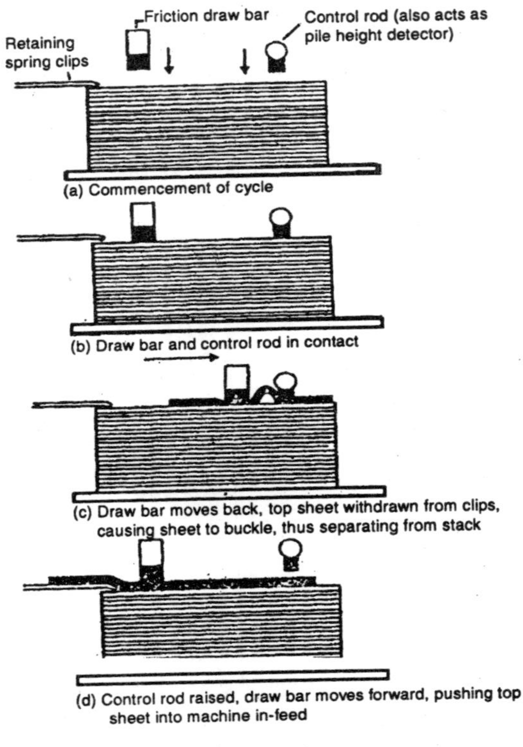

Figure : Friction feed action

Sheet control in the feeder stack is obtained by means of the loading gauges referred to earlier. This method does not permit sufficient stability at the top of the stack, in particular because the sheets are subject to air blast. So a variety of fittings are provided, including clips or claws, flip springs, stripping brushes, front flaps and back stops. Not all of these are required simultaneously; they must be adjusted, removed for interchanged to suit the stock. Small, simple feeders will have the minimum necessary for handling average types of stock. Most single-sheet feeders will have some form of front-edge retaining clip to act as a separating aid. These must be related to pile height and sucker angle for different stocks, and may need removing or changing to a lighter or heavier spring (e.g. it may not be possible to induce the curl in a heavy, rigid material).

Flip springs can be used with similar effect on back separation units. This may be independently adjustable or combined with the back stops, although this tends to restrict positioning. Both back stops and flip springs can be moved sideways to suit sheet size. Other aids include stripping brushes and blowers with nozzles angled downwards. Most back separation systems used for offset machine feeders operate at the centre of the sheet, and do not provide for blower separation along the full length of the stock's back edge. This is probably because it would create too much disturbance, and so the blowers-acting in two stages, as outlined in fig, but sometimes three-are used; hence the angled blower mentioned above.

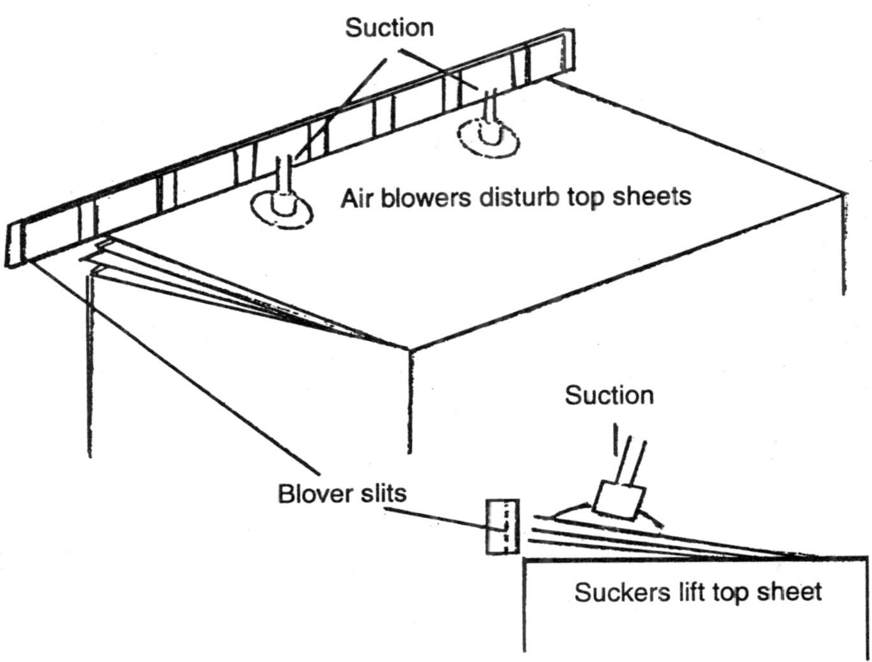

Figure : Sheet separation front separation unit

Not all units use the blower foot but provide the forwarding blast from separate directional nozzles which can be set to direct the air angled outwards towards the front edge corners, and also reduces the possibility of air being funnelled under the sheets along the centre of the feedboard, with consequent disturbance of register and risk of creasing. A foot is still used for holding the lower sheet, to prevent it being dragged forward by the upper sheet's movement. As with the blower foot, it moves down when the upper sheet is lifted clear.

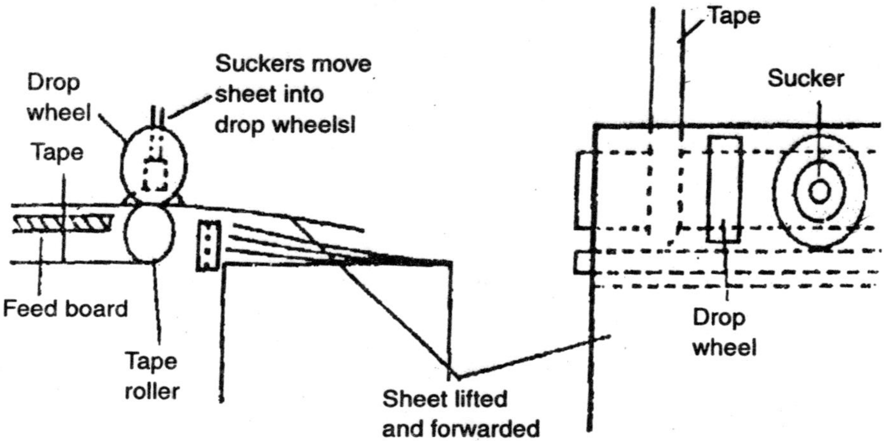

Figure : Initial forwarding, front sepration unit

Pile height is relative to the separator unit's action, and needs to be maintained during printing. The sensing or detection system operates through some feature which contacts the top of the pile once every cycle of operation. This feature will usually serve at least one other purpose; e.g. control foot or rod in a single-sheet feeder, or blower foot in the back separator mechanism of a stream-feeder. In both cases the item acts as a holding or steadying agent to permit another operation which might otherwise disturb the top sheets. Because it rises and falls it will be sensitive to the load's changing height, and can therefore be adapted through mechanical or electrical linkages to actuate the pile hoist mechanism. Unfortunately, the resistance of the stack to the pressure of the sensing device tends to vary with the type of stock and with the amount of material in the load. In other words, there will be "springiness" in the load varying with its thickness (top to bottom) and with the air in and between the sheets. Provision is accordingly made for adjustment, and this may be fitted to the control feature itself or be a separate control on the feeder side-frame. Initial adjustment is made at the appropriate stage of make-ready but constant checking during the run-down of a load is necessary.

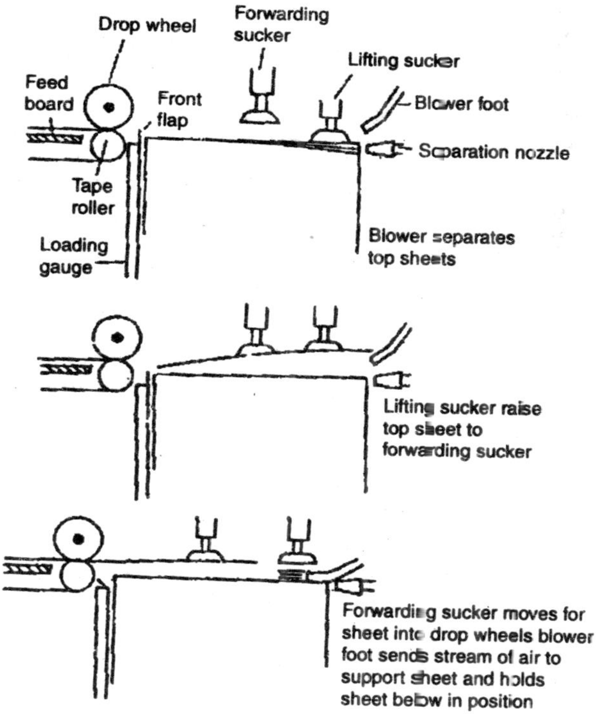

Figure : Sheet separation and initial forwarding, back separation unit

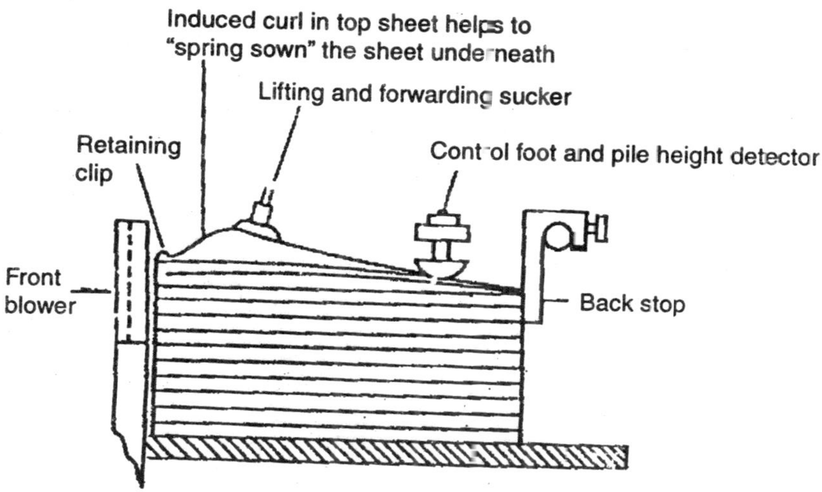

Figure : Use of front edge retaining clip in a single sheet feeder

Preparing a stack of sheets for an automatic feeder requires care and skill, when completed, the sides of the load must present smooth faces on all four sides with the top as level as possible. The sides should be perpendicular to the loading plateform, with no tendency for learning in any

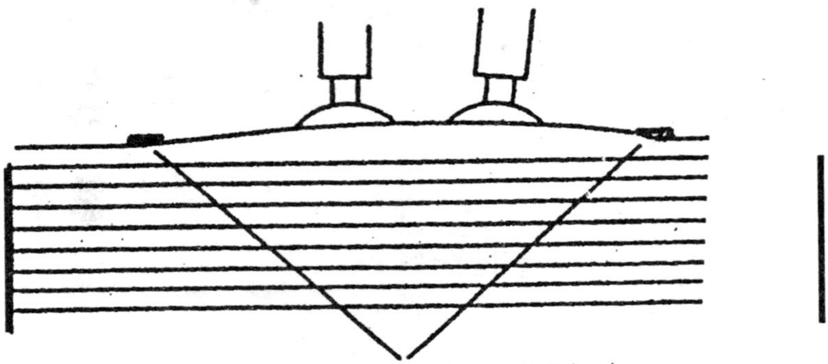

Figure: Use of flip springs on back separation units

direction. Each batch of sheets should be "winded" as it is lifted and placed in position; i.e. gripped between forefinger and thumb at diagonally opposed corners and slightly "fanned". This is necessary to break the tendency for sheets to cling together or "stic", and assists in the straightening operation ("Knocking up"). Sheets tend to cling for several reasons;

(1) pressure in the stack

(2) edges "burred" by cutting

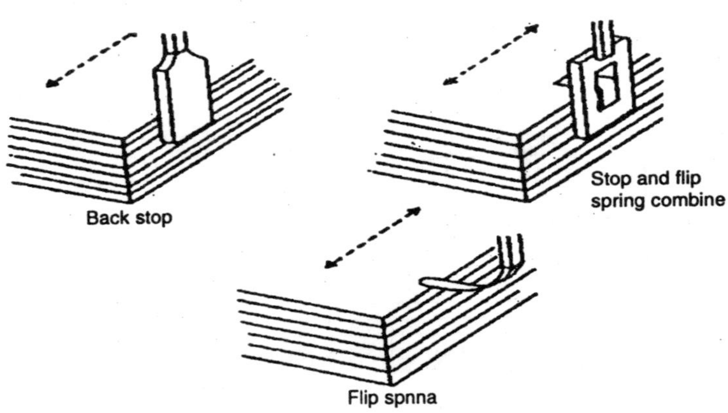

Figure : Types of back stops and flip springs

(3) static electricity

(4) set-off, if sheets have one or more colour already printed.

Concerning point (4), sticking may still be due to ink particles on the backs of sheets, resulting from contact with wet in during an earlier printing, even if no set-off is obvious. Great care is needed when handling large-size sheets, to prevent kinks or marks in the material's surface, especially with coated stock.

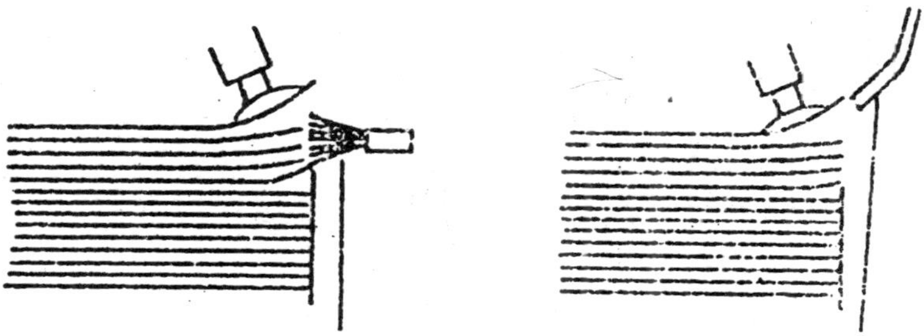

Figure : Use of stripping brushes and angled blowers

The operating difference between single-sheet feeders and stream-feeders will be obvious to those acquainted with both types. On the former, each sheet travels singly along the feedboard from separation to register points, whereas in stream-feeding the sheets travel overlapping (or underlapping, depending upon the viewpoint). Streaming is used almost exclusively for large sheet size machines where the advantages are more important. Operation of a sheet-feeder mechanism must for obvious reasons, be synchronised with the printing unit. But the speed of travel between separation and registration need not be the same as the machine cylinder surface speed if sheets can be put into a "pipeline", i.e. a stream. Sheets move from a stationary position on top of the load to a stationary position at the register lays (no matter what the rate of machine revolution, sheets always stop to be registered). Accordingly, in single-sheet feeding they must accelerate and decelerate within a distance determined by the circumference of the machine impression cylinder. In stream feeding, the distance is only approximately one-third of the circumference, assuming three is sheet underlap(the most usual system).

For comparison, a machine with a cylinder diameter of 30 cm. revolving at a rate of 6,000 impressions per hour would have a surface speed of

$30 \times 3.14 \times 6000/60 = 9420$ cm per minute.

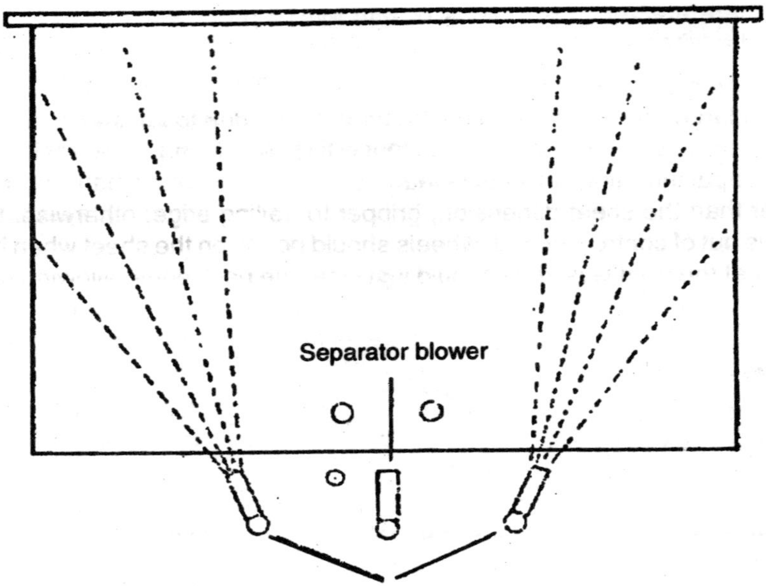

Figure : Forwarding blowers

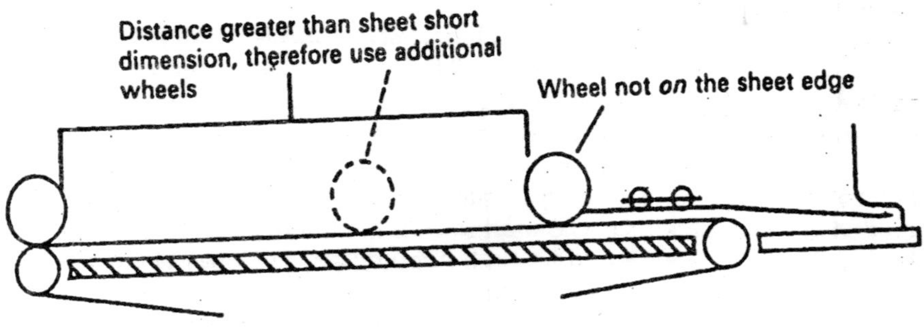

Figure : Controlling Sheets along the feed-board and when positioned at the front lays

There are versions of feeders with four or more wheels, and some where the wheels rise and fall (hence the term "drop").

A variety of small wheels, brushes and ball smoothers are supplied with the larger and more versatile feeders. The wheels may be adjustable for pressure, although for many purpose they need exert only their own weight. Spacing between wheels in the direction of travel should never be greater than the sheet dimension, gripper to trailing edge, otherwise, the sheet is out of control second, wheels should not be on the sheet when it is exactly at the front lays; they should just clear the back edge, allowing free sideways movement for the side-lay pull-up. Ball smoothers should be on the sheet at this point, to give it the necessary control without hindering sideways movement.

The conveyance method with a travelling gripper system appears to offer more positive sheet control during movement, but in its usual form must be coupled to a singlesheet feeder. The front separation suckers lift the sheet straight into the travelling grippers, which operate in slots cut in the feed-board. These grippers then take the sheet down the feed-board and pull it into the front lays, releasing it in time for side-lay registration. It is then taken into the printing unit by the machine's insertiom system, after which the feed-board grippers return to the separator to receive the next sheet. The grippers may be fitted with detection devices for operating the machine trip systems in the event of a mis-feed or the absence of a sheet.

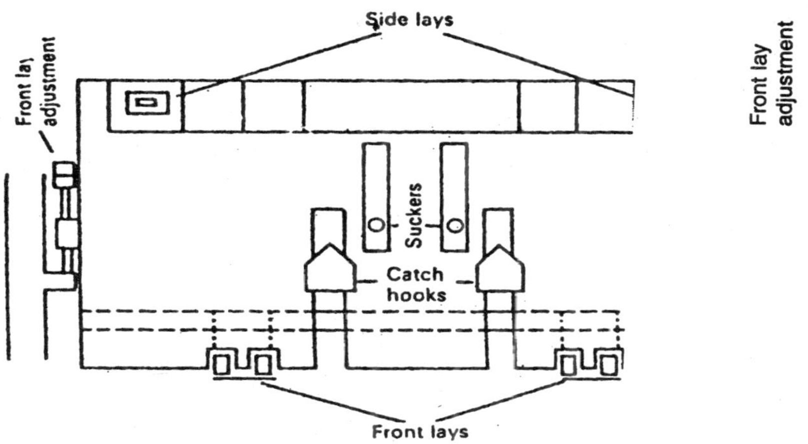

Figure : Front lay adjustment embodying whole register system and pre-alignment mechanism. Side lay is mentioned at 90° to front lays

The feed-board on certain types of large, high-speed offset presses are fitted with pre-alignment (pre-register) slow-down devices. These are fitted at the end of the tape run, usually in and combining with the feed-plate, which is the extension of the feed-board used for the side-lay and front-lay systems. The pre-alignment device normally consists of two or more "catch hooks" which operate in longitudinal slots cut in the feed plate. The hooks move forward in the raised position, and are so timed that the sheet gripper edge enters the catch hook on its forward travel. Two or more vacuum suckers below the feed plate, working in time with the hooks, guide the sheet edge into the catches. The hooks continue forward movement with a slightly retarding action, thus permitting sheet alignment during travel, and ensuring that very little if any movement of the sheet for lining up to the main front-lays is required.

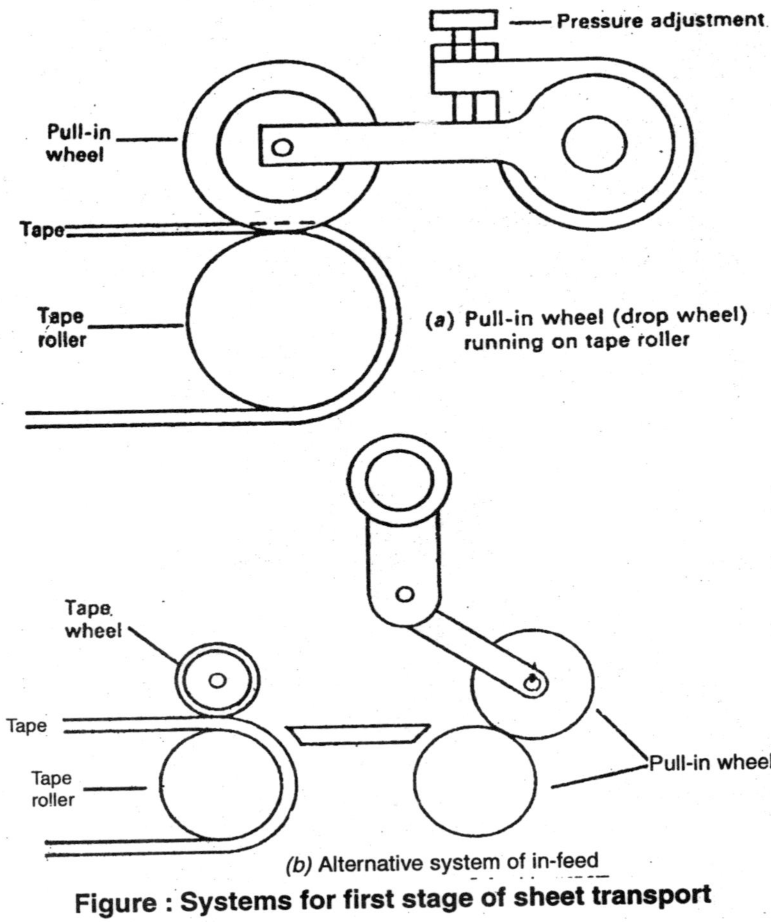

(a) Pull-in wheel (drop wheel) running on tape roller

(b) Alternative system of in-feed

Figure : Systems for first stage of sheet transport

Any required adjustment to the front-lay alignment automatically adjusts alignment of the catch hooks, because it is arranged for the adjusting screws to operate on the feed-plate assembly on which is fitted the whole registration system. After the catch hooks have deposited the sheet in the front lays they continue slightly forward and drop, returning in the lowered position until raised in time to meet the leading edge of the next sheet.

Sheet-fed offset presses, in common with most other types, use the three-point register system, i.e. two points for the long dimension of the sheet (front-lays) and one for the short dimension (side-lay). Front-lays, sometimes referred to as head-lays or front-stops, are found in varied shapes and sizes and with different methods of adjustment. All will have a common feature, viz.the part against which the sheet edge is placed, i.e. the register point. This can be referred to as the lay piece, which normally requires little maintenance but should be inspected periodically for cleanliness and possible surface cuts. The lay away from the register position, permitting free passage of the sheet once the feed grippers have closed and pull the sheet into the printing unit. Timing of this movement is critical, especially when grippers are moving forward as they close,which is the case with the "tumbler" - operated grippers fitted to many small machines with a direct feed on to the impression cylinder. The lays may swing from above the feed plate or from below. This will be a design feature of the machine, but will depend to some extent on the type of in-feed. The tumbler system already mentioned will require the lay to swing from above.

The lay mechanism may incorporate the adjusting screws used for altering the amount of gripper margin on the sheet. Alternatively, the adjustment can be fitted ends of the operating shaft outside the feed plate side frames, or to the whole feed plate assembly. In the later case, an adjustment to one side only, which will put the sheet gripper edge out of alignment with the machine axis, will maintain the side lay at 90 degree to the front lays. Adjustments to individual lays can make each lay piece out of square with the sheet-edge.

The lay smoother will be fitted to the overhead type of lay or will be integral with its design. Adjustment to suit stock thickness will be either individual or through the operating shaft, in which case a single adjusting screw can set all smoothers together. Lay which pivot below the feed plate are usually provided with separate smoothers, which can be either individually set or adjusted from one point.

Front lay systems can consist of two or more sets of lays. Except for the smallest presses, alternative positions for the two register points must be provided in order to facilitate registration of the different-sized sheets for

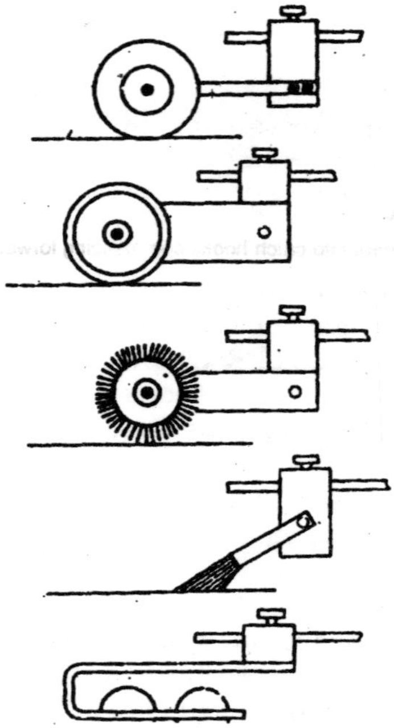

(a) Metal wheel

(b) Rubber tyred wheel

(c) Brush wheel

(d) brush

(e) Bath smoother

(f) Heavy bath smoother

Figure : Types of feed board sheet control devices

which the press is specified. Where only two lays are fitted there will be a means for moving them to alternative position, usually determined by slots cut in the feed plate's leading edge. Another method employs four, six or more lays, each with adjustment enabling those not required to be moved clear of the sheet. A third possible arrangement uses a similar number of lays, but only two provided with lay pieces which are detachable and can be refitted to whichever pair of lays is required. Almost all types are fitted to the feed plate, and positioning can take place only in a lateral direction, i.e. the necessary slots, bars or channels run from side to side of the feed plate. Many roller side-lays operate in a slot covered by sectionalised plates over all parts left open after moving the side-lay.

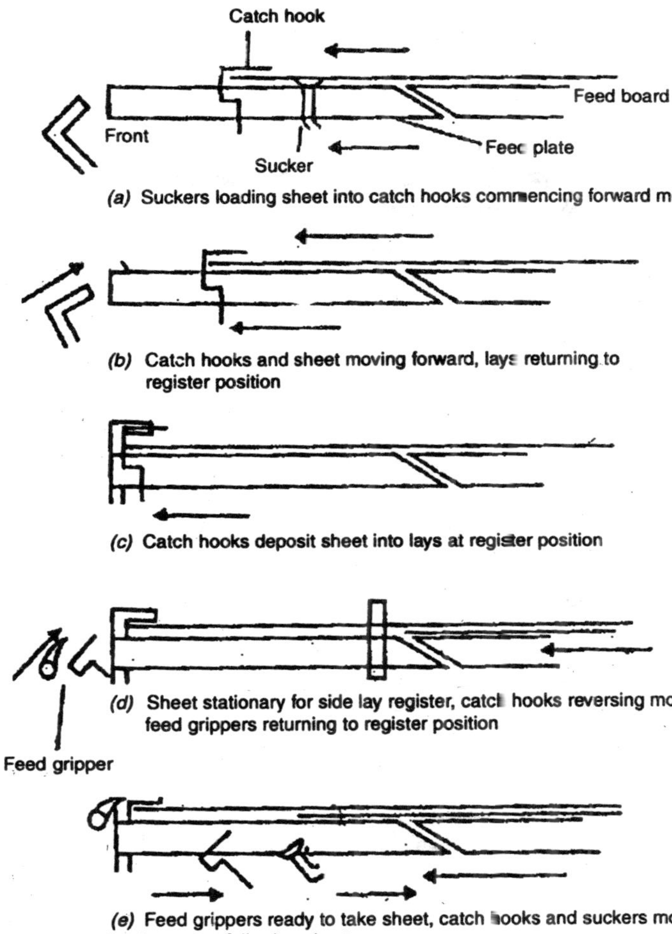

(a) Suckers loading sheet into catch hooks commencing forward mention

(b) Catch hooks and sheet moving forward, lays returning to register position

(c) Catch hooks deposit sheet into lays at register position

(d) Sheet stationary for side lay register, catch hooks reversing motion, feed grippers returning to register position

(e) Feed grippers ready to take sheet, catch hooks and suckers moving back to meet following sheet

Figure : Operation of pre-alignment mechanism

Positioning of the side-lay can be regarded as the coarse setting. A micrometer adjustment is provided for the fine setting, either on the lay itself or by remote control from the feeder or machine side-frame. Other adjustment on the side-lay are for the smoother setting and for the spring tension of the drop roller, which must be sufficient to draw the sheet-edge up to the lay piece without buckling and marking. Another small lever or adjusting pin will be fitted to enable the drop roller action to be stopped.

Most presses will have two side-lays (one termed the off-lay), but it will be obvious that only one is required to operate to provide three point register.

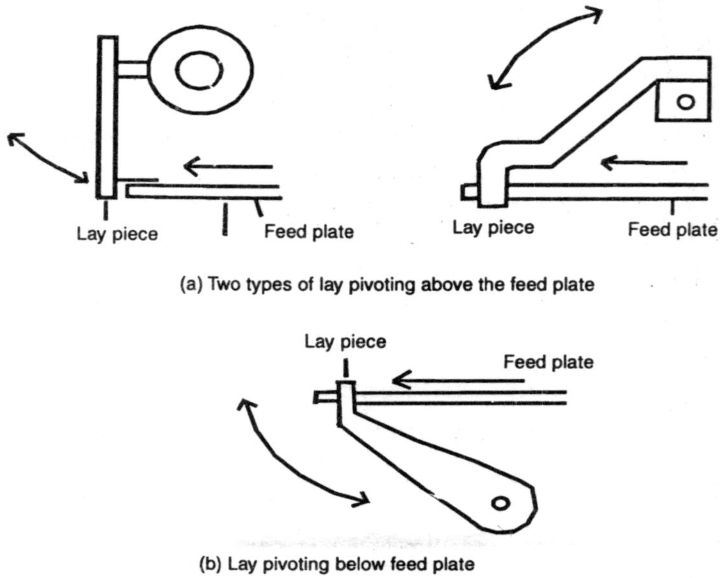

Figure : Type of front lay movement

Stroker side-lays have features similar to the roller side-lays and, in general, require similar treatment. If a separate adjustment is needed to position the stroker relative to the lay, this should be carried out as accurately as possible (following instruction manual directions). Faulty setting can result in mis-register and, in some cases, damage to the mechanism. Both roller and stroker movements require regular maintenance, particularly to prevent accumulation of paper dust and dried in. Lubrication is required where indicated by manufacturer, but must be very carefully carried out otherwise excess oil finds its way onto the stock. The point applies to all sheet-handling mechanism.

Push-type side-lays are found on small offset presses, usually those of the simplest type, where sheets are fed with the short dimension to the gripper. But they can be succesfully used on medium-size sheets if provision is made for lack of rigidity in the long dimension of thin stock. In addition, an opposing compensation system is needed in order to prevent thicker sheets bouncing away from the lay (this applies to all sizes). Sheets are more rigid

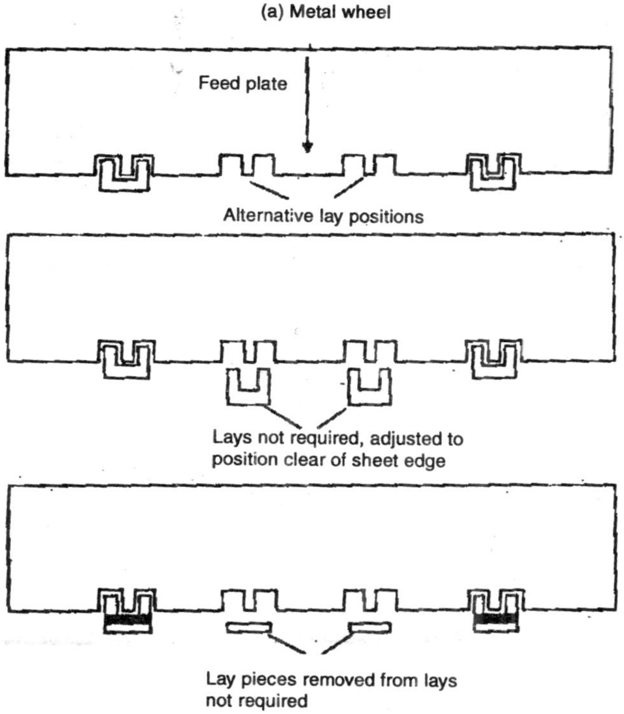

Figure: Alternative front lay arrangements

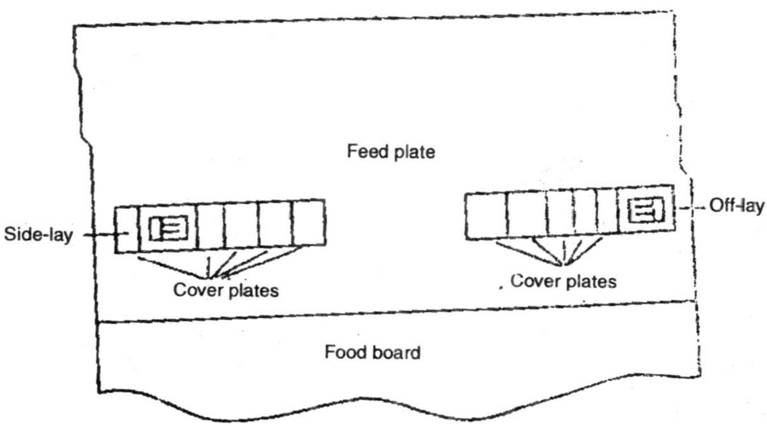

Figure: Cover plates fill gaps when side lays are moved

in their length if curved slightly in the short dimension, hence the reason for the use of convex curved feed-board on some machines.

The compensation system can consist of a light spring fitted to the opposing side-lay with provision made for changing the spring from one side to the other when changing from side-lay to off-lay and vice versa.

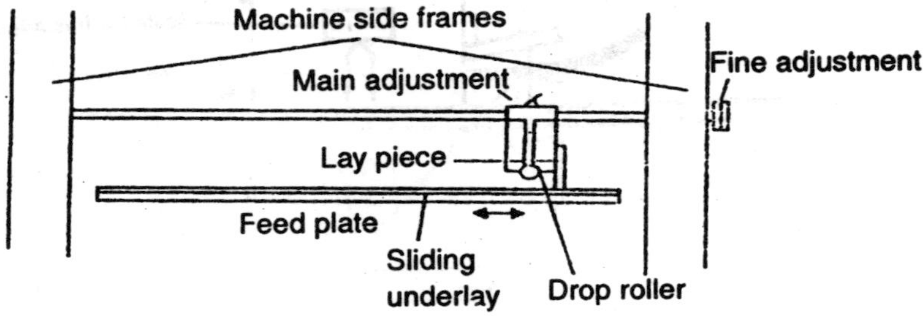

Figure : End elevation, view from feeder-settings for pull type side-lay with fine adjustment outside machine frame.

The machine trip systems are operated by sheet detection or sensor devices fitted on the feed-board or the feed plate and/or related apparatus. Each device will have two main features; the actual detector and the linkage system transmitting the signal to the machine trips. Detectors can be mechanical, electro-mechanical, photo-electric or pneumatic. Older machines and some small offset presses will use mechanical devices throughout, but the larger, high-speed machines have various combinations, depending on the position of the detection points. For example, photo-electric sensor may be used for the feed-lay position, but electro-machanical devices may be found at other points.

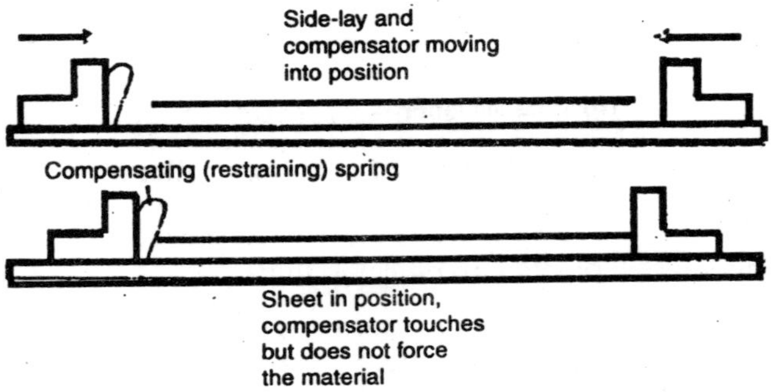

Figure : Setting of push side lay

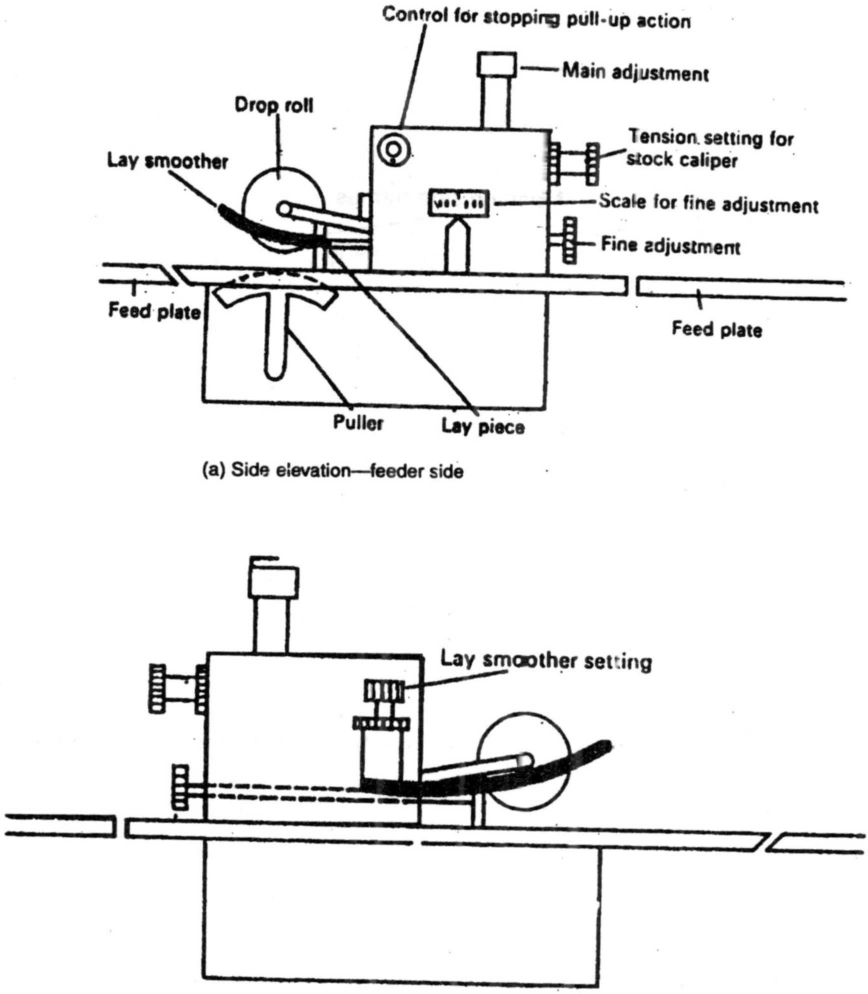

(a) Side elevation—feeder side

(b) Side elevation—gripper side

Figure : Setting of pull-type roller side lay

The various points can be listed thus:

(1) Late sheet detection, i.e. a sheet does not arrive at one or both front-lays at the correct time (also detecting, of course, the absence of a sheet); a few very sophisticated machines have similar sensors fitted to the side-lay.

(2) Early or fast sheet detection, i.e. a sheet which arrives at the front-lay position before the lays have returned to this point.

(3) Two-sheet detection, usually fitted at some position on the feed-board which will enable the trips to operate before two or more sheets are fed simultaneously into the register system; a rejection system is fitted to some machines which diverts the additional sheets into a separate receptacle.

(4) Twisted sheet detection of sheets fed out of square from the feeder pile; these are used for erratic sheet movement, sheets only slightly out of line being detected by the late sheet devices.

MABEG FEEDER

Setting Instructions

This type of feeders is designed for continuous loading when on long runs, it being possible to load a pile outside the machine and truck into a pre-determined position either from the back or the side of the feeder.

Pile Loading

Place a folded sheet in the feeder and position as in fig.2 having checked centre. Position the side guide (5) against the side of the sheet, on the side to be side layed. The same guide is interchangeable for each side of the pile.

If the feeder is to be loaded by hand, load as per. If a preluded board is to be put into the machine, it should be positioned against side lay guide by adjustment to.

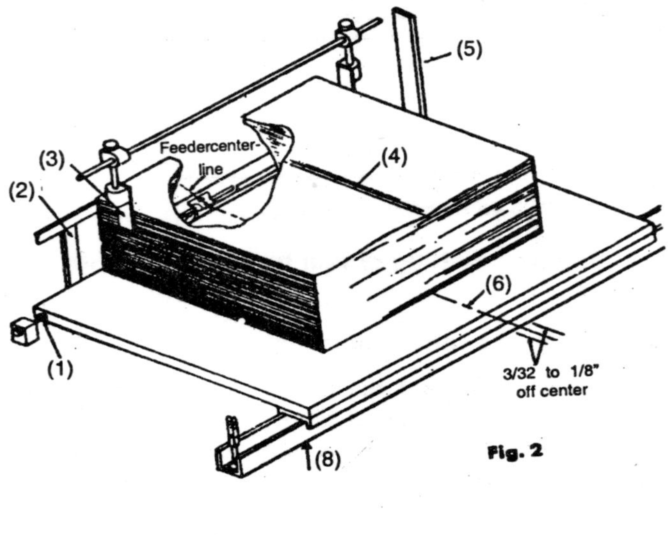

Fig. 2

Pile Raising

The raising and lowering of the pile is done by electric motor using the operating buttons shown in the Fig.4 and Fig.5. Fig.4 the position for lowering the pile -Fig.5 for raising the pile.

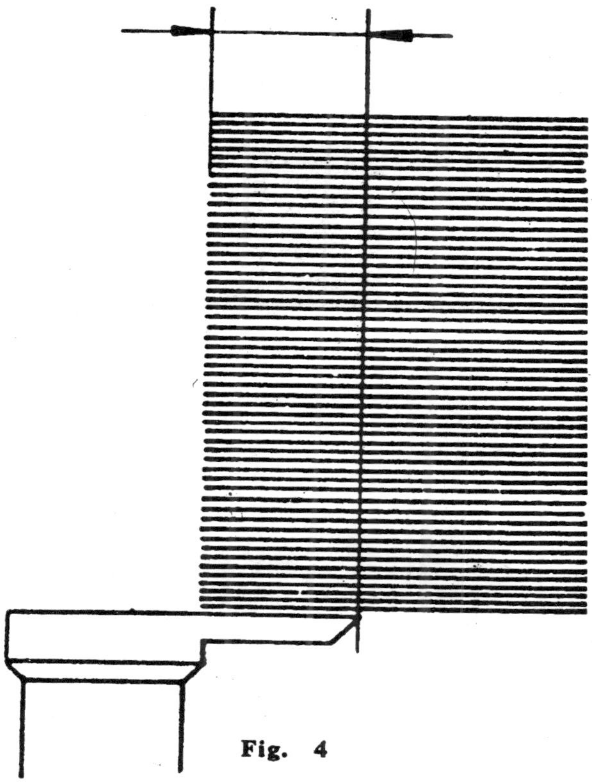

Fig. 4

Sheet Separator Unit

Move the sheet separator unit by the handle provided, until the rear pile guides (9)Fig. 6 just touch the back of the pile. Turn the feeder by hand until the pressure foot is (5) Fig. 6 in its lowest position. The foot should not extend more than 9/16 in, over the rear of the pile. Readjust pile guide (9) Fig. 6 to achieve this position.

When the pile raising motor automatically cuts, out the top of the stack is just below correct operating height to bring to correct height set controls according to (II) Fig.19, run machine with feeder moving without air, then adjust back pressure foot, when in its lowest position, to get the top of the stacks as per front edge as per (3) Fig.6.

With the stack at the correct operating height, position forwarding suckers (1) Fig. 6 agreeably to sheet, and adjust knurled screw (2) Fig. 6 so that the suckers is 1/16 in clear of the top of the pile when in its lowest position.

Set guide rod (4) as per diagram when lifting sucker (11) is in contact with the top of the pile.

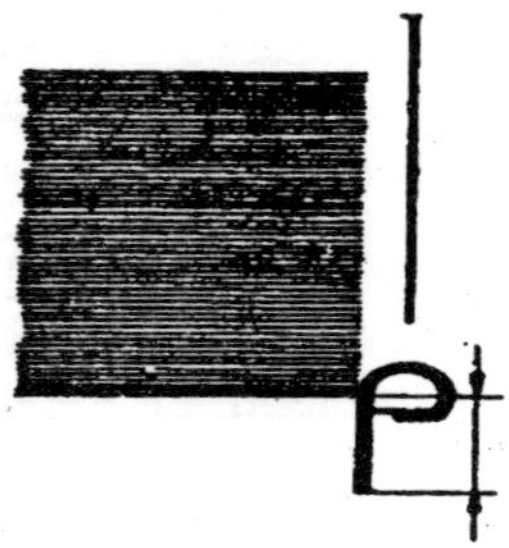

Fig. 5

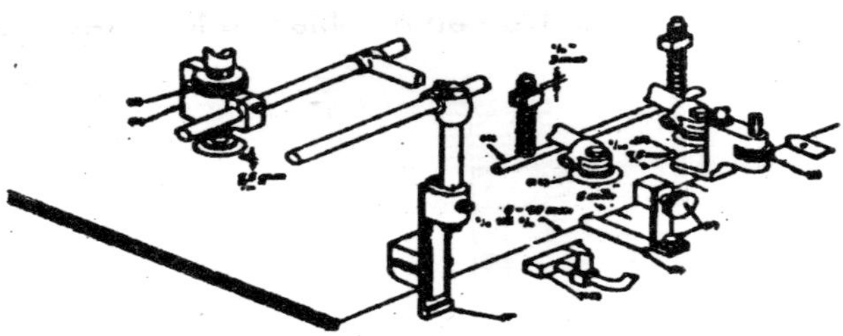

Fig. 6

Pick-up Suckers

Pick-up suckers Fig.7 should be set to a clearance of approximately 3/16 in above the top sheet when the pile is in its correct operating height. This can be increased to 1/4 in when on card.

This position can normally be achieved by adjustment to pressure foot. If this is not possible, adjustment can be made by adjusting (a) Fig.8.

Different sucker feet and a variety of rubber suckers are provided to be used with different paper stocks as in Fig 14.

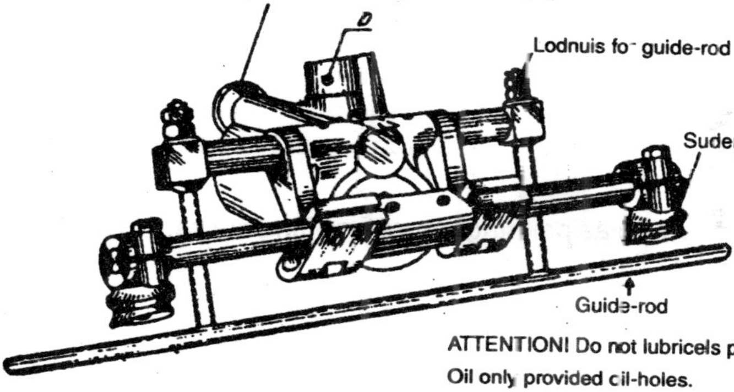

Figure : 7 PICK-UP SUCKER

To ensure trouble free operation, the pick-up suckers should always be set parallel to the top of the feeder pile. The amount of suction air is adjusted by turning knob on side frame, and should be set with the minimum air to achieve good sheet pick-up.

It should be born in mind that wooded wedges can also be used to obtain a level top of stack. It shows the settings of the guide rod, which should rest on the pile when the lifting suckers are in their lowest position. It shows an extra adjustment that can be obtained to increase the tilting mechanism.

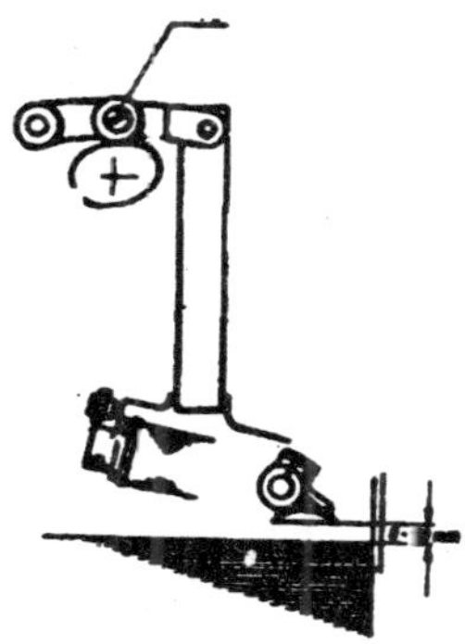

Fig. 8

Rear Air-blast Nozzles— Fig. 15 illustrates the position of the air blast nozzle in relation to the top of the feeder pile, set approximately 3/8 in, behind pile.

Two types of nozzle are provided, one for paper, one for card.

Fig. 9

Fig. 10

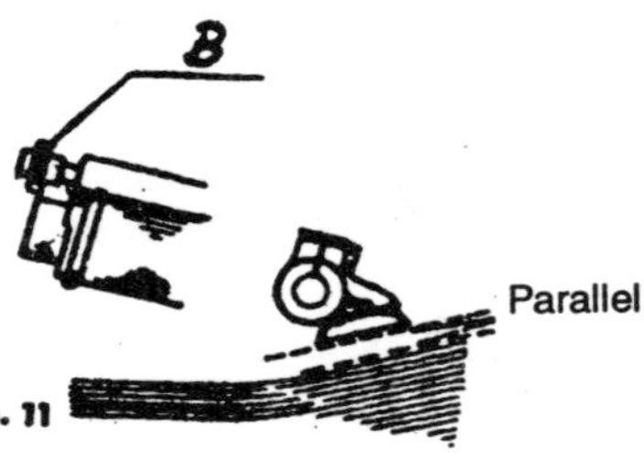

Fig. 11

Fig. 12

Fig. 13

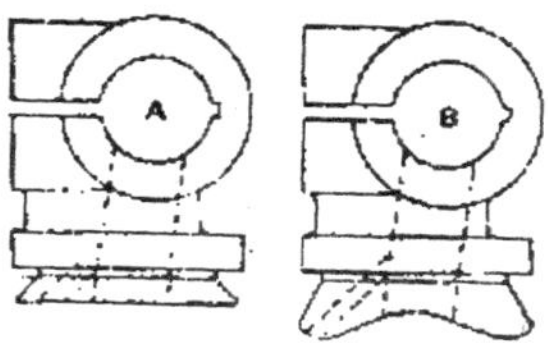

Fig. 14
Illustrates the two different sucker feet. (A) paper. (B) card.

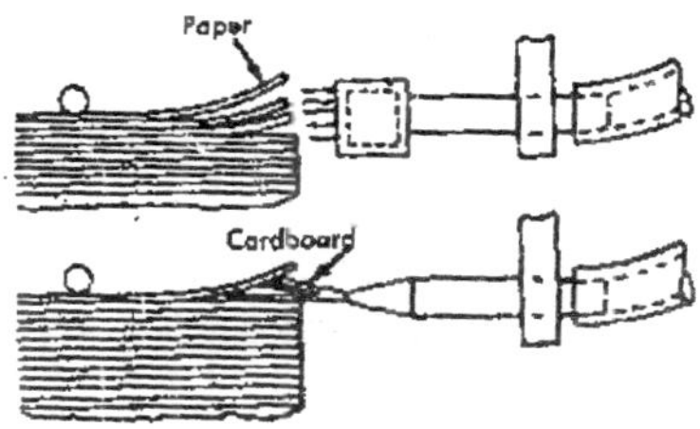

Fig. 15

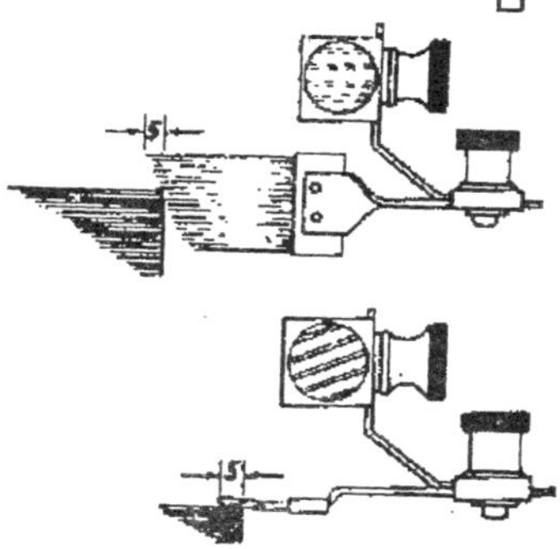

Fig. 16

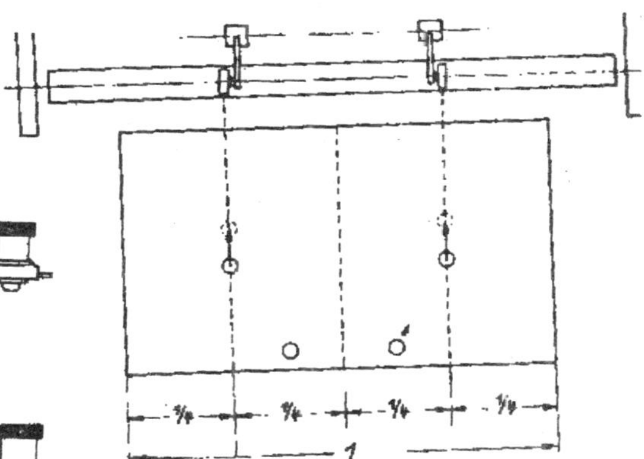

Fig. 17

Rear Flip Brushes— The two sets of brushes should be set as illustrated in Fig. 16 When working on heavy stocks spring flaps are provided, to be used instead of flat brushes.

Setting Transport-Rollers— The setting of the transport rollers so that they are to in line with the forwarding suckers, Fig.17 is the best position to ensure accurate for warding of sheet. These rollers should never be set to run on feeder tapes.

The Adjustment of the Transport Rollers

(a) The cam performing the swing is brought in the position shown below so that on the minimum point the follower is just running free, (A).

(b) The set-screw (B) now absorbs the spring pressure (C).

(c) Turn the set screw marked (D) counterclockwise, until the roller (E) touches the tape roller shaft, marked (F).

(d) The set-screw (D) is adjusted until there is a 0.012 in, clearance to (H) Lock up at (G).

(e) The spring pressure for the transport roller is regulated by turning screw marked (J).

(f) Level setting of both transport rollers is essential.

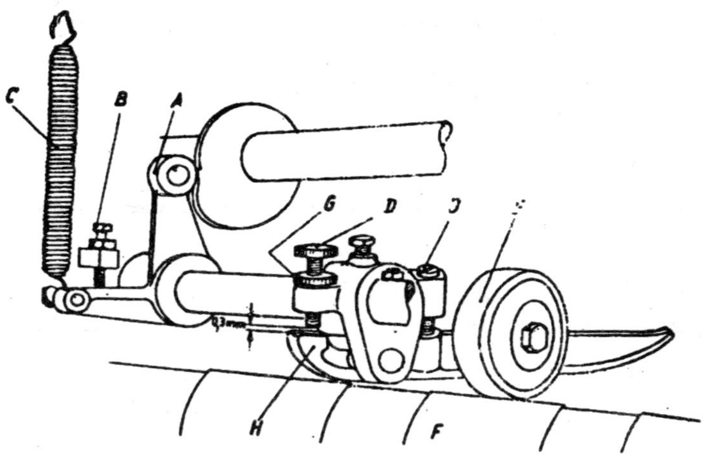

Figure : 18

Adjusting the Hold-Down Rollers

The rollers are set to hold the sheet in contact with the tapes during its travel on the feed board. To increase or decrease the tension the sheet adjustments are made with knurled screw(1), movement along the tape is made with after releasing knurled screw (2). Each sheet should be under the control of two rollers, each side of the board, during its travel down the feed board.

Setting Plastic Transport Rollers

The plastic rollers are set to keep the sheets towards the front-guides. These are positioned just at the back edge of the sheet when the sheet, is in the front lays. Adjustment is the same as above.

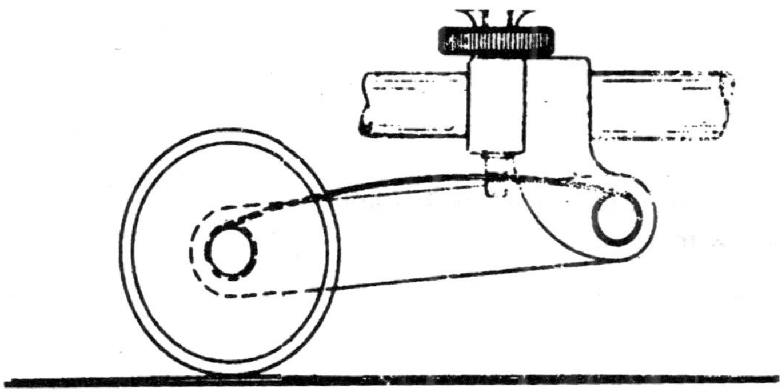

Figure : Adjusting the Hold-Down Rollers

Adjustment of Rotary Brushes

The rotary brushes are set the same way as the rollers. These brushers are set to hold the tail and of sheet down after it has arrived at the front - guides (A). If, on thin papers, there is a tendency for the sheet to buckle, the brushers should be moved back about 2 inches to (B). Pressure should be adjusted so that side laying is not impeded.

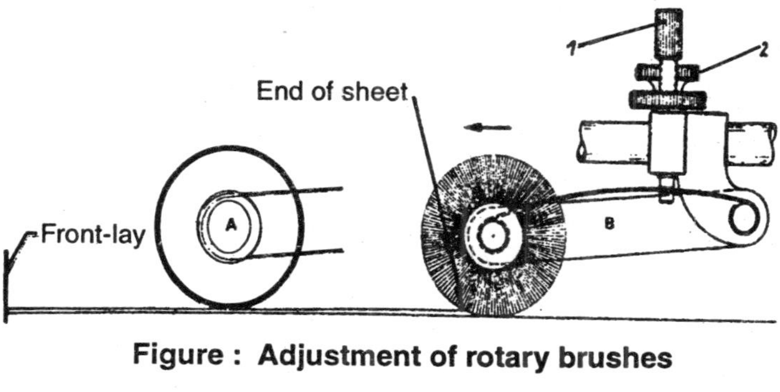

Figure : Adjustment of rotary brushes

Adjusting the Sheet Hold-down Balls

The sheet hold-down balls hold the sheet to the front lays during the side registering of the sheet. Setting and adjustment is made by setting screws 1 and 2.

Setting Flat-Brushes

The flat brushes are set to hold the sheet after it has arrived at the front-lays. This prevents the sheet from bounching back.

Adjusting the Double-Sheet Control

(a) Feed three sheets under the caliper roller (1)(small sheets only feed two sheets.)

(b) Turn feeder by hand until line on cam (2) is in line with the centre of the roller (3) of the double sheet control.

(c) Loosen lock nut (4) and turn screw (5) until caliper roller (1) lightly contacts the sheet.

(d) Tighten nut (4) and re-check caliper setting.

(e) Loosen nut (6) and adjust turn screw (7) until the conductor gap is about equal to the thickness of the sheet.

(f) Tighten nut (6) and re-check gap.

(g) Timing the adjustment is as follows : as son as the sheet is controlled by the transport rollers, the ball bearing (3) should be at the lowest point of the cam (2) this is marked with a line and (0).

The whole of the double sheet control mechanism can be adjusted sideways to avoid running over wet work.

Recheck of Double-Sheet Control

(a) Lift caliper roller (1) and place one extra sheet under it.

(b) The contacts should now be firmly closed.

(c) Remove the extra sheet and see that there is a gap in between contacts.

(d) Another test is to run the machine and feeder and hold a strip of paper so that it engages under double sheet roller, this should cut out feeder.

Front and Side Lays

The front lays are adjustable towards the swing-arm gripper by loosening the thumb nut (1) with the knurled screw (2). After making the adjustment, re-tighten the thumb nut. By turning the hexagonal and nut (3) inside the frame on the operating side, the front lays can be adjusted together height-wise, this is necessary when changing from paper to carton or vice versa. When printing carton, height is adjusted so that the sheet may be moved under the lays without meeting resistance. When printing paper height is adjusted to two or three times the thickness of the stock. This also applies to height adjustment of the covering lay. (4) on the side lay, which is affected by means of screw (5). The spring pressure of the pull roll (6) is set according to the stock by means of screws (7) is screwed out completely and the next stronger spring is inserted. When handling heavy carton the pull element (8) may be replaced by a roughened partby loosening nut (9). The roll (10) rests on an eccentric halt and by turning this bolt the distance between roll (1) and cam (11) is altered. This distance should be 0.012 in or 0.03 mm when the roll (6) contacts the segment (8). The pulling distance should amount to 5 in to 3 in or 8 to 10 mm. When changing the sheet size the side lay can be adjusted laterally by lessening the thumb screw (12). Fine adjustment is effected by the knurled nut (13) and the amount of adjustment can be read on the scale (14). The side lay may be inactivated by pulling out the pin (15) and turning it by 180 degrees until it slots in again.

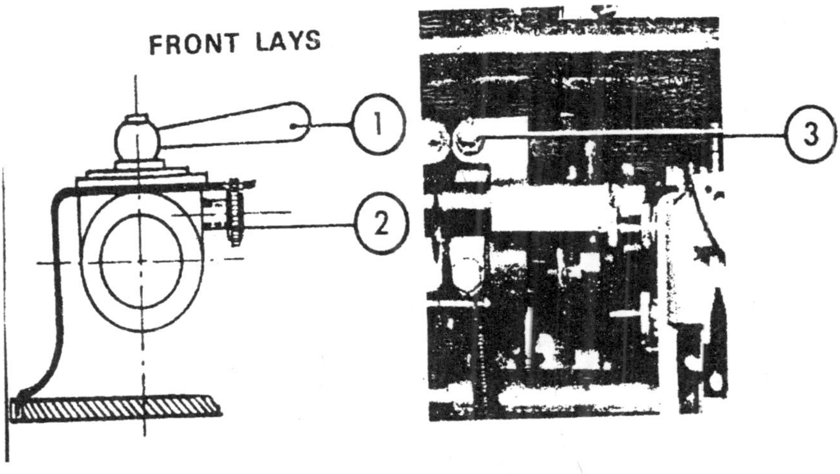

Figure : Front Lays

The detector in the centre of the lay position is designed to prevent sheets which arrive too early from the feed table to enter the machine. if a sheet travels under the front lays, the detector (19) cannot dip into the corresponding space in the plate and is lifted by the sheet, the contacts (2) touch each other and will effect throw-off.

The detector increases the operational safety of the machine, and the minder is advised to check correct functioning periodically by placing a sheet under the detector (19). The contacts (2) must touch in this condition normal dip of the detector into the plate is about 2.5 mm.

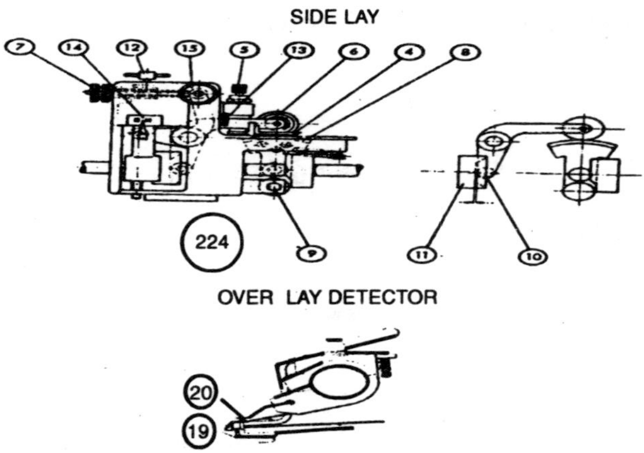

Figure : Side lay

Swing-Arm Grippers

The swing-arm grippers take the sheet laid by the lays and transfer it to the impression cylinder. After prolonged machine stoppages, e.g. when resuming work in the morning, etc. it is recommended to move the swing arm gripper springs towards the feed board to relieve the swing-arm gripper springs. The gripper pads (1) are adjusted according to thickness of the stock and the distance (a) should be two to three times the thickness of the paper. When printing heavier carton the height of the prepare pad is set so that the sheets can not 'jump' from the front lays.

Before making adjustments, loosen the yellow-head screws (2) eight, left and centre. After loosening the nut (3) adjustment is made by means of kunrled nut (4) and the scaled rings of the kunrled screw show the amount of the adjustment. The figures indicate the distance (a) in steps of 0.004 in or 1/10 mm. After adjustment the nuts (3) and screws (2) are re-tightened.

When making new adjustments or when checking the swing arm grippers, the gripper pads are set to the distance of 0.012 in or 0.3 mm (scale value 3) and a shim of 0.004 in or 1/10mm thickness is placed under the stops (5) on right and left. The adjustment screws (6) on the gripper (7) from the shaft, lightly press the grippers against the pads and re-tighten the clamping pieces with their screws on the shaft. When shims of 0.004 in or 1/10 mm thickness are removed from the stop, the grippers will lie on the supports exerting a certain pressure.

Pressure of the individual grippers can be checked for simultaneous opening by using a strip of tissue paper.

Additional adjustment may be made by means of adjusting screws (6).

Figure : "Swing-Arm Grippers"

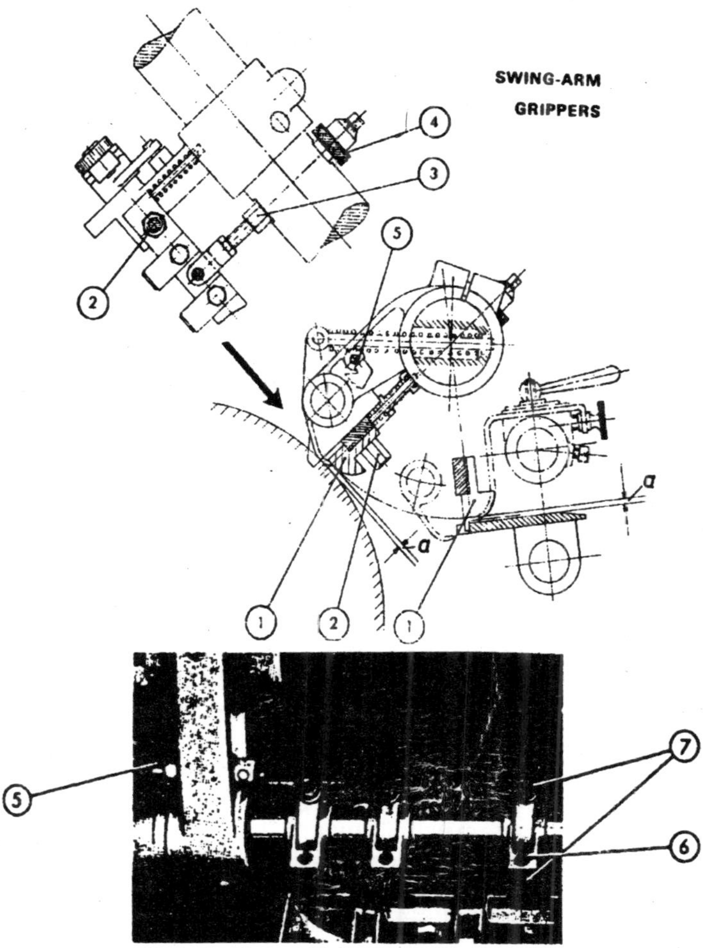

Swing- arm grippers seen in arrow direction

Sheet Detection.

The sheet detector mechanism works mechanically, that is two detector needles run over the front edge of the sheet after it has been adjusted by the front lays. If the whole or part of the sheet lies back, the pins fall into the holds provided in the feed board and the detector rod puts the check mechanism into operation; impression is thrown off, sheet feeding is interrupted, the machine is switched to slow speed, the front lays remain down and the swing-arm grippers stay open, the supply of ink and water is interrupted. The same happens when the double sheet detector in the feeder operates.

An air blast serves to lift the top sheets from the paper pile. A suction food (sucker) then picks up the uppermost sheet by means of vacuum and transfers it to a set of pull out rolls. The pull-out rolls then roll the sheet onto the register table.

On the register table the paper first passes through a multiple-sheet detector. If more than one sheet is detected, the defector plate will open and the paper directed into the loosed of the press. A properly set multiple sheet detector serves to ensure that only one sheet of paper will be printed at a time. It also prevents gain-ups from occurring on press. Single sheets that pass under the detector are conveyed down the register table to a front stop by means of conveyor tapes. Here, the paper is jogged into a preset position before being into printing system.

SHEET TRANSPORT– There are two basic mechanical methods of transferring the sheets from one part of the press to the next. The first is by gripper trolleys in which the sheet is held in one set of grippers between transfers, and the second is cylinder to cylinder transfer in which the sheet is passed from one cylinder to the next.

The chain and grippers method has the merit of lower cost, simplicity and fewer transfers, however, the sheet support is not always positive and flutter could take place which would be a particular problem with stock. The cylinder to cylinder method is more expensive from an engineering point of view, but it does ensure that the sheets are positively supported at all stages. However this system usually means more gripper change over and where the wet face of the work is presented to a cylinder it will be necessary to ensure that either the ink is properly set or that contact with the cylinder is not so firm as to cause ink transfer with subsequently marking up and set-off problems.

For thicker and more rigid stock such as board, it is important that impression and transfer cylinders are not so small in circumference that sheet are distorted as they press around them. For this reason several manufacturers provides "jumbo" size cylinders on their presses, which make them particularly suitable for carton printing and similar work.

THE DELIVERY SYSTEM - Sheet are finally deposited on to a delivery board, which is a mirror image of the feeder board. As the feeder board rises, so the delivery board lowers. After image transfer, the sheet is transported from the printing system to a tray or table. Either of chain delivery techniques may be used for this purpose.

The delivery system is known as the sheet exists from between the blanket and impression cylinders, it is simply ejected into a delivery tray. The texture rolls serve to propel the sheet into the tray.

In the chain delivery technique, gripper bars are used to transport the printed sheet from the impression cylinder to the delivery table (receding stacker). These gripper film of ink. If the ink tacks were not strictly measured and controlled on multicolour presses than ink transfers to the blanket from the previous printed film causing in back trapping and a muddying of colour values.

3. Register tolerances- As there may be more than 100 dots of varying colours per square centimeter. It is obvious that the margin can be spared for error is very small.

4. Paper/substrate stability-As we all know moist paper and board stocks are gyroscopic, that means, they will take in moisture from the atmosphere and this will affect their physical dimensions. Therefore, it is to be understood that printing stocks are not subject too undue atmospheric variations and further more install an air-conditioning system by which atmospheric condition can be kept stable.

Then the question does arise for the facility of a coating unit which will protect the final printed sheets with a liquid polymer film for additional protection and better finish. Radiation drying by means of infrared, ultra violet etc. will need to be installed between the final coating units and the delivery in order to make sure that sheets are fully dry before being let one upon another.

PANELS CONTROL– on single-colour offset presses making adjustment manually is quite reasonable but in the case of multicolour it is essential to have a centralized control system where by the operator can initiate all the electro-mechanical functions such a sheets feeding, damping inking and impression and also control our print variables such as register, in film thickness, damp film feed and so an for each variable in each press unit. The best position for centralized press control panel is at the delivery end were the machine operator can examine a sheet immediately it has come at the need of its travel and can make any corrections if required.

Press Makeready and Operation

Preparing the press for printing is known as 'makeready'. It begins when the last printed job is completed and the press is set up for the new production job which follows.

These following steps are essential to good makeready:

1. Checking the work instruction
2. Preparing the pile
3. Feeder setting
4. Feedboard setting
5. Setting the sheet path
6. Delivery setting
7. Fitting the plate
8. Preparing the dampers
9. Ink preparation
10. Washing off plate
11. Printing
12. Setting register controls
13. Obtaining the pass sheet and running on
14. Wash up

These points are considered in detail throughout this chapter.

Checking the work instruction

Careful scrutiny of the work instruction is necessary if efficient use of time is to be made. Much time can be wasted if a plate fault or wrong imposition is not discovered until sheets are printed. The plate should be checked methodically for image, text and imposition quality and accuracy. The paper for the job must be checked by name, size and weight; the same must be done for the ink specified for the job in the work instruction.

Preparing the pile

Paper is delivered in sealed packs and if the temperature of the pressroom is different from the pack temperature it must be allowed to adjust its temperature with the room before being opened and stacked in a pile on the feeder platform. If paper of a different temperature is exposed to the pressroom atmosphere it will change the temperature of the air surrounding the stack and alter the surrounding humidity. If the relative humidity surrounding the stack is higher than the moisture content of the paper the paper will absorb moisture until it attains a level of equilibrium.

Since the edges of the stack are exposed to the surrounding air they will absorb moisture but there stack centre will not. The paper will expand at the edges to produce wavy edges thus making sheet feeding a problem.

If the relative humidity surrounding the stack is lower than the moisture content of the paper, the edges of the stack will desorb moisture to the atmosphere producing tight-edged paper.

The paper is checked for squareness by bringing the leading and back edges of the sheet together in the centre. This will quickly reveal any un-squareness. Stock for printing close register, multicolour work and jobs which must be printing to size should be paper with its grain running parallel to the leading edge of the sheet (Long grain paper). Changes in sheet dimension due to absorption of moisture are easier to remedy with long grain paper than with short grain paper. Paper is knocked up carefully to break guillotine edge burrs and encourage air to circulate between the sheets. Coated stock may be damaged at this stage if not carefully handled.

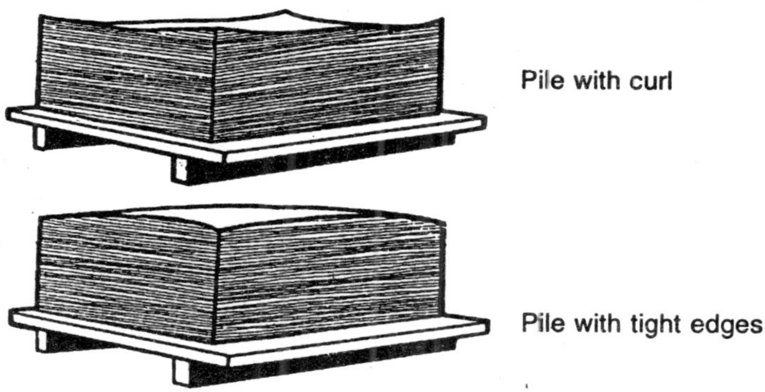

Piles

During makeready, sheets are drawn off the pile to make press adjustments. These makeready adjustments fall into three categories:

(a) The sheet must be guided through the printing unit so that is printed image falls in the correct place on the sheet. This is sheet registration and requires that the sheet controls on the feedboard and the position of the plate on the cylinder are adjusted accordingly.

(b) The amount of ink printed onto the sheet has to be adjusted to the required standard. This operation may required many sheets to pass through the printing unit to allow the ink supply to become uniform and of consistent thickness.

(c) The makeready operation culminates in the printing of a sheet which in every respect passes the scrutiny of manager of overseer. This is called the 'pass sheet' and is used during the print run to assess the running quality of the job.

The makeready book is prepared to achieve these aims without constant recourse to placing additional sheets on the pile during makeready. Waste sheets (spoils) from previous jobs which are of the same weight and size are used to makeready the printing unit. The following list gives an outline of a typical makeready book on the top of the pile.

100 sheets of waste	This provides an initial run through the press to settle the damping and inking systems.
5 sheets of stock	One sheet is sent to the reader, two to the finishing department, one to planning and one used for press adjustments.
10 sheets of waste	
2 sheets of stock	
10 sheets of waste	These sheets are used for sheet registration
2 sheets of stock	
10 sheets of waste	
2 sheets of stock	
50 sheets of waste	
2 sheets of stock	These sheets are used for settling the ink weight and uniformity
50 sheets of waste	
2 sheets of stock	
100 sheets of waste	Final adjustments to produce a pass sheet
2 sheets of stock	

The makeready book is prepared from the bottom upwards and then placed onto the top of the pile.

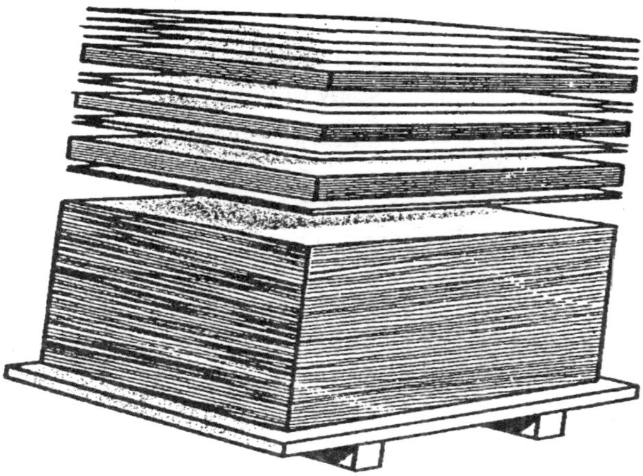

Make ready Book

Feeder setting

The pile of stock is centered in the feeder bay by folding one sheet in half and using the centre fold to relate the pile to the correct position. A scale on the face place of the feeder is also used to position paper correctly. An allowance of about 8 mm should be allowed for the sheet to be moved sideways by the side lay mechanism to bring the sheet into the dead centre of the printing unit.

The correct working height of the pile is ascertained by trial. The pile height governor senses the top of the pile and must be adjusted, until the top sheets are high enough for efficient separation.

The suckers must be positioned to lift the sheet clear of the pile. Multiple suckers on a common bar used in single sheet feeders will require that suckers outside the sheet width are isolated.

Heavy stock may require that the sucker out is changed for a softer material such as rubber.

Sucker vacuum must be set to a minimum consistent with good separating. If the vacuum is too great, double sheets will be picked up.

The flip assemblies must be positioned close to the suckers so that double sheets are knocked off the suckers.

Once the sheet has been lifted by the suckers it is advanced between the forwarding rollers which then take over the advancing role. If the sheet enters the feedboard slightly twisted this can be remedied by altering the pressure between the forwarding rollers. The greater the pressure, the faster the sheet moves between the rollers.

Position of sucker and flip device

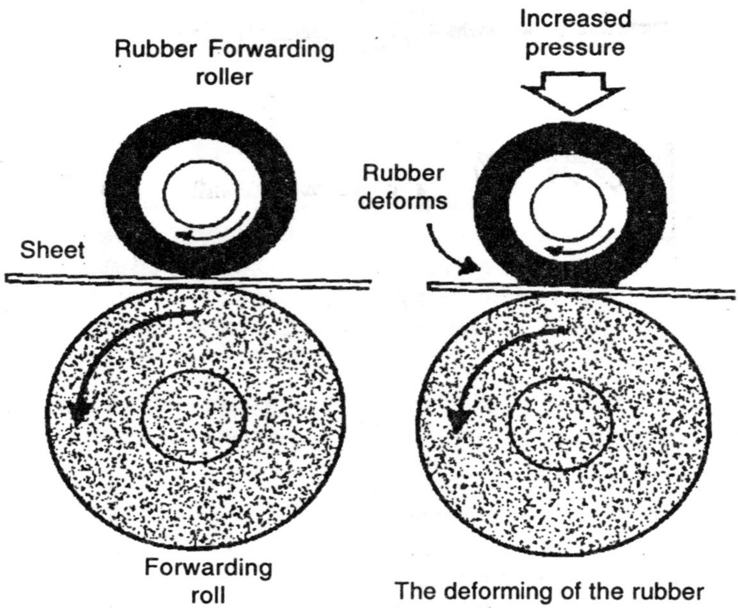

The deforming of the rubber forwarding roller has the effect of increasing its surface speed at the point of contact. This causes the sheet to move forward with increased speed

Forwarding wheel pressure

Feedboard setting

The running-in wheels on the tapes are placed so that there is adequate sheet control before the sheet leaves the forwarding rollers. When the sheet settles at the front lays the running-in wheels are placed just off the back edge of the sheet but not on it. At this point the side lay will move the sheet and if there is any impediment to side lay movement the sheet will register badly. The sheet lying at the front lays is usually controlled by rotary brush wheels or ball assemblies.

The two sheet detector will require setting. This is usually done by tearing a strip of paper (or two with stream feeders) and inserting the strip under the detector with the press running on crawl. The adjuster is altered until it checks the feeder out on two strips but not on one.

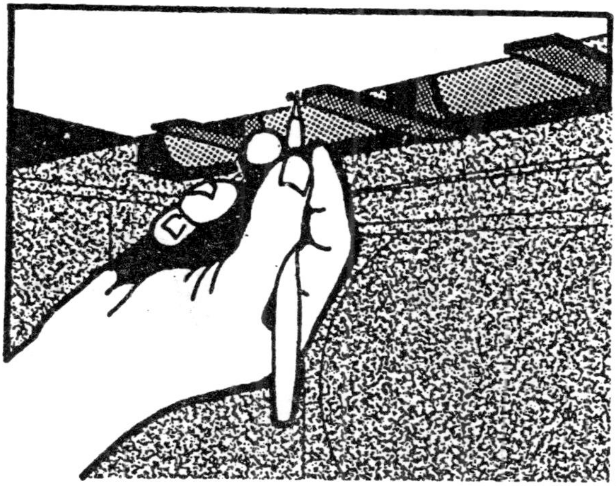

Checking grip on sheet

Setting the sheet path

Very little setting is required on most presses to effect good delivery of the sheet. Skeleton wheels or discs and star wheel assemblies should be positioned to contact the sheet in the non-image areas. Pneumatic sheet control devices may require adjustment to give sheet support without fluttering.

Delivery setting

The side joggers must be set to contact the sheet and move it slightly (no more movement than 5 mm) to platform centre. If the joggers buckle the sheet, marking will occur. The same is true for back edge jogging.

The sheet must be dropped vertically from the grippers so that air is trapped between it and the previous sheet. If the sheet drops at an angle onto the pile it is likely to expel air from the surface of the previous sheet and may mark the sheet as it settles into place.

Slow down devices are usually set when the press is on the run.

If the press is fitted with anti-set off spray it may be necessary to adjust the position of the spray nozzles to give adequate coverage of the sheet.

Once the settings have been made, the feeder and delivery of the sheet should be checked by running fifty sheets through the press at speed, making final adjustments until satisfied.

Time taken to set up the sheet transport in this way will allow the pressman to concentrate this attention on the printing process during makeready.

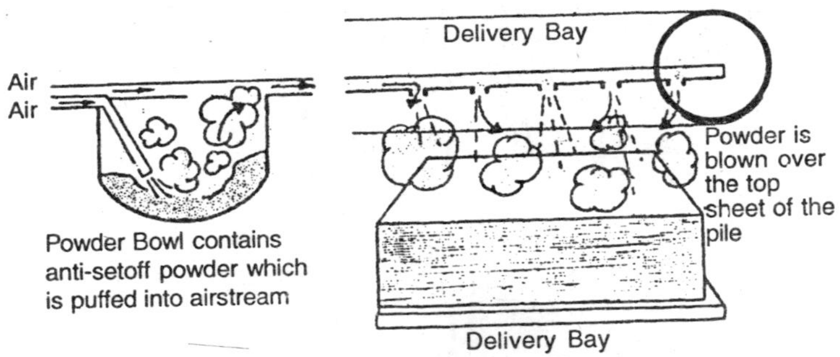

The principal of anti-setoff spray

Fitting the plate

Before mounting the plate on the cylinder the following procedure is be followed:

(a) The plate cylinder is wiped.

(b) The back of the plate is wiped and pieces of tape are removed.

(c) Some pressmen like to re-gum the plate at this point before putting it on the press.

(d) With many presses it is an advantage to bend the back edge of the plate slightly to make fitting easier.

(e) The plate clamps are centralised laterally.

The plate and its packing are placed into the leading clamp and on one. A fully Place registration to the cylinder may be achieved by using a punch register system in which case the leading place clamp is pre-registered by pins before the plate is fitted.

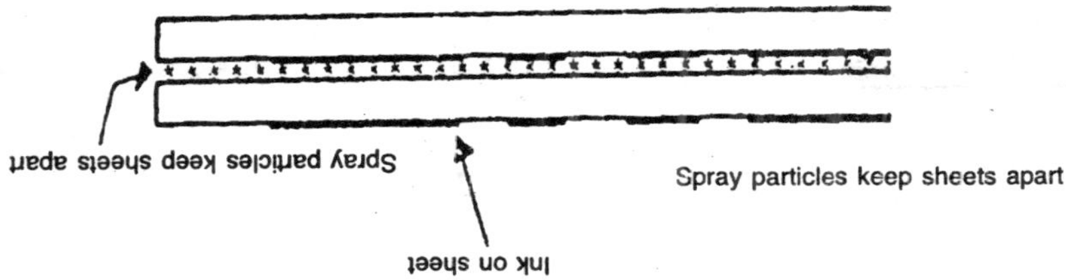

Another common method is to align the plate with marks engraved on the cylinder (pitch lines). In this case register linens are printed down onto the edges of the plate at the same time as the image, the lines always in true parallel corresponding with the image. When this is done with multicolour plates the plates can be positioned accurately on the cylinder with very little discrepancy between one plate and another.

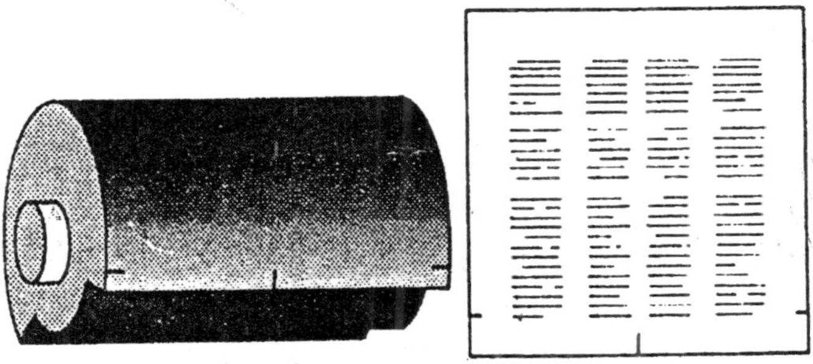

Plate Cylinder alignment mark

After aligning the front edge of the plate, the cylinder is roated and the back edge of the plate fitted into the rear clamps and tensioned. Tension is applied to both rear and black edge clamps, care taken to maintain the register position of the plate to the cylinder. With punch register systems plate tensioning must be performed with a torque wrench to ensure that every plate fitted is pulled down with the same degree of effort.

The standard method for testing correct tension is to rotate the cylinder slowly and to tap the plate surface with the knuckles. A hollow sound indicates slackness in the plate. Over-tensioning invariably results in the plate cracking along the clamp line, so care is taken with thin plates on small diameter cylinder.

Preparing the dampers

Dampers should be clean and slightly damped hand before fitting into the unit. When freshly fitted, the end bearings of the dampers should be lubricated as all lubrication is removed from these bearings when the dampers are cleaned.

Usually dampers are slightly drier than required so that additional moisture supplied to the damper will bring it to its working level. If damper too wet at the start the moisture level may be difficult to reduce and water may become entrained in the ink during makeready.

Ink preparation

Litho inks are generally thixotropic. This characteristic means that the ink must be agitated to make it flow correctly and agitation can be performed on the ink slab before placing it in the duct or agitated in the ink duct by palette knife or automatically with an ink duct agitator device. This latter device traverses the ink duct and continually stirs the ink to maintain its fluidity. All thixotropic inks flow when agitated but revert to non-flow mode when left for a short time. "Hanging back' is produced in the ink duct when the ink ceases to flow and ink supply dries up because the ink has not been agitated regularly. Before placing ink into the ink duct the duct blade must be firmly locked in position.

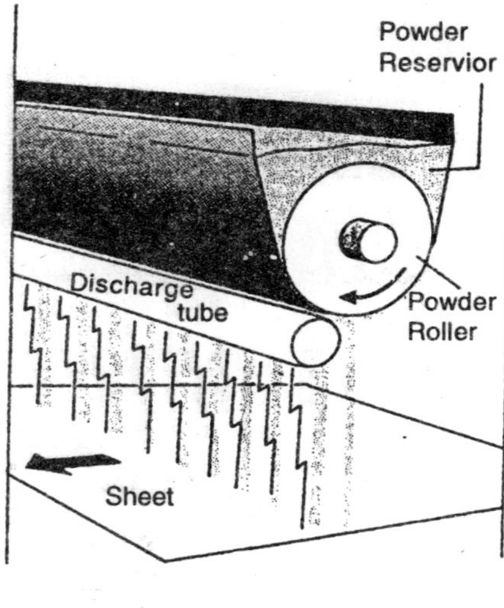

Electrostatic powder application. The oxy-dri anti-setoff powder applicator produces an electrostatic field which carries the powder from the engraved surface of the roller onto the sheet. The electrically charged powder evenly distributed over the entire width of the sheet also neutralises static in the stock.

Hanging back

The amount of ink carried by the duct roller is controlled by the blade setting.

The blade is adjusted by bringing the ink feed roller (vibrator) into contact with the ink duct roller and rotating the two together by hand for several revolutions. By allowing both rollers to rest momentarily, the ink will flow into the roller nip and produce an ink line to indicate the level of ink along the length of the roller. To increase the ink supply at any point the duct screws are turned anticlock wise and the rollers rotated by hand for several revolutions until the foreign ink level is evenly distributed on both rollers. When halted, the ink will flow into the roller nip and the ink line will show a peak where ink supply has been increased.

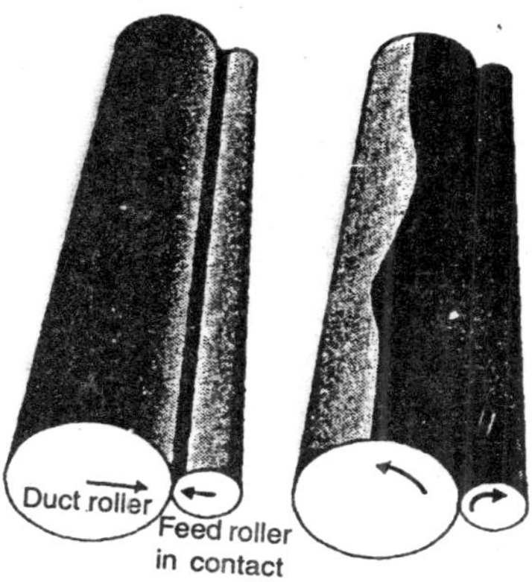

Checking the adjustment of the duct blade by producing an ink contact line on the roller

To reduce the ink the same procedure is followed, turning the screws clockwise to produce a dip in the ink line. A straight line indicates uniform setting across the duct.

A dew minutes spent setting the duct in this fashion, assessing ink supply to the image requirements of the plate, reduces makeready time.

The pressure runs up a minimum quantity of ink onto the rollers. Ink supply may always be increased during makeready with rapid response from the inking system but excess ink on the rollers is difficult to reduce.

The pressman should aim at setting the ink so that he has capacity to increase or decrease the overall supply during makeready without having to readjust the duct screws.

Washing off the plate

Before printing can commence the plate gum must be removed with water. Some pressmen prefer to remove the image ink at this stage by giving the plate a wet washout'. This method requires that the plate is first washed over with water to dissolve the gum, and while the plate is wet a cloth containing ink solvent or washout solution is rubbed over the inked image. Care must be taken with the wet-washout technique to ensure that the plate is kept wet with water while the ink solvent is used to remove image ink. The plate is finally cleaned with water before proceeding to print.

Printing

The press is now started up and the dampers lowered onto the plate for two or three revolutions until the moisture on the plate surface is even. This can be checked visually by looking across the plate at an angle while it is rotating. An excess of plate moisture will produce droplets at the back edge which will quickly become entrained in the ink pyramid.

The inkers are dropped onto the plate and by observation allowed to run for several revolutions until the image is uniformly inked. During this operation the moisture supply may not be adequate and 'catch up' occur. Catch up is the taking of ink to the non-image areas. Catch up usually begins at the leading edge of the plate along the clamp line and when moisture supply is increased the catch up disappears. It is easier to balance the ink and moisture supply when catch up occurs because catch up indicates low damping and if this is increased gradually, a point will be reached where the minimum of moisture is carried on the plate consistent with clean working.

Once the damping and inking are working correctly the feeder is started and the first sheets of the makeready book passed through the press. When plain sheets are passed through the feeder is cut out and the sheets allowed to pass into the delivery befcre the press is halted. If possible, the dampers should be lifted from the plate and the plate cylinder parked with the depression gap behind the dampers. If several adjustments are to be made, many minutes may elapse in which the plate dries out uniformly. If the dampers are close to the plate surface during a halt the damp air surrounding the dampers may give rise to oxidation of the plate surface. This is not a problem with many anodized plates although it could occur and can be avoided by careful parking of the plate cylinder.

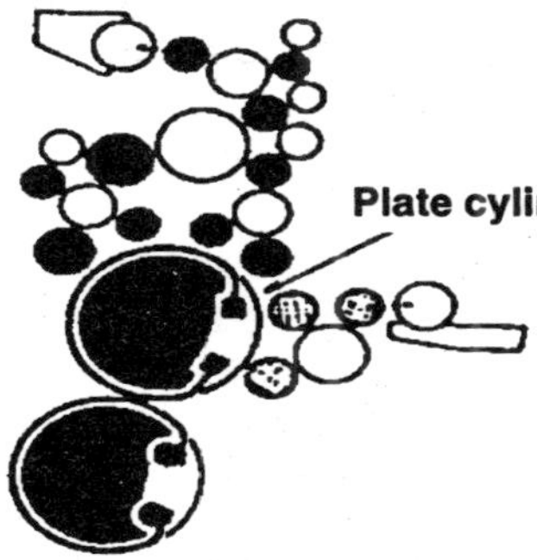

Parking the dampers

Setting Register controls

The sheet is first adjusted to bring the image into its correct position on the sheet. Various methods are used to do this. The sheet may be folded in half and the image related to the folds. If may be necessary to place the sheet onto a ruling-up table and lines drawn in to align the image squarely. A rapid method is to place an acetate gid over the printed sheet and to ascertain squareness visually.

With perfected work the outer side is aligned to the inner by holding the sheet up to the light and viewing the register between both sides.

If the image is tilted this can be remedied by altering the front lays but movement is limited here and on the high speed press tilted front lays can give register problems. The recommended method for squaring the printed image with the sheet is to tilt the plate.

Lay marks are placed on the plate to check register on the run. After the Pass sheet is obtained lay marks are made in the following manner:

(a) A sheet printed in good register is selected and a corner torn away on the side lay edge. The torn portion should be large enough to cover the point where the side lay contacts the sheet edge.

(b) The plate is gummed up and dried. The torn sheet is placed onto the plate and the printed image lined up with the plate image carefully.

(c) A side lay mark is scored into the plate at the point where the side lay contacts the sheet edge. The lay mark ideally should be about 10 mm long and print right on the edge of the sheet.

(d) A front lay mark is also made on the plate on both sides of the leading edge. This should be about 5 mm long and placed to run off the edge of the printed sheet.

Once the run begins the side lay mark will be seen clearly on the edge of the pile as a thick line. The front lay marks shown as a running line down the edge. When examined, a batch of printed sheets are taken from the delivery during the run at intervals land knocked up against the gripper edge then fanned out to show the side lay and front lay marks. Variations in the positions of the lay marks will indicate faulty register.

Obtaining the pass sheet and running on

When the pressman has finished makeready the sheet is usually scrutinised by a supervisor for quality. Printing faults such as poor quality image, incorrect colour, wrong inking level land blanket defects should be discovered before the run commences.

When a Pass sheet is obtained, a final check of sheet transport controls must be made before the run on. During the early stages of the run repeated examination of the sheets should be made to ensure that the Pass sheet standard is maintained. Within a short time the ink/water balance will settle down, the pressman endeavouring to keep the moisture supply to a minimum.

Quality control aids on the printed sheet are evaluated at the Pass sheet stage and standards set then are evaluated as the run continues.

Densitometer readings of the ink film thickness printed on the sheet may be taken from the Pass sheet and these used for reference during the run. When a density reading is taken from the Pass sheet, the time between impression and reading must be noted and all subsequent readings taken from production sheet at the same time interval. The longer the period between impression and reading, the greater the absorption of the ink into the stock. Density readings will vary therefore with time, and to ensure that readings taken -during the run are compatible with the Pass sheet, this time interval must be carefully considered.

Wash up

At the completion of the plate and its pickings are first removed. The plate is gummed up and prepared for storage by washing off the press ink and rubbing the image up with inking-in solution. This may not be necessary with plates in which the image is dark enough to make identification of the plate easy when it is in store. Most negative working plates do not require inking-in unless it is for identification purposes. Positive working plates may require inking-in to protect the image during storage.

Roller washup is achieved with the automatic washup device. The washup blade is usually made of rubber, fibre or metal and is placed in firm contact with one of the hard distributor rollers of the ink pyramid.

Wash up solvent such as white spirit is usually used. High evaporation solvents should be avoided as the ink must be removed from the rollers into the washup tray before the solvent evaporates.

For good cleaning of the rollers prior to changing colour, a two or three phase cleaning method such as Pira red, white and blue solution should be used. Each coloured solution is used in order, one to remove ink, another to remove lint, and water-based debris, the third to keep clean and dry the rollers.

Press speed should be increased for washup. This ensures that solvents used do not evaporate before they have removed the ink to the washup tray.

One half of the rollers should be saturated with solvent at a time. This ensures that the dry side of the rollers gives traction to the wet rollers and the rollers do not skid or stop rotating.

The surface of the blanket is not improved by frequent washing with ink solvents. The surface pores of the rubber are opened by the solvent and if the ink particles are not removed rapidly they penetrate the surface and dry to produce a hard shiny surface (glaze) which has poor ink transfer properties.

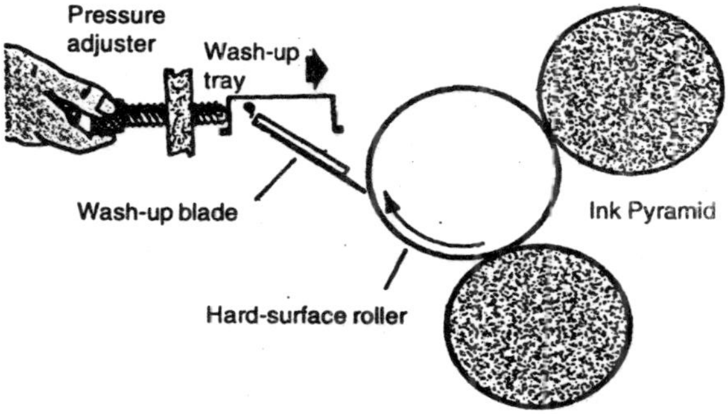

Principal of the wash-up device

If the blanket requires alienation during the run it should be washed over with water to remove accumulated lint and fluff. The final wash should be performed using a recommended solvent which is used on a saturated cloth. One cloth is used for wiping the surface and another dry cloth used for drying immediately to remove the ink-laden spirit before it evaporates. A small area of the blanket should be cleaned in this way, working progressively over the entire surface. When the blanket is dry the surface should be dusted with french chalk to reduce surface tackiness.

Plate cylinder packing

Large offset presses are designed with a plate cylinder which carries a standard thickness plate plus a packing sheet. Packing sheets are not normally used on small offset and web offset presses.

Plate packing is necessary when printing multicolour and close register work because it allows the pressman the facility of changing the cylinder diameter when the printed image size must be changed.

Material for plate packing may consist of manilla paper or stock paper or a more stable material like I.C.I. Melinex plastic sheet. Stock paper used for packing is generally unsatisfactory for long runs because it compresses and the edges soak up moisture. When this happens the edges of the paper swell giving light tints a heavier appearance at the plate edges. Manilla paper and oiled manilla paper are commonly used because they do not

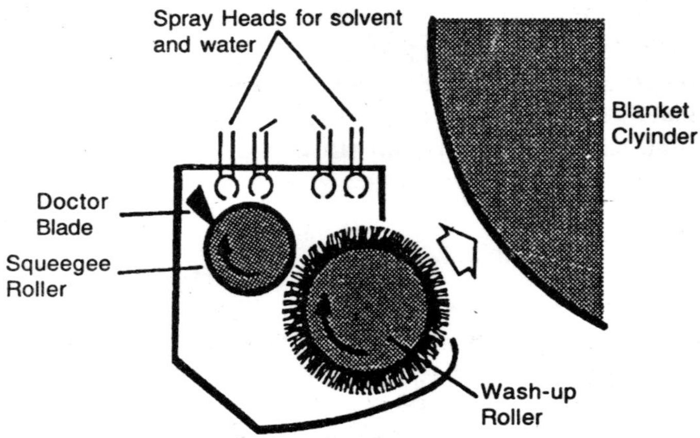

The automatic blanket washing device is situated close to the press blanket where it can be moved contact with the blanket for washing. The cloth covered roller (or belt) scrubs the blanket surface while being saturated with ink solvent and water from the spray pipes inside the unit. As it rotates the cloth roller is squeegeed to remove the accumulating sludge and this squeegee roller is scraped with a doctor blade while being doused with solvent.

compress significantly, but they are not as durable as plastic sheet and must be replaced at each plate change.

Plastic packing sheets may be obtained in a variety of thicknesses to give the pressman a choice of increasing of decreasing packing by increments of 0.05 mm.

Change of image size

When the printing unit is set up according to the manufacturer's specification with correct thickness plate, blanket and packings, the printed image should be the same size as the plate image.

A change in the printed image size can be affected if the plate cylinder packing is reduced or increased in thickness to change the overall diameter of the cylinder. This facility has been used by lithographers for many years to counter stretch or shrinks problems which occur with paper after it has passed through the press for its first printing. In the majority of cases, paper absorbs moisture from the blanket which is carried there because of its contact with the plate. A high proportion of the damping solution applied to the plate ends up in the paper. It is for this reason that the level of moisture on the plate surface should be kept to a minimum.

When damped evenly, the cellulose fibres in paper swell slightly and the overall dimensions of the paper change. This dimensional change is more pronounced in the circumferential direction (gripper to back edge of the sheet) when using Long Grain paper. Fortunately, the change in image size effected by changing plate packing thickness is also circumferential, so the pressman can remedy dimensional changes in the sheet by altering his plate packings.

The principles governing the change of the printed image size are as follows :

(a) When all cylinders are dressed to the manufacturer's specification the press will print to size.

(b) By increasing the plate packing the plate cylinder diameter is increased thus giving the plate a greater surface speed on rotation than the blanket cylinder. Because the plate image is therefore transferred to the blanket at a higher speed, the image size is reduced slightly in the front to back edge dimension.

$$P \times 2\ 3.142 \times F = N$$

where P = packing thickness change

F = fraction of the cylinder occupied by the image

N = new image size

(d) A reduction of the plate packing thickness has a reverse effect, the printed image is slightly larger (gripper to back edge only).

(e) An increase or decrease in the impression cylinder diameter also has the effect of changing the printed image size but few offset presses have the facility for altering the impression cylinder diameter.

(f) A change in the diameter of the blanket cylinder by increasing or decreasing its packing thickness has no effect on image size.

When the plate cylinder diameter is increased the surface speed of the plate will be increased and this will produce a slight slur to the printed image. The increase or decrease of packing thickness must be kept to tolerable limits otherwise image quality will deteriorate and plate wear become excessive.

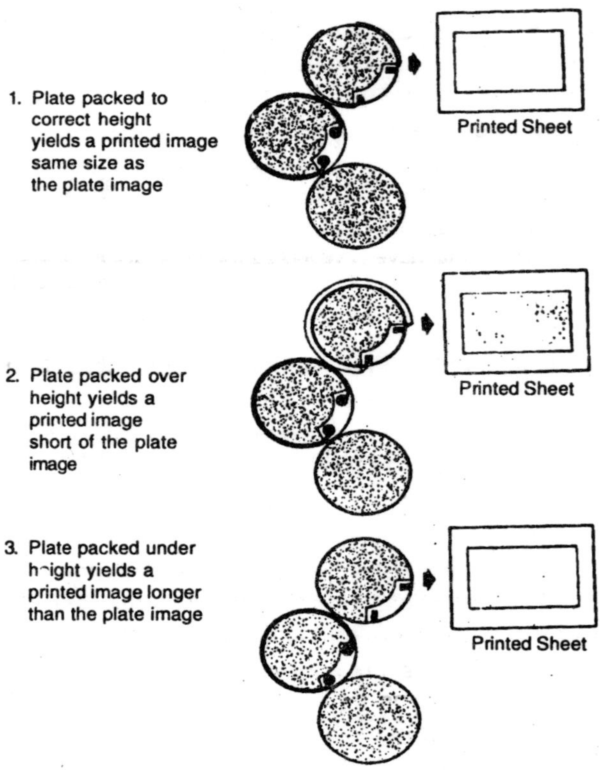

Figure : Principal of changing image size

Impression cylinder pressure

Pressure between the blanket and the impression cylinder is adjusted to accommodate the thickness of the printing stock. Coarse-surfaced papers require bore impression pressure than the same thickness of smooth surfaced paper to ensure that the image 'bottoms' in the paper surface.

The normal method for setting the impression cylinder pressure is to move the cylinder out of printing contact with the sheet when the impression is on and then to gradually adjust the pressure during makeready until a good impression of the image on the sheet is obtained.

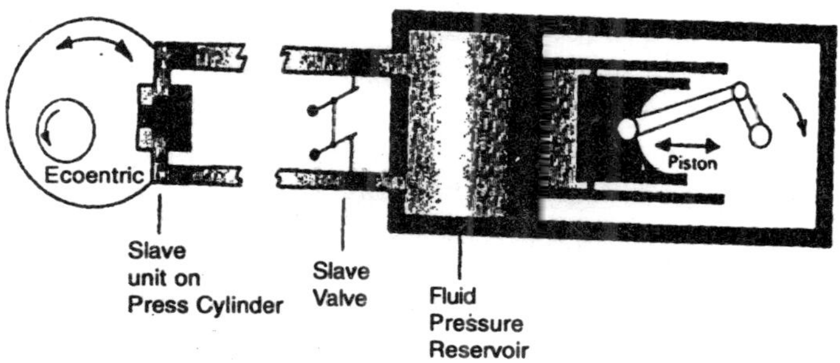

Schematis of the hydrolic ram. The moving of the slide valve allows hydraulic pressure to move in one direction. When the valve is reversed the pressure is reversed.

Hydraulic triping

On the conventional press, cylinders and rollers are tripped in and out of contact with each other by means of a system of levers, springs, cams, etc. At the moment of tripping, considerable power is drained from the press drive and in the case of large presses, and multi-unit presses, this power drain is unacceptable.

The hydraulic control of tripping eliminates the connection systems of links and cams and also makes it possible to trip components when the press is halted. No power for triping is derived from the press drive.

Ram power. A pump, pressure reservoir and pressure switch are used to produce constant hydraulic pressure in the system. A simple slide valve moved by electrical solenoid opens the system and power is applied to salve units which trip the press. In most cases the slave ram moves an eccentric cam or bearing to move cylinders in and out of contact.

Blanket maintenance

Priming

When a new blanket is used it may absorb a proportion of the ink solvent and cause the ink tack to rise and produce poor sheet release from the blanket. When this occurs, especially with large solid images, blanket priming may be an advantage. Priming a new blanket is performed in the following manner

(i) An ink solvent or distillate reducer obtained from the ink supplier is wiped over the new blanket surface and allowed to soak into the surface for two or three hours. This may require repeated replenishment from time to time.

(ii) When the surface saturation has been completed the excess solvent is removed with a dry cloth and the blanket surface dusted with french chalk.

Priming may not be necessary with many of today's blankets, although inks with a volatile solvent can be troublesome. Inks with such a solvent base are absorbed into the blanket during the run but evaporate quickly when the press is halted to leave a high tack ink on the blanket. When the press is started again, the increased tack may tear sheets or web. This type of problem is remedied either by washing the blanket with distillate or as is more common to spray distillate over the blanket while the press is on crawl just before printing.

Repairs

Low spots on the blanket surface are usually cause by folded sheets or foreign objects passing through the printing unit (blanket smash). If the surface rubber is undamaged the low spot is probably caused by the compression of the carcase fibres. These fibres can be revitalised by removing the blanket and soaking the back with water for about an hour and then allowing it to dry slowly. A recommended method is to plunge the blanket in a water trough for an hour and then hang the blanket up to dry in a dark place.

Rapid blanket repairs are often called for and local swelling of the compressed fibres can be achieved by injecting solvent, water or oil into the fibres with the use of a hypodermic syringe.

Local swelling of the surface rubber may help to keep things going to the end of the run and several proprietary swelling agents can be used, although such methods affect the rubber for a limited period.

If the surface rubber is damaged, very little can be done to effect a good repair. With some jobs it may be possible to lift the blanket and turn it around through 180° leaving the underpacking as before. This can move the damaged area away from blanket image.

Another remedy is to move the blanket forwards or backwards on the cylinder body to move the blanket forwards or backwards on the cylinder body to move the damaged spot out of the image line. These remedies are temporary but may enable a particular job to be finished before a new blanket is fitted.

Washing the blanket

Most blankets are washed too often with ink solvents. In many cases a wash with water is all that is required to remove lint and fluff from the surface and this will improve the printing impression greatly. Solvents which swell the blanket rubber in general do no good because they open the surface pores and remove anti-oxidants, plasticisers, waxes and other ingredients necessary to long blanket life.

Most blanket washes are based on aliphatic hydrocarbons such as white spirit. Other ingredients in the wash are selected to give it properties such as low flash point, pleasant odour, miscibility with water control of toxicity.

Blanket washing if not performed correctly will reduce the life of the blanket.

Reviving the blanket

Restoring the qualities of a blanket after use will prolong its working life. The elasticity of the blanket may be regained if it is released from tension overnight and at weekends. Periodically the blanket may be soaked in water for a few hours to allow the compressed carcase to swell and regain its former bulk. After soaking, the blanket should be hung up to dry slowly before refitting it to the cylinder.

Rejuvenation of the surface rubber requires the removal of dried ink from the surface pores. Poor blanket washing allows particles of ink and driers to penetrate the surface rubber to produce a hardened film called 'glaze'. This glaze prevents good ink transfer and will affect image quality.

Special glaze remover solvents are recommended by blanket makers to deep clean and restore the surface to a velvet texture. If abrasives such as pumice powder are used for removing glaze they should be used with care.

MULTICOLOUR PRINTING

Multicolour printing on the single colour press requires careful planning. In many cases a second colour on a job may not require close register (spot colour) but if close register is involved a number of factors have to be taken into account.

Image coverage

If one of the two colours has a larger image area this should be printed last. This general rule may be reversed if one of the two colours is an opaque ink and the lighter weight ink has to be printed over it.

If the heaviest image is printed first it may be difficult to print the second colour onto it. When the first ink film dries hard the second printing lays the liquid ink on the surface very much like putting water over the surface of glass. The gloubles of ink reticulate and refuse to flow to an even film giving the print a mottled appearance. The phenomenon is called 'poor traping' or 'refusal' (see Troubleshooting for remedies).

If the heaviest image is printed first and the ink is not perfectly dry before the second colour is applied, the first print may set off onto the blanket cylinder to produce image doubling or contamination of the second colour.

As a general rule when using transparent inks, the lightest image is printed first followed by the heavier.

Ink transparency

When two inks are printed, one on top of the other, the final colour ink will produce a dominant cast to the combined hue. For example, a green hue produce from yellow and blue inks will be yellower if the yellow is printed last. This occurs because the light reflected from the surface yellow predominates over the underlying colour.

This must be taken into account when attempting to match a third colour made from two other colours. Opaque inks tend to obliterate the surface on which they are printed. A transparent blue printed over an opaque yellow will give a more satisfactory result than opaque yellow printed over the blue.

Ink colour

Short print runs with numerous colour changes should be planned to make washup easy. It is often difficult to wash up the press and remove all traces of the previous colour if it is a dark colour like black. Short run jobs therefore should be planned to move from the lightest hue inks to the darkest.

The same can be said about damper contamination. Ink impregnated in the dampers may contaminate a lighter colour printed later. To avoid repeated damper washing the jobs must be planned to print lighter colours first.

Process colour printing

Full colour printing uses four ink colours of Cyan, yellow, Magenta and Black. These inks are transparent and can be printed in any order but the recommended sequence is Cyan first to giver clear visual prints on the first run. Yellow next to allow the weight of yellow to be determined by the green produced in combination with Cyan. Magenta usually follows next with Black to finish. This sequence may have to be altered however by the following considerations.

1. Many colour jobs are proofed before a run and the progressive book may indicate a different sequence of printing the colours.
2. The last colour printed may impart an overall cast to the final print. For example, the order Cyan, Yellow, Magenta and Black will produce a slightly reddish cast to the final print while a greenish cast will be produced if the order Magenta, Yellow, Cyan and Black is used. In most cases this will not be important but it can have a significant effect when, for example, an attempt is made to match an advertiser's colour of lipstick.
3. The trapping of one ink upon another has to be considered. If trapping is likely to be a problem then the colour with the heaviest coverage should be printed third or last in the sequence.
4. The traditional order or printing yellow first has the drawback of printing sheets of white paper on which Yellow is difficult to evaluate. This is especially true with gloss papers. A densitometer is necessary for this order.

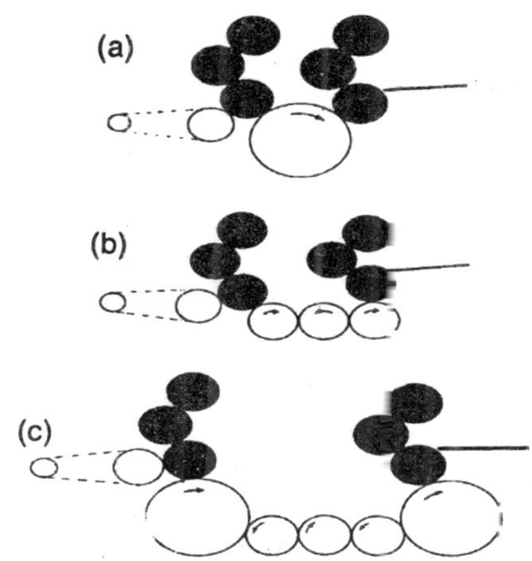

Figure : Drum Transfer

(a) One drum **(b) Three drums** **(c) Five drums**

The multicolour press

Modern multicolour press design falls into two categories. The modular unit press is based on single printing units which are coupled in-line to form two, three, four, five and six unit machines. The second category is based on a two colour press design (which may have a common impression cylinder) which is coupled in tandem to form a two, four or six unit press.

The modular unit design offers greater combination flexibility and lower productions costs, while the tandem press offers higher standards of registration between units.

Each separate printing unit on the multicolour press is based on the same working principles as the single colour press. Differences of design occur in the method of sheet transfer between each printing unit. The prime object here is to avoid sheet marking and to maintain good register as the sheet is passed from one set or gripers to another. This is generally achieved by drawing the sheet over a number of transfer drums (each complete with gripers) which may number five, three or single drums. Sophisticated machines make use of high velocity air to give sheet support without marking.

The convertible press utilises a special transfer drum mechanism which turns the sheet over for perfecting. A simple configuration uses the two colour press to print two colours on one side of the sheet or when converted, prints one colour on each side of the sheet. A four colour press can be converted to print two colours on each side of the sheet.

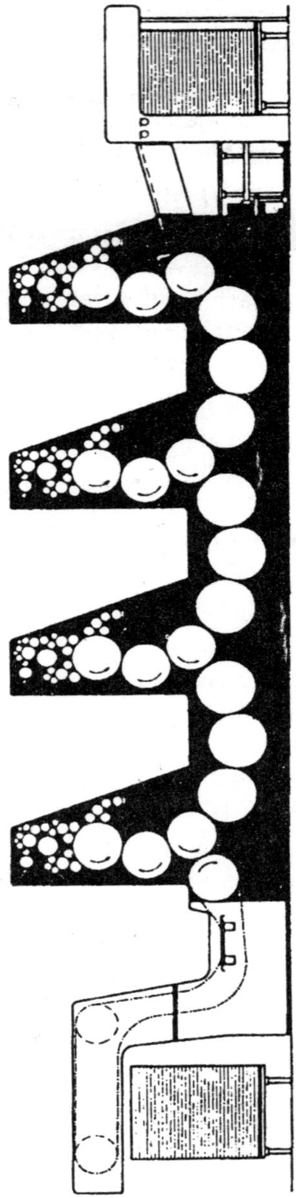

The makeready procedure used on the

single press in applicable to the multicolour press with the following additional considerations :

1. The front lays on the press must be set to feed the sheets squarely into the first unit of the machine. The practice of adjusting the front lays to obtain register must not be followed on the multicolour press.
2. Image register between one unit and another is achieved by adjusting the position of the plates. Pin register plate clamps are commonly used to assist in plate register.
3. Each printing unit incorporates mechanisms for adjusting the lateral and circumferential register of the plate cylinders. Many press designs allow this adjustment to be performed while the press is running.
4. The multicolour press prints 'wet-on-wet', that is, each successive unit impresses a wet film of ink upon wet ink from the previous unit. This may produce ink trapping problems and is remedied by lowering the ink tack on each of the following units.
5. Remote control and computerised systems for automatic adjustment of the ink ducts are of great value in reducing makeready times on the multicolour press.

Process colour printing on the four colour press is a simpler operation than that on the single colour press because the final full colour print can be assessed for colour balance after one pass through the machine. The imposition of four films of ink onto the sheet may lead to ink set-off. Accelerated ink drying systems are an answer to this problem, and permit rapid turn around of printed work for perfecting.

A new development in colour separation techniques known as Achromatic Separation produces four colour separation sets with a greatly reduced level of ink in the colours cyan, magenta and yellow. This is turn lowers the possibility of set-off in wet-on-wet printing.

Quality control

The standard of quality for the job is set when the pass sheet is obtained. This establishes the image position, register, image quality, ink density and half tone reproduction quality.

During a run of many thousands of sheets perhaps occupying several days, the pass sheet standard must be maintained while room lighting changes from daylight to artificial lighting, and often when there is a change of pressmen on working shifts. Quality control therefore should not be left to the estimation of the individual pressman whose visual judgement over a long period may alter with eye fatigue and a change in room lighting.

A number of quality control aids are produced to assist the pressman in keeping to pass sheet standards:

Star target devices

The Star Target is a device designed by the Graphic Arts Technical Foundation of America but it has its counterparts throughout the world. The Target is a wheel-shaped design printed down into the trim areas of the sheet. Its main function on the press is to indicate dot gain or dot loss in half tones, but it also indicates press slur and doubling. When dot gain occurs during the run the centre of the Target fills in and indicates gain before it becomes visible in the half tone subjects on the plate. Dot loss produces an open area in the centre of the Target. Press slur is indicated by a figure eof eight appearing in the Target centre, doubling produces a moire pattern in the Target.

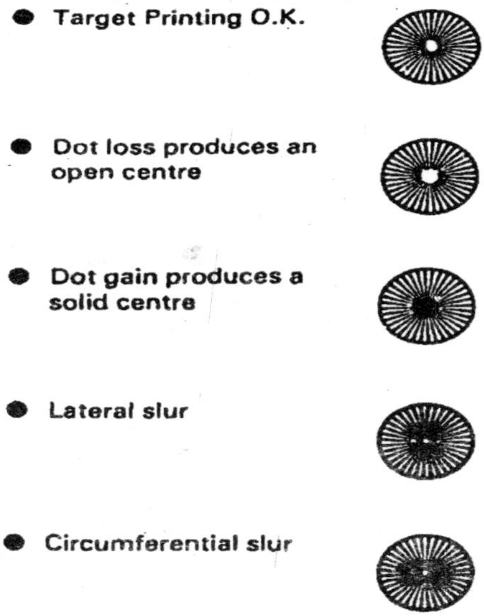

Figure : The Star Target

Dot gain scales

This device is printed down in the trim areas of the plate. The scale contains a numerical scale on a tone background. At the pass sheet stage the number which corresponds in visual uniformity with the background tone is noted and thereafter dot gain or loss is indicated when this tone uniformity changes up or down on numerical scale.

Figure : The GATF dot gain scale

Slur bars

These devices are printed in the trim areas of the sheet at an angle of 90° to the gripper edge. Press slur may be caused by mechanical defect or by faults in the passing of the stock through the press.

Lateral slur causes the vertical lines in the slur bars to join together thus indicating the defect visually.

Circumferential slur causes the horizontal lines of the bar to join together. There is a slight variation in the design of these devices but they all function on similar principles.

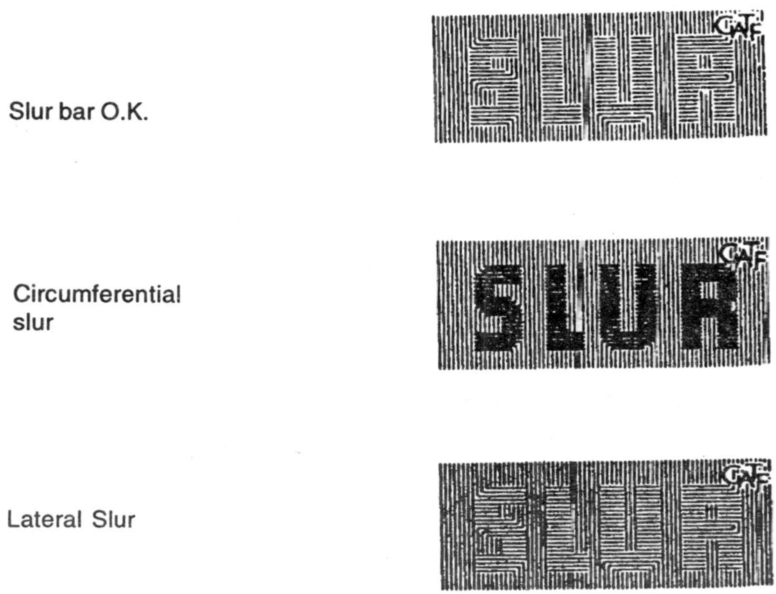

Figure : The GATF slur bar

Register marks

These devices are printed down in the trim areas of the plate. Usually the register mark consists of one horizontal and one vertical intersecting line. Circles or ellipses are sometimes added to aid registration between one colour and another.

Sheet registration consists of placing the register marks into their correct positions related to the sheet in single colour work. With multicolour work the register mark must fall perfectly over the previous printed mark to give good register.

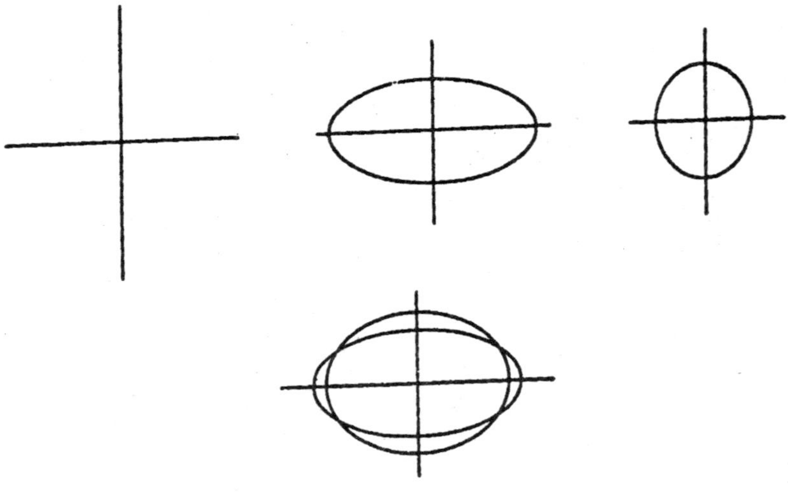

Register marks

Colour control bars

These are devices printed down in the trim areas of the sheet. The colour control bar consists of a number of devices such as those considered above plus special areas to control colour printing quality.

The bar is made up of patches of about 3 mm square (some are circular) and large enough to be evaluated with the reflection densitometer.

Solid patches in the bar give solid density readings are taken and these are used during the press run to maintain the ink level. The printed ink density changes as the ink penetrates the stock and dries, therefore densitometer readings of the solid patches are usually taken after a timed interval, say of one minute, and this interval is observed throughout the run.

Trapping patches in the bar are solid areas which when printed in two or more colours are superimposed on one another. The Cyan is first printed and the density of the trapping patch evaluated with the densitometer. When this is overprinted by the Yellow the densitometer is used to compare the density of the Yellow on its own against the density of the Yellow over the Cyan. Any discrepancy between the two readings indicates trapping error.

Percentage tints are placed in the bar and are used to check the degree of dot. gain due to squeeze between plate, blanket and stock. Most litho presses produce a gain of approximately 5% so that a tint patch of 25% printed onto the plate will produce a tint of 30% on the printed sheet. This type of gain may not be uniform however, with heavier tints showing more gain than lighter things. Percentage tints incorporated in control bars therefore range from 15% to 80%. Dot gain may be lowered by experimenting with different grades of blanket, change in printing pressures and the viscosity of the ink. Percentage tints will aid in the process of using any of these remedies.

Grey balance patches consist of half tone patches in Magenta, Cyan and Yellow which when superimposed produce a grey. This grey is usually compared to a grey patch alongside in the bar which is produced by a black tint, or may be compared to a printed scale of greys held in the pressroom. Any shift away from neutral grey will indicate a predominance of one of the printed colours.

The reflection densitometer

This is an electrical instrument used for taking density readings from colour control patches. The instrument contains a light source which is focussed on the printed control patch and the reflected light passes through a filter into a photocell to give a dial or digital reading of the density of the ink film. A complementary filter to the colour being evaluated is used so that the ink colour is changed black to the photocell.

Polarising filters in the densitometer minimise the differences between highly reflective and matt surfaces.

Sophisticated densitometers may also offer facilities for measuring the surface reflectivity of the stock, the press contrast factor and half tone density.

A quality control chart is often used with the reflection densitometer. The pass sheer density readings are taken and recorded onto the chart. Subsequent readings are taken during the run at predetermined intervals and recorded not he chart to indicate clearly the movements of ink density. If the printed in density goes beyond the tolerance limits (the limits where density changes become obvious to the eye) the parts of the job where this occurred can be determined thus making stock sorting an easier task.

15

Web Offset

A single unit may print, say, eight pages newspaper and the more pages required, the more printing units will need to be installed in the press line.

National dailies of large circulations may require bigger printing units, to accommodate larger plates carrying more paper and these are more likely to be arranged with the web running vertically between the blanket cylinders and the printing units arranged in an area from either side of the web for convenience and accessibility. These are generally three-tier installations with the webs coming up from reel room, in the basement.

Colour printing may be undertaken by running the web from one printing unit to the next unit. A four unit, in-line, blanket to blanket press, could print four colours on each side of a single web. Alternatively, the four units may take two webs and print two colours on each side of both webs and so on.

Satellite units may be used where multicolour printing is a standard feature of the publication. In this, a number of blanket cylinders are groped around a common central impression cylinders. If colour is required on both sides of the web, then two satellite units may be employed.

Combination presses will contain a mix of blanket to blanket perfectors. For the monochrome work, plus a colour satellite unit. This is an ideal situation where the majority of pages are monochrome & colour is required only on a limited number of pages.

The basic principles of offset lithography are similar for both sheet and web printing. Plates are produced in much the same way, although the high volume requirements of web production are more likely to require direct camera-to-plate system and automated plate making lines.

The designer of the printing units will also need to be modified, for web printing as the intermittent pattern of sheet fed production is replaced by the continuous flow of a web, and this will influence practically every element in the printing unit.

Technical data for WEB-OFFSET
Speed max. (copies/h)
Cut-off (mm)
Web width (mm)
Thickness of printing plates (mm)
Thickness of rubber blanket (mm)
Printing paper g/m^2
Max. diameter of paper reel (mm)
Core diameter (mm)

Cylinder dimensions are usually of a fixed diameter, therefore at every revolution a certain amount of substrate will be fed through the press, corresponding to the cylinder circumference, and this will relate to the final end product size.

This is completely different to the sheet-fed situation where various sizes of printing stock can be fed into a press, upto the maximum sheet size. There are no recognized international standards for web press circumstances, although there are certain recommended cut-off size for periodicals. This can give some problems where publication are to be printed simultaneously in more than one centre on presses with different cylinder diameters.

Plate cylinders will be designed with the minimum possible gap, to hold the ends of the plate. This will enable the maximum surface area of the cylinder to be available for printing. This does give technical problems, as the ends of the plates, which enter the gap must be accurately formed to enter the limited space available. The plate locking system must be simple, quick and ensure that the plate conforms, accurately to the cylinders circumference, without imposing undue stresses, Poor design of plate lock-up will lead to edge cracking, which is probably the most common cause of plate failure on web-offset press.

Blanket cylinders will also need to be designed with a minimum gap for gripping both ends of the blanket. A bulge in this area will give rise to a number of printing and web-control problems, even the quite possibility of web breaks. On some simpler web offset presses, sticky back blanket are used, where by the blanket is held on the cylinder by an adhesive backing, However, this technique is not recommended for high-speed publication presses.

Impression Cylinders do not need to have a provision for grippers to hold the substrate, as in sheet-fed production and these may in their simplest form be continuous steel cylinders geared into the press drive, which in turn drive the other cylinders in the printing unit.

Inking system on web offset presses will be designed on the same basic principles as for sheet-fed presses, except that a more continuous flow of ink will be required as against the intermittent flow on sheet fed machines. A bulk supply of ink is also a common feature, pumped to the appropriate station, in order to cope with the high volume demand. Remote control of inking is standard in most web-offset presses together with automatic monitoring and correction.

Damping will also need to be continuous and precautions must be taken against the feed-back of ink into the damping system, conventional damping techniques, similar to those on sheet-fed presses are often employed, plus non-contract methods such as sprays or brush damping, which apply moisture without direct roller contact to either system is used. There should be continuous circulation from bulk storage, with automatic filtering etc. to maintain clean and consistent solution.

Photo Polymer Relief Printing - This is a modern version of the original letterpress process, in which the image to be reproduced is raised in relief above the base material. Metal type plates have long since been replaced by photopolymer material, which can use photographic input to produce flexible, one-piece relief plates, which can be wrapped around a printing cylinder in much the same way as a litho plate.

Relief printing has several advantages when compared with lithography:-

1. The ink/water balance, essential to the lithographic process no longer applies. The moisture variable is eliminated, only the ink level is to be determined. This makes for faster start ups, quicker make readies, lower wastage, less maintenance and lower material costs.
2. Direct printing from plate to substrate, is possible without the need for an intermediate cylinder, thus simplifying press design and lowering costs.
3. Image quality is consistent. Once established, the image is less likely to fluctuate than litho with its ink/water balance problems and Three-cylinder system.
4. Durability of the photopolymer material, is superior to most litho printing surfaces. Runs in the order of millions are quite common, compared with the tens or hundreds of thousands expected from a typical lithoplate.

THE FLEXIBILITY OF WEB-FED MODULAR CONFIGURATION

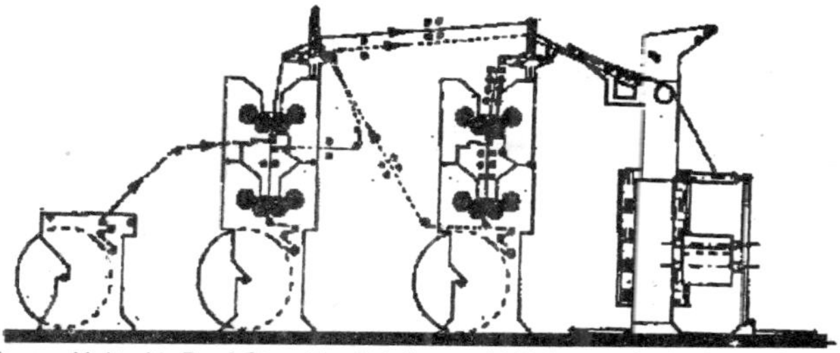

Standard/super Unit with Reel Stand for Printing multi-Colour on both sides of the Web.

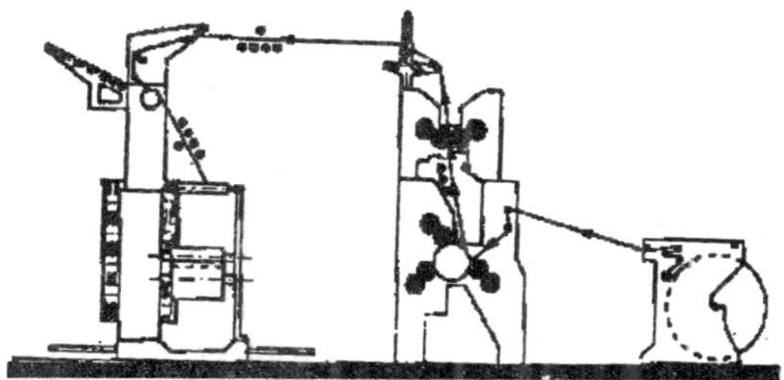

3-Colour Satellite-stack Unit with Reel Stand for Printing 4-3 Colours on one side of the Web and Single Colour on the other

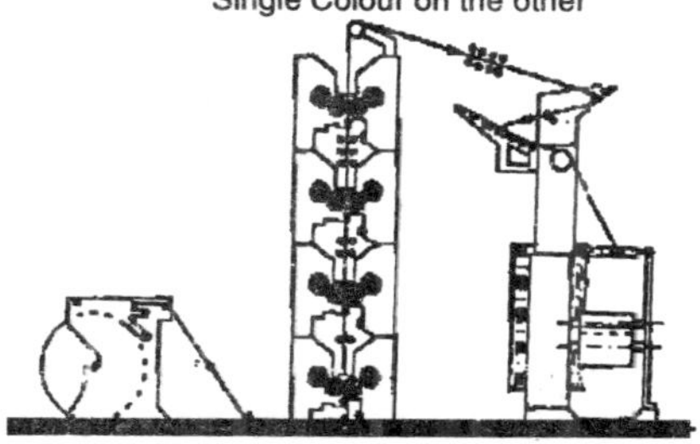

4-III with Reel Stand for Printing 4-Colour on both sides of the Web

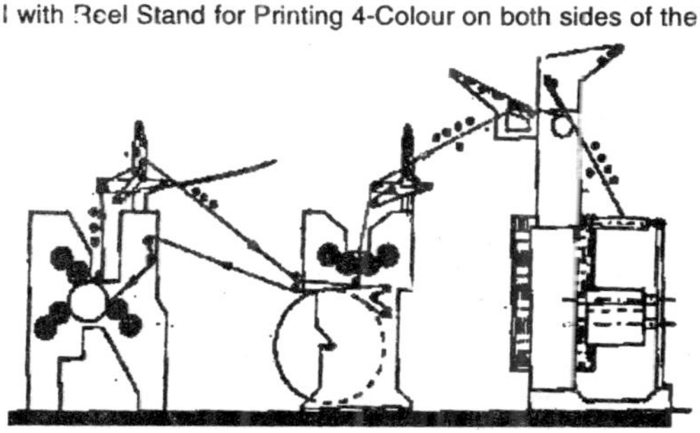

2-Colour Satellite + Standard/Super with Integral Reel Stand for Printing 4-Colour on one side of the Web and Colour on the other Saving the space of Reel S

5. Plate size will correspond to the printing area, A one centimeter square patch will only need a one centimeter square plate and not a full cylinder size plate as with Litho.

Photopolymer relief plates- photopolymer platemaking systems can be quite varied in their applications. Small desktop types can be purchased for producing the rubber stamp type of plate, or for making relief plates, for the imprinting units of litho printing presses. And in-line production system are available for high-volume production such as books, magazines and newspapers.

The hardness of photopolymer plates may be varied to suit the needs of the work in hand. Hardness is measured in degrees shore, on a scale which runs from zero hardness to soft glass hardness at 100 S(soft). A soft plate could be less than 40. Stand may be used for run-of the mill work such as paper back book work or business forms. Harder plates of 60-70 S(soft) will be used where fine line work and halftone printing work are concerned.

Generally there are two types of photopolymer relief plates:-

1. The solid plate, which is provided by the manufacturer in a range of basic sizes, calipers and hardness for subsequently processing.
2. The liquid plate which initially comprises a molten resin mixture, which can be extended from its reservoir to a required size and caliper prior to processing.

Both types may be used either as unsupported plates, which can be secured to the printing cylinder by adhesive, or fused on to plastic or metal based to provide better stability and rigidity and mounted via plate tensioning and champing systems.

Solid Plates- Solid plates usually comprise a stable metal or plastic base with the light-sensitive photopolymer material keyed firmly to it, plus a thin protective film to avoid premature exposure.

A flash exposure may be given in the first instance, in order to wake up or presensitized the material prior to the main exposure.

The protective film is then removed and the main exposure, made in a vacuum frame, with the plate material in firm contact with a negative.

Exposure to ultra-violet light hardens the photopolymer and determines the printing image areas.

The plate is then passed to a procession unit, in which it is washed out or sprayed with a solvent to remove the unexposed polymers, thus forming the relief image. Some systems employ an aqueous-based wash-out solution which may be an advantage where environment and pollution problems are of consequence.

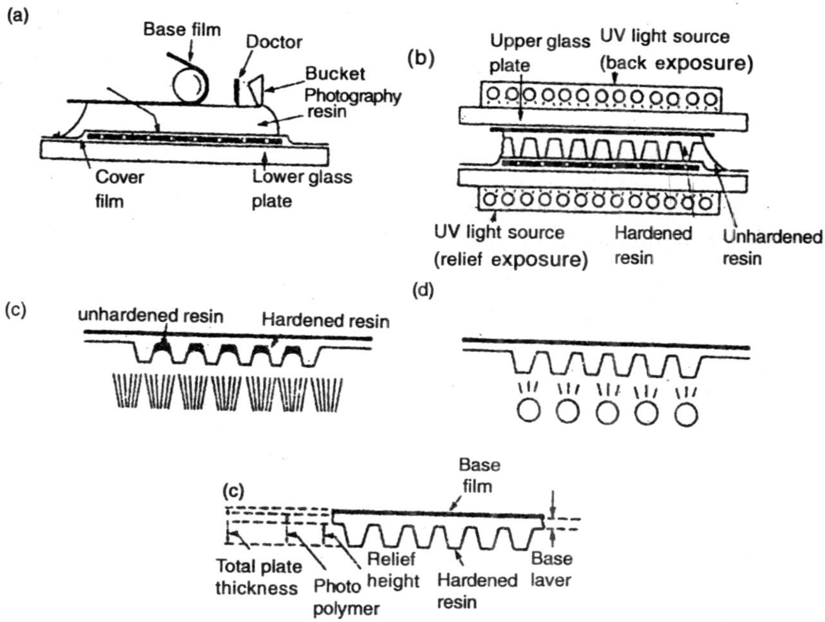

Figure : Photopolymer relief plates - the liquid resin system (a) Preparation, (b) Exposure, (c) Wash-out, (d) Post exposure, (e) Finished APR plate.

A final drying and curing unit will solidify and finalize the plate ready for forwarding to the printing press.

Liquid plates- Liquid plates function on similar principles to the solid version, except that the polymer is kept in molten form, in resin tank, until it is required to make a plate. The polymer is then extruded on to a light table holding the negative, which is protected by a thin cover film and doctored to the required thickness. The resin sets rapidly and at this stage can be exposed to ultra-violet light, to establish the image areas.

Washout is similar to that of a solid plate plus final at drying, curing and hardening. In most liquid systems surplus resin may be turned to the bulk supply and recycled.

Proofing

Hard copy proofs- Hard copy proofs may be obtained, prior to platemaking, such as a laser printer linked directly to a page planning system, or as contact proofs from the negatives, which will be used to make the printing plates, or the paste-up copy prepared at the page make up stage.

Soft copy proofs- Soft proofs may be viewed on VDU terminals, at the make up stage. This mainly occurs where text or monochrome (copy to print time is short) as in newspaper production, very time bound job is concerned.

Press proofs- Unlike majority of litho press proofs, these will be printed on rotary proof presses, where the plates are curved around a cylinder of a similar circumference to that of the production press.

Rotary proofing- Rotary proofing is preferred owing to the fact that photopolymer plates are significantly thicker than litho plates and will distort more when curved around a cylinder. It means the print length will be shorter, if the plate is proofed flat, compared with printing on the curve. The significant factors in determining the final print length are plate caliper and cylinder circumferences.

Photopolymer plates may be less than a millimeter in thickness or more than 5mm for certain purposes. A typical 2mm plate wrapped around a 60 cm, cylinder may print more than 5mm longer than it would if printed on a flat bed proof press.

Ideally, whenever accurate print length is of importance, photopolymer plates should be exposed, processed, proofed and printed all on cylinders of the same circumference.

There are three main component areas in any relief printing unit and these are: plate cylinder, impression cylinder and inking system.

Plater Cylinders- Plate cylinders are designed to take the wrap round plates and hold them firmly and accurately in the cylinder either under tension, or by magnetic force.

Tension systems requires the plates to be gripped head and foot and put under stress to ensure accurate conformation to the cylinder. the clamps gripping either side of the plate must be below the cylinder surface and the gap or slot into which the ends of the plate are inserted, should occupy the least possible areas of the cylinder, in order to allow the maximum possible circumference to be available for printing. An automatic plate locking mechanism should be employed, particularly on multi-unit, mulitplate, multiweb publications presses.

The printing cylinders will have either built permanent magnets or electro magnets which can be switched on and off as required.

Pin or stud registering will also aid quick and accurate mounting of the plates on the cylinder. In this technique the leading edge of the plate is punched with precision holes at the planning plate making stage.

Impression Cylinders- Impression cylinders may be plain, continuous steel cylinders, which serve to support the web and act as a counter force for the plate. This is acceptable where the softer grades are employed for more precise image. It is more common to use an impression blanket on the cylinder to

1. Soften the effect of printing pressure
2. Reduce wear on the plate and
3. Allow an element of tolerance in the impression Zone. Blanket generally comprise a number of piles of fabric and rubber or plastic material. The idea structure is to have a resilient base, which will compress under pressure from the printing plates, but a firm surface layer to ensure accurate of the relief image.

As with plates, the blanket need to be secured at both ends and tensioned around the cylinder, with the least possible loss of printing surface. Over tensioning means that it will risk stretching and damaging the blanket.

On the other hand, under tensioning will permit the blanket to move under printing stress and cause bulges, mainly in the areas of the blanket gap, which will affect the printing quality and increase the chance of web breaks.

Pressure between the plate and impression cylinders can normally be finally adjusted to allow for different caliper and types of web material. The cylinders are provided with hardened steel bearer rings at either end, as an integral part of the cylinder. The hight of the bearers will be equivalent to either the plate caliper or the blanket caliper respectively, plus any allowance for under-packing

Non-bearer cylinder are also used, where the need for really precise cylinder circumferences and minute pressure control is not so critical.

Inking System- Inking of photopolymer plates is, by means of roller trains, not dissimilar to those on litho presses. The nature of the ink varies from paste to semi-liquid consistency and is generally somewhat less viscous than litho ink.

The ink is metered to the roller from a duct where the ink flow is controlled by keys acting upon a flexible blade, making near contact with a rotating steel roller, as in sheet-fed. For very high volume production the ink may be held in a bulk store and pumped to the individual units, then metered to the roller train. For simpler presses, manual control of ink metering may be employed. But remote control via a central control panel will be a standard feature on high volume presses.

There are few types of arrangement of the printing units but will depend on the nature of the work and end product.

1. **Arch-type perfecting units-** may be used for multiweb and publication work, where the webs are fed in from below the units. These will differ from the web-offset versions in that they do not operate on the blanket-to-blanket principle as there are no offset cylinders. The webs are therefore fed from one printing plate and impression cylinder to the next. Good control of the webs is ensured as they wrap around the impression cylinders but it is also to ensure that ink from the first side printed does not transfer to the second impression cylinder and cause set-off on subsequent impressions. To minimize this problem, blankets may be coasted with a silicon layer which will refuse to accept ink from the web.

Extra spot colours may be obtained either by additional printing plate and impression cylinder, possibly located above the main units or by adding a second plate cylinder with its own inking system to impress against a common impression cylinder.

The another technique, simply by reversing the direction of travel of the units and re-threading the web in a different route may be used, for perfecting in black and white or for adding extra colours.

2. **In-line units -** In line units will be used where the web travels horizontally and several colours are to be printed on one side of the material, as with the majority of packaging work. The plate cylinders and their inking system will usually be above the web for convenient accessibility and the impression cylinders below the web.

Printing on the underside of the web may be, by means of specially designed reverse printing units or by the reversible unit technique, where the plate and impression cylinders is mechanically reversed and the web re-routed.

3. **Multi colour units-** Multicolour units may be used, where precise register is essential, as with colour half tone work. Four or more plate cylinders, each with its own inking system will be arranged around a central impression cylinder, to provide a compact but efficient unit. This type of arrangement is actually suitable, where the majority of a publication is monochrome and a colour section is required. Care must be taken in ink formulation to ensure first, that the inks by or trap rightly one on top of the other and second that the superimposed ink film are sufficiently dry before being subjected to the stress of folding, cutting, rewinding etc.

Web offset - The arrangement of the printing units in a web offset press can vary depending on the requirements.

Main production possibilities : to blanket perfectors for the monochrome work, plus a colour satellite unit. This is an ideal situation where the majority of paper are monochrome and colour is required only on a limited number of paper.

Folding possibilities

T Former fold (first longitudinal fold)

Z Cylinder fold (first cross fold)

K Third fold (cross fold, second longitudinal fold

P Parallel fold (double parallel fold second cross fold)

Figure : Folding Possibilities

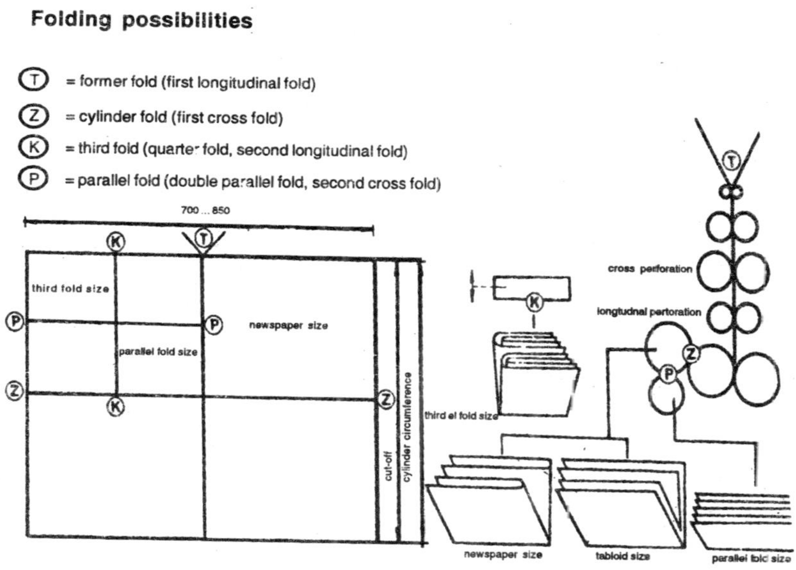

NEWSPAPER PRINTING

With the introduction of offset process, extensive automation entered the pressroom. Because they must compete with T.V, the newspaper publishers feel compelled to take full advantage of the strength of offset, i.e. colour printing, which in turn shall bring about a greater demand for higher quality colour reproductions. This orientation towards the changing media landscape inevitably results in an enhanced newspaper printing quality, which means that the press installations must become accordingly larger,

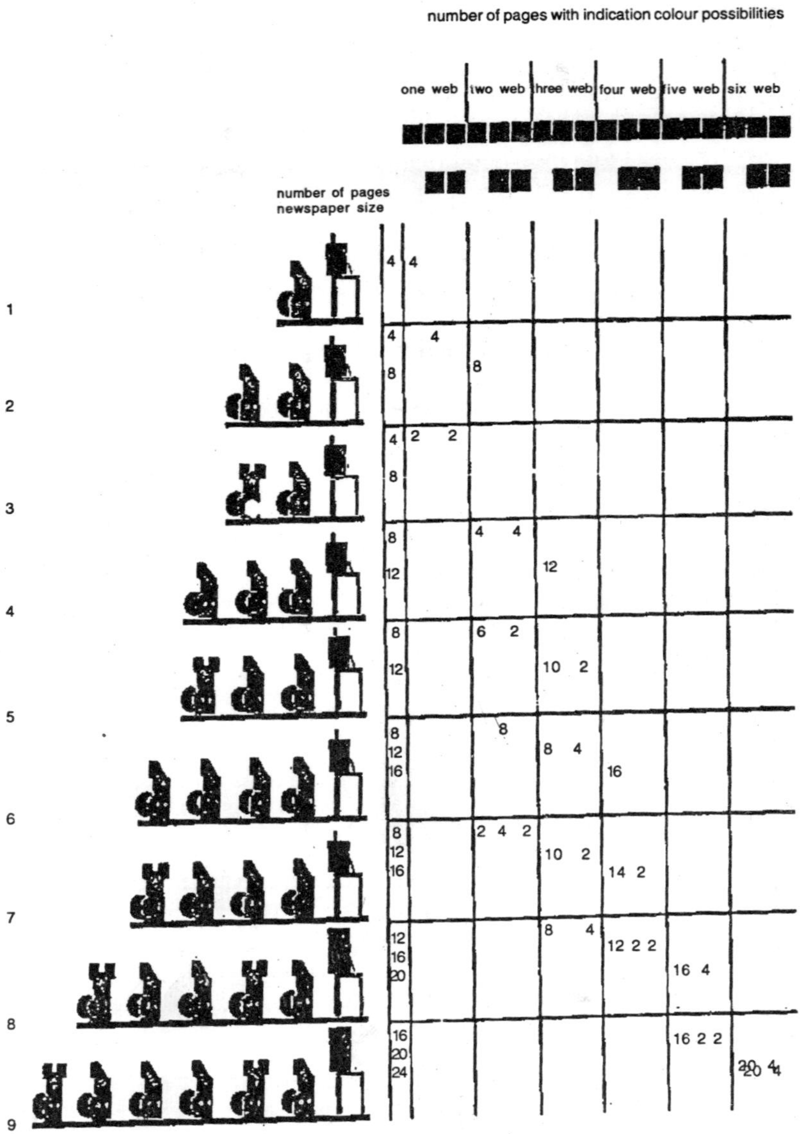

Figure B : Machine Configuration

more powerful, offer a greater flexibility of application, and, therefore become more difficult to operate.

Diversification and product differentiation are becoming increasingly important for newspaper printers. The competition of the print media, especially the daily newspaper, with local and regional information providers from the new media sector is bringing about.

Newspapers must become even more up to the minute. With colour supplements and advertising free sheets they are fighting for share of the advertising market, always in the hope of increasing circulations should readers and the advertising industry be animated by the high standard of printing quality and colour also in the editorial section. The newspapers can meet the exacting demands resulting from this development, provided that their production equipment can guarantee both short change over times, as well as consistently high production performance. The level of efficiency required must be achieved by short non-productive times and shift operation.

The modern offset process, that is being used increasingly for newspaper production, continues to set the standards for quality and production safety. This enables the newspaper to visibly enhance its attractiveness and its appearances. However, neither the quality of the printed pictures, whether editorial or advertising pictures, nor of colour printing, have yet to fulfill their full potential.

Since the start of euphoria surrounding offset, the press designers have constantly presented new printing unit configurations. The development went from the Y-type unit to the various satellite presses upto the blanket to blanket printing units.

The press superstructure is usually equipped with a hot-air or infrared drier, thus enabling also better paper qualities to be run without smearing problems and consequently a better printing quality to be achieved.

The good result that is achieved both in monochrome and four-colour printing is obtained with relatively simple, and therefore favorably- priced technology, combined with less start -up waste and a consistent printing quality throughout the production run. A greater operating comfort by simple holding and reduction of the maintenance costs, confirm the correct approach of the anilox technology.

Anilox is not a printing process, but in fact is only an old type of plate inking system, adopted from gravure printing and used in flexo to obtain a thin ink film. We the students of this subject know that both in letterpress and in offset, to obtain an absolutely clean and high quality print image it is necessary to have a consistently thin ink film on the ink from rollers, and thus achieved by multiple splitting and distribution by a large number of rollers. During this process, image- dependent resettings frequently cannot be avoided, depending on the ink taking. Usually it was attempted to eliminate the resulting irregularities in the print-out by means of a complicated inking unit. Moreover, ensure a consistent ink supply, a large number of ink screws is necessary that today are preset and remotely controlled by electronic means. Where ink screws are now no longer required, the plate scanner also become superfluous. When with, ink anilox - letter press, hard plates

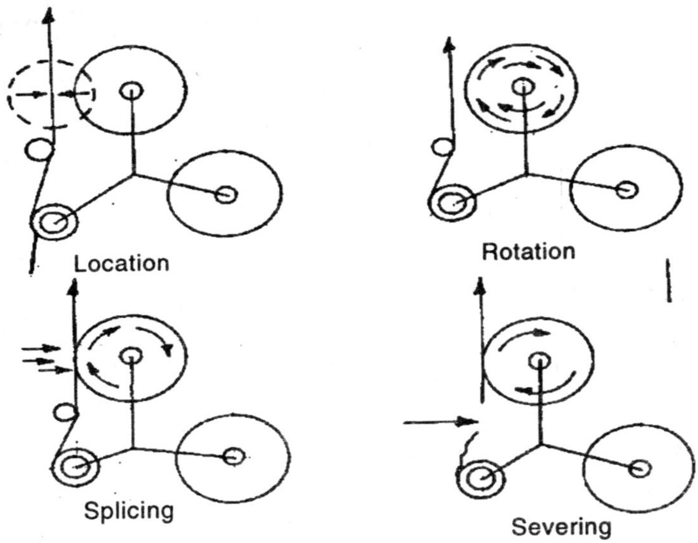

Figure : High speed reel changing

and oil-based inks, after the elimination of the unavoidable initial problems, the simplicity of the process was confirmed, no bearer rings, no throw on and throw off impression, no damping unit, no auxiliary drive for anilox rollers were needed, no drying - in of the ink was observed, and even less waste was produced, efforts began to apply this principle also to the offset process. A certain disadvantage of anilox- letter press compared to offset is the expensive plates.

The keyless anilox newspapers offset short inking system of the commander consists of the anilox engraved rollers, the chamber doctor blades, the ink pump, the ink rough, the cylinder-sized forme rollers, a small blanket roller, an oscillating ink roller between the anilox and the forme rollers. A roller washing device is not neccesary in the anilox press.

In anilox offset, the ink is transferred from the anilox roller to the plate, via an ink roller and a forme roller. It has been established in developed countries that ghosting can be eliminated when the forme roller has the same diameter as the plate cylinder. The chamber doctor blade has three main functions. It ensures that the anilox roller always receives a consistent supply of ink, removes the excess ink from the surface of the engraved roller, and re-introduces it into the ink circulation via a return system.In this way, the plates continually receive a new ink film without any electronic controls. Because the ink Keys of the conventional offset process are eliminated, the usual inking corrections during the printing run can be carried out only across the full width by tempering the anilox rollers via a change in the viscosity that is kept within narrow tolerances, In this case, the quality of the printed results depends mainly on an the standardized, uniform reproduction.

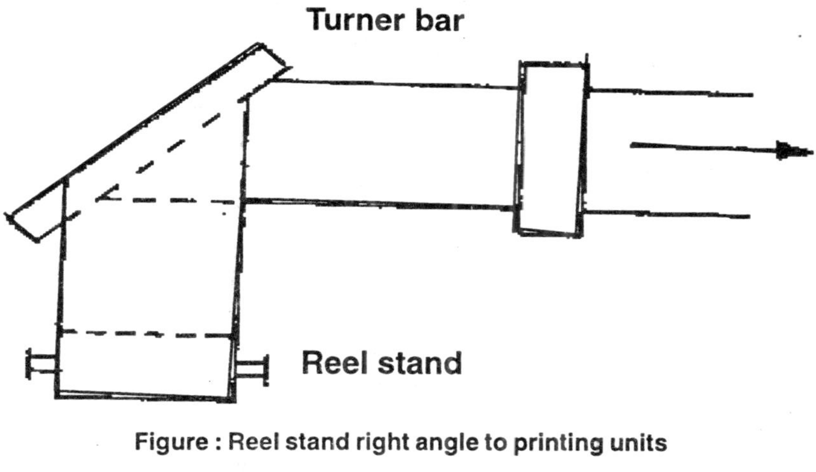

Figure : Reel stand right angle to printing units

Because damping of the plate continues to be necessary in offset, there is the danger of ink emulsifying. Anilox newspaper presses work successfully with a non-contact brush damping unit that guarantees a fast adaptation to changes in the ink supply, a quick plate release at the time of reproduction starts as well as an extremely thin, uniform moist film on the plate.

But again small runs do not require the super technology of a high performance press with the latest automation level. This development is intended to show the way into the future, especially for developing countries and their markets where there is no shortage of manpower and it is not deemed necessary to achieve a maximum performance with high-tech equipment.

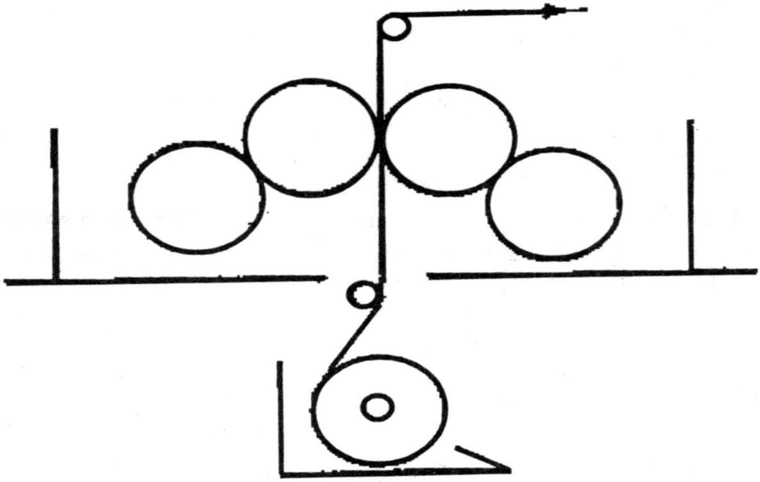

Figure : Reel stand in basement below printing units
POSITION FOR REEL STANDS

Folding : Folding may be either along or across the web.

Length folding

Length folding, as typified in broadsheet newspaper production is most commonly performed by a vertical, triangular 'former' plate, around which the web is wrapped and pulled by 'drawing' rollers to the cross fold and cutting sections. Several webs can be wrapped around one former, and the technique has the merit that, apart from the feed rollers there are no moving parts.

An alternative method of length folding is the 'plough' folder, so called for its superficial resemblance to a plough share. This technique makes the fold around a horizontal former is used more in packaging work than for publications.

Cross folding

Cross folding usually requires the web to be pushed in to the nip of a pair of rollers by a metal blade. In most instances the web will by this stage have been cut in to its final print length. More than one cross fold may be made in order to achieve magazine, pocket book, digest or diary page sizes.

Concertina folding

Concertina or zigzag folding is a technique for cross folding uncut webs for such items as business forms (continuous stationery) pack labels and computer listing papers. The web is usually pre-perforated laterally as an aid to precise and positive folding.

Combination folding

Combinations of these techniques are often necessary to produce a final completed product such as a newspaper or magazine, and to lend versatility to the end-product provision.

A typical newspaper folder, for instance, may start with a set of feed rollers which propel the combined webs to the former plate. A second set of feed rollers, at right angles to the first, will 'draw' the webs around the former to make the length fold on a 'broadsheet' paper.

The webs may then be cut to length (hence the term' cut-off) and passed to one or more sets of folding cylinders or rollers for cross folding. Finally the completed papers are delivered via an endless belt delivery system for counting, batching and despatching to their destinations.

Tabloid newspapers are produced in the same manner except that the webs are slit before folding in order to produce the final page size.

An additional 'quarter page' fold may be provided to produce a magazine-size product.

Ribbon folding

'Ribbon' folders employ a different approach in that the full width web is first slit in to page-width strips, which are then diverted around angle bars and associated before cutting to length and cross folding. This is a particularly versatile method for publications in which page sizes and numbers of pages may vary from one run to the next.

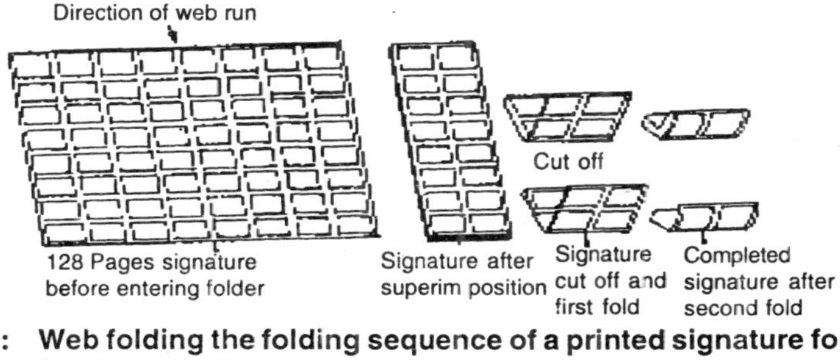

Figure : Web folding the folding sequence of a printed signature for pocket book production

Cutting :

Cutting the web to its final size is usually by means of a rotating blade cutting against either a counter -cylinder or another blade. Where a 'wad' of paper is to be severed and the quality of cut is not too important, a saw-toothed blade may be used, cutting into a resilient 'buffer' on the counter cylinder. This produces the typical newspaper-type cut-off being efficient and economical and requiring simple maintenance. In some instances such as magazine and sectional production, a saw-tooth cut may be used on press for fast delivery from the folder, with the final trimming taking place at the finishing stages.

More precise cutting will be by two finely-set and sharpened steel blades, usually one to rating and one fixed. To avoid a heavy load on the two blades as they make contact a slight 'scissor' action is made so that the rotating blade passes across the fixed blade from one side to the other. This requires the cutting section to be set at a slight angle to the web to ensure that the final cut is at a true right angle to the edge of the web.

Rewinding :

There are many occasions where the web is required to be rewound rather than cut to length and folded. For instance the end product may be required in roll form, such as adhesive tapes, wallpapers, snap-off bags, wrapping papers, etc.; or it may be that rewinding is an intermediate stage for further operations as in the case of preprinted colour work for newspapers, overprinting, varnishing, laminating, bag and sack making, etc. There are two basic approaches to rewinding.

The hand-filled or typed individual instructions are usually contained within a background 'tint' area, often comprising the name of the issuing institution printed in minute lettering and repeated in closely spaced lines. The ink used for this feature is normally of a 'figitive' nature in that it is water or solvent based and easily disturbed if attempts are made to erase, obliterate or otherwise modify the information typed or written upon it.

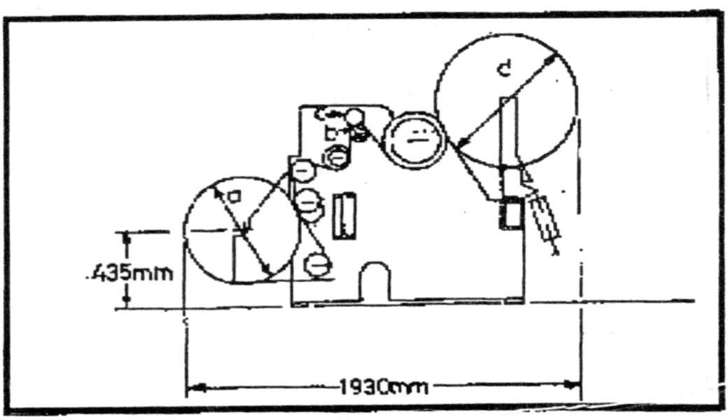

Figure : Basic arrangement of a typical slitter/rewinder

The individual cheque number and other coded information may be printed with an in containing minute ferrous particles which can be detected by a magnetic ink character recognition (MICR) system. This is not so much a security precaution as a means of identifying and sorting cheques for re-routing information to individual branches and accounts. To aid rapid recognition the characters are printed in a distinctive, stylized form so that the magnetic reading head can distinguish each characters pass below the reader. Ink film thickness, character size and spacing, and the position of the coded information on the cheque must be precisely controlled to ensure that MICR cheque sorting and data recording function efficiently.

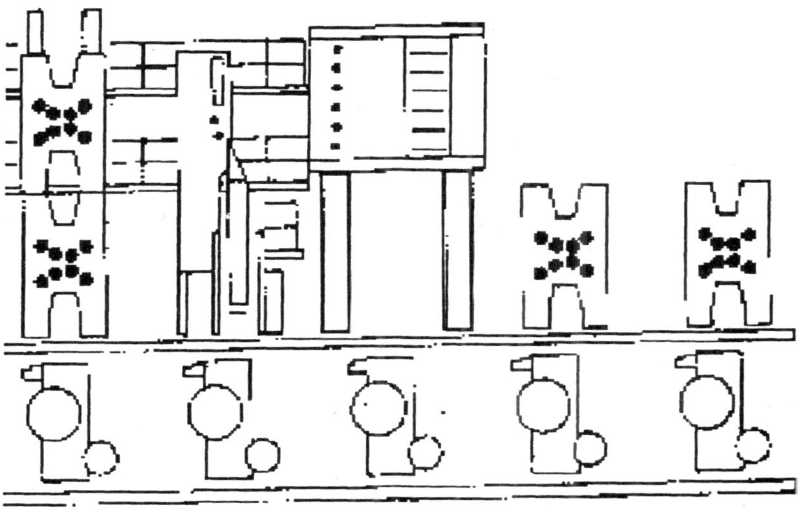

Geoman Weboffset (This configurations) installed in Times of India (Mumbai & New Delhi).

INFRA-RED DRYING ON WEB-OFFSET

The demand for assisted drying comes mostly from newspaper printers. Tests with infra-red (IR)- drying, carried out abroad, have already proved that this drying process could provide the possibility of improving quality and broadening the usefulness of presses. Initially, printers in foreign countries used IR-drying on their machines, but after R&D teams formulated the straight web technique, all the intermediary rollers were rendered superfluous. It also made it possible to top produce multi-colour mailings without smudges. The only disadvantage was the slow running speed of the press.

There have been problems with the planning and fitting of IR-drying units, to the already existing machines because the space available in the original construction of the press is usually insufficient to obtain length required for drying. Moreover, feeding in more energy does not achieve the expected result.

This means all Indian machine manufacturers and their respective R & D divisions have to start thinking about how a new rotary can be developed and installed with sufficient space left for IR-drying.

A possible solution could be positioning the IR-drying units shortly after the last print unit. The fewer guide rollers touch the printed paper web between the last print unit and the IR-emitter, the better- because any marks left by the guide rollers cannot be reduced or eliminated through IR-drying. To make the guide rollers ink-repellent, there is a whole range of rollers, from grained chromium rollers to hedgehog belts, available, as also different types of coating for each of these rollers. We can get acceptable results and running speeds, if a stretch of 3000 mm for the distance between the last print unit, or before the folder, can be provided. In this case, an IR-drying unit can then be reasonably planned for, and running speeds of 8.5 meters per second can be obtained under good working conditions.

The print carrier is also significant when IR-drying is used. The paper should, to a certain extent, still be absorbent, Even light-weight papers and satinised papers can be printed upon and dried. But the elements of the folding apparatus can easily become soiled if the ink balances are not altered immediately after the IR-units are switched off, this might result in a web break. So we have to be very careful about "radiation drying", and the "reaction of inks to radiation."

Different inks require different temperatures. The best results may be obtained with temperatures ranging between 80° C and 95° C. In higher temperatures, it is possible that the oils in the inks will evaporate or burn out. This will result in undesirable vapour formation.

Some inks react far quicker at low temperatures. But, upon operation, IR-dryers can get heat up to temperature of 140° C. It is always better to cool the web if it gets heated. Several options are available for this purpose, and if new rotary installations are made in the planning stage, a driven cooling tower can be provided for. Wherever plants are to be modernised, cooled guide rollers which are dragged by the web, can be used as this is considerably more cost effective.

The printed paper web is then dried out. This means adding humidity which had been removed during the drying process. Re-moisturising units can be supplied in the cooled or the un-cooled versions. Adding silica reduces any problems of static electricity which has a negative effect on the environment. However, warmth dispersed by the IR-driver to the surrounding layers of air should be removed, as this opens up the potential for heat recovery systems.

OFFSET PRINTING MACHINES

Sheet fed : single, two, four, and five colour machines are being produced in the country. Though there are about twenty five single colour machines manufacturers yet only 5-6 manufacturers are quality conscious e.g. Dominant Offset Ltd., J. Mahabeer & Co. Ltd., Manugraph Industries Ltd.,HMT, Max Machinery Mfg Co. & Speed-O-Graphics (India), Two colour quality sheetfed machines printing on one or both sides of the sheet are being manufactured in the country by four manufacturers Dominant Offset Ltd., J. Mahabeer & Co. Ltd., Manugraph Industries Ltd.,HMT and Max Machinery Pvt. Ltd.. When talk about four and five co our sheetfed there are only two manufacturers the total range of sheet fed offset machines right from single colour to five colour are being produced in the country by 2 manufacturers who account for 90 % of the requirements Dominant Offset Ltd. J. Mahabeer & Co. Ltd., Manugraph Industries Ltd.. However high speed operational machines i.e. over 10,000 iph with sophisticated C P C controls printing four or five colours or more are being imported.

Webfed : there are 4 or 5 reputed manufacturers of these machines in the country and they have a good share of this market specially for machines with operational speeds upto 40,000 i.p.h. like the Printers House Ltd., Raghbeer Machinery Ltd., J. Mahabeer & Co. Ltd. & Manugraph Industries Ltd.Machines with higher speeds and sophisticated controls continue to be imported. Manugraph Industries Ltd., Dominant Offset Ltd., J. Mahabeer & Co. Ltd. and HMT are exporting sheet fed offset machines to developed, developing and undeveloped countries. The Printers House Ltd. Raghbeer Machinery Pvt. Ltd., J. Mahabeer & Co. Ltd. and Manugraph Industries Ltd. are exporting web offset machines.

Binding and print finishing equipment : Except sophisticated machinery like total binding systems, flow line systems and special purpose machines, most of the other types of equipment including CNC. paper cutting machine are being manufactured in the country by a number of small-scale units. HMT Ltd. & Speed-O-Graphics (india) are exporting CNC cutting machines.

Quality control in sheetfed & web press : The importance of quality control has become very crucial along with the advancement in technology. The lack of quality control is the highest intangible cost variable in many printing companies. The lack of knowledge or implementation or neglegency represents hidden costs and are deducted from profits.

The very concept of quality control has touched a great height and as a result most high quality cautions, profitable printing presses have a quality comfort department, run by professionally qualified personnel. By virtue of which they are capable to prevent problems before they occur and thus help production without interruption. It minimizes downtime and waste. With advent of hi-tech, hourly cost are very high if we compare with oil mechanical presses.

DELTA J MAHABEER & CO. LTD

DELTA 662

Features	DELTA 661	DELTA 662	DELTA 664	DELTA 665
1. Pre - Loaded Feeder	S	S	S	S
2. Feeder Head : One - System Feeder Head	S	S	S	S
Two - System Feeder Head	O	O	O	O
3. Double Sheet Detector	S	S	S	S
4. Rotary Gripper Infeed System	S	S	S	S
5. Automatic Impression Engagement and Disengagement	S	S	S	S
6. Front & Side Lays Micro Adjustments On The Run	S	S	S	S
7. LED Monitoring Display	S	S	S	S
8. Film Dampening	-	-	S	S
9. Recirculating Water Pump System	-	S	S	S
10. Bearer To Bearer Contact Between Plate and Blanket Cylinder	S	S	S	S
11. Quick Action Plate Clamps	S	S	S	S
12. Powerful 19 - Roller Ink Train	S	S	S	S

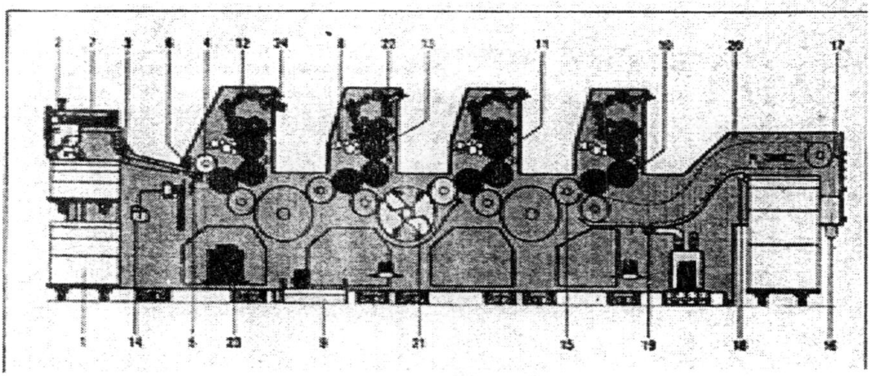

Features	DELTA 661	DELTA 662	DELTA 664	DELTA 665
13. Micro Adjustment Of Circumferential And Lateral Running Register	-	S	S	S
14. Central Lubrication Systems	S	S	S	S
15. Specially Treated Cylinders To Prevent Setting Off	-	S	S	S
16. Powerful Spray Unit	S	S	S	S
17. Delivery Jam Detector	S	S	S	S
18. Powerful Vaccum Brake	S	S	S	S

19.	Sheet Decurler	O	S	S	S
20.	Delivery: Low Pile	S	S	S	S
	High Pile	-	O	O	O
21.	Quick Change Perfecting Unit	-	O	O	O
22.	LED Changeover Display	-	O	O	O
23.	DC Main Drive With Controller	S	S	S	S
24.	Remote control Of Inking Unit And Register	-	O	O	O
25.	PROTOCOL/BACHER Pre-register Systems	-	O	O	O
26.	Alcohol Dampening	O	O	O	O
27.	Additional Rollers For Inking Unit	O	O	O	O
28.	GE FANUC Freely Programmable Control System	-	-	O	O
29.	Non-stop Delivery	O	O	O	O
30.	NPI Attachment	O	O	O	O

<u>Densitormeter</u> :This can measure density, dot gain, print contrast, apparent trap, hue error and grayness changes during a pressrun and indicate differences between production press sheets and prepress proofs. Modern developed electronic presses have on-board densitometers. Scanning densitometers appropriate for any press provide density readings on solid patches of colour bars and also as critical halftone or solid image areas.

Optical density tolerance should be determined from density measurements taken as an approved press sheet before the press run begins. These measurement can either be manually recorded or stored in computerised press densitometers or in stand alone electronic scanning densitometer datalases.

A formed density measurement program will result in minimum colour fluctuations within a press run and keep the press operator aware of any changes in ink balance in the press.

Computerized presses allow density tolerance to be maintained within ± 0.04 density units. Tolerances for mechanical presses may be between ±0.05 and ± 0.06 but density of 0.04 are often achieved when other press conditions are properly maintained.

Dampening solution ph : To achieve the optimum ph, dampening solution be mixed to á ph between 4.5 and 5.5 . This acidic tolerance seems to work best for most plants and provides the latitude to increase acidity to 4.0 if necessary . Remember dampening solutions with a high acid concentration (below 4.0 ph) frequently cause a number of predictable problems. The most common are retarded ink drying and ink set off which reduces print sharpness.

Once optimum ph is achieved, it is essential to maintain ph consistency. Dampening solutions ph should be measured at least once a shift and even more often on high-speed well presses where acid extracated from paper often contaminates the dampening solution, increasing its capacity.

Printers now are conductivity meters to monitor the tampering solution's concentration. Such type of devices can be an alternative to conventional ph measurements. Acidity should be measured with a ph meter or with a reliable ph strips.

Alcohol in dampening solutions : when alcohol is used in dampening solutions, ph measurement showed be made before and after adding alcohol. Though theoretically alcohol should not affect ph, but impurities in the alcohol some times change the acidity of dampening solutions.

Water treatment for fountain solutions : As we all know that water is a major component of fountain solution for offset printing. Erratic dampening fluctuation plate problems and slow ink drying are just a few problem that can be solved by water treatments. There are many different types and methods of water treatment e.g. filters, water softeners, DI (deionization) and RO (reverse osmosis). Printers can contact a water treatment expert to help them.

Filtration for fountain solution : This systems are designed to increase the life of fountain solution and reduce waste water discharge from a printing plant.

The filtration system will pick up containers, that accumulate in the fountain solution allowing for better control of ph and conductivity for an extended time.

Micrometer measurements : Ideally the pressroom should be, equipped with a table model deed with micrometer to measure the caliper of plates, blankets and packing sheets. Such a micrometer has the advantage of allowing measurement at the centre and both edges of blanket and paper. Because the device works on the principle of dead weight, human error is eliminated.

Measurement of ink film thickness : Ink film thickness tolerance limits for sheetfed offset presses are normally 0.2 m.m. (0.0002 inch,

0.005m.m.) and 0.4 m.m. (0.0004 inch, 0.010 m.m.) as measured on the oscillating roller above the form roller farthest from the dampening system.

Web presses : which use less viscous inks usually require a slightly higher ink film thickness-ranging from 0.3 mil (0.0003 in, 0.008mm) to 0.6 mil (0.0006 in, 0.015 mm)

Fluorescent and metallic inks may require higher ink film thickness to print properly, but even with such ink the high-end tolerance rarely exceeds 0.8 mil (0.0008 in,0.02 mm).

If an ink film thickness exceeding the high-end tolerance is needed to achieve a required ink density, it is likely that ink and fountain solution are not properly balanced or that the ink being used does not have enough tinctorial strength for the form being printed.

At least one wet film: thickness gauge, calibrated from 0 to 1 mil, should be available in all offset pressrooms.

Form roller settings : If form rollers are set too highly against the plate, the printer is likely to encounter plate wear & poor ink transfer from roller to plate. If the rollers do not contact the plate ink may not transfer at all. This will prompt the press operator to increase ink feed to achieve what appears to be adequate ink film thickness. As a result of increased ink feed, the ink film thickness on the oscillating roller above the form rollers probably exceeds the recommended 0.4 mile maximum for sheetfed printing and the 0.6 mil, maximum for web printing, thereby creating problem caused by excessive ink flow.

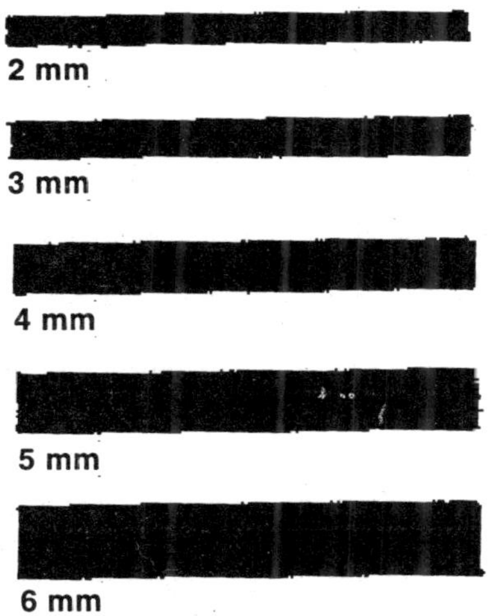

Figure : Ink stripe gauge

If the setting between the forme rollers, and oscillating rollers is less than between the form rollers and the plate, the form rollers may be driven by the plate, which could result in plate wear, slur, improper inking of the plate, and a loss of control of the inking system . In sufficient pressure may also result in roller slip due to ink lubrication, causing printing problem and possible roller damage. Form roller settings should be checked at least once a week by dropping the ink form rollers on to the plate and measuring the width of the resulting stripe.

Durometer - Roller hardness is measured with a durometer. Rollers on the press should be checked for hardness on a regular basis to avoid poor ink distribution and premature plate wear. 'The hardness specifications of a newly manufactured roller vary considerably depending on where the roller will be used and how it was made.

Upto 25% of the weight of the rubber covering of a roller is platicizer. Since plasticizer is leached out by washup solvents and solvents in the ink vehicle, some of the plasticizer in the roller is inevitably lost. In addition, plasticizer is driven off by the heat generated as the roller rotates and flexes on the printing press. As a result the roller diameter decreases slightly and the roller becomes harder.

Anti set-off spray powder control- Many advances have been made in developing rapid ink drying systems, but most printers will have to be contended with spray powder to prevent setoff.

If excessive amounts of spray powder are applied to overcome ink setoff in the delivery pile, the press operator may be overlooking the real cause of the problem such as an excessively acidic dampening solution that retards ink drying or an excessively thick ink film.

A bigger problem is the removal of excess spray powder from the pressroom air, especially at the delivery end of the press. One common solution is to exhaust the air away from the delivery.

Pressroom lighting:- In addition to lighting around the press, the booths or tables where press sheets are viewed must have lighting that conforms to recommended standard viewing conditions. Viewing at 5000 k is the current standard for assessing art work and proofs as specified by the American National Standard Institute.

Moisture content of paper - Paper for printing usually contains about 6% moisture. Paper is hygroscopic meaning it will take on or give off moisture until equilibrium is reached with the relative humidity of the room.

Sheetfed printers can use a sword hygroscope to measure moisture content of paper.

Excessive moisture content say above 7% generally causes register problems because of the increased dimensional instability of the paper, inadequate moisture, say, below 5% often causes folding problems.

Paper picks up moisture front dampening solution on the blanket of sheetfed and web offset presses. It has been estimated that upto 7% of the moisture that dampens the plate can be absorbed by paper. If paper moisture is more than 50% the moisture content of the paper will become particularly excessive, because the paper will absorb additional moisture as it passes through the first printing unit of the press.

A sword hygroscope cannot be used effectively for rolls of paper, so to provide accurate moisture readings for web paper, the web pressroom should be equipped with a moisture meter that can be clamped onto the lead of the web.

Most web press can be equipped with preconditioning devices, that either reduce or increase the moisture content of the web of paper. An equally distributed increase or decrease of *more* than 1% is difficult to achieve. Web presses are also equipped with remoisturizers to replace the moisture removed during the drying process. The web temperature should not exceed 350° F (180° C) other wise paper can cause blistering & folding problems. If IR (Infrared) dryers are used on sheetfed presses, the pile temperature of the sheet should not exceed 100° F.

BUSINESS FORM PRODUCTION

Web-fed presses are most commonly used especially for the production of continuous form for computers and data processing purposes. Offset litho is the main printing process although relief offset may be used for longer runs and direct relief printing for imprinting and numbering.

The basic construction of form press would be an in-line arrangement consisting a reelstand, infeed, printing units, ancillary units for perforating, punching numbering etc. and finishing stations for either rewinding, cutting or pack folding. Multiweb presses may be used for high-volume production of multipart sets, but it is more common to print individual webs and collate the parts at a later stage.

Fixed-size presses are the simplest. Economic and efficient permitting continues production and rapid change over from one job to another. This requires a measure of standardization, on a limited number of form sizes, as they have to correspond to the circumferences of the printing cylinders plus a volume of work to keep the press occupied.

Variable-size presses are built to cater *for a* number of form sizes by changing the main printing cylinders for cylinders of different diameters.

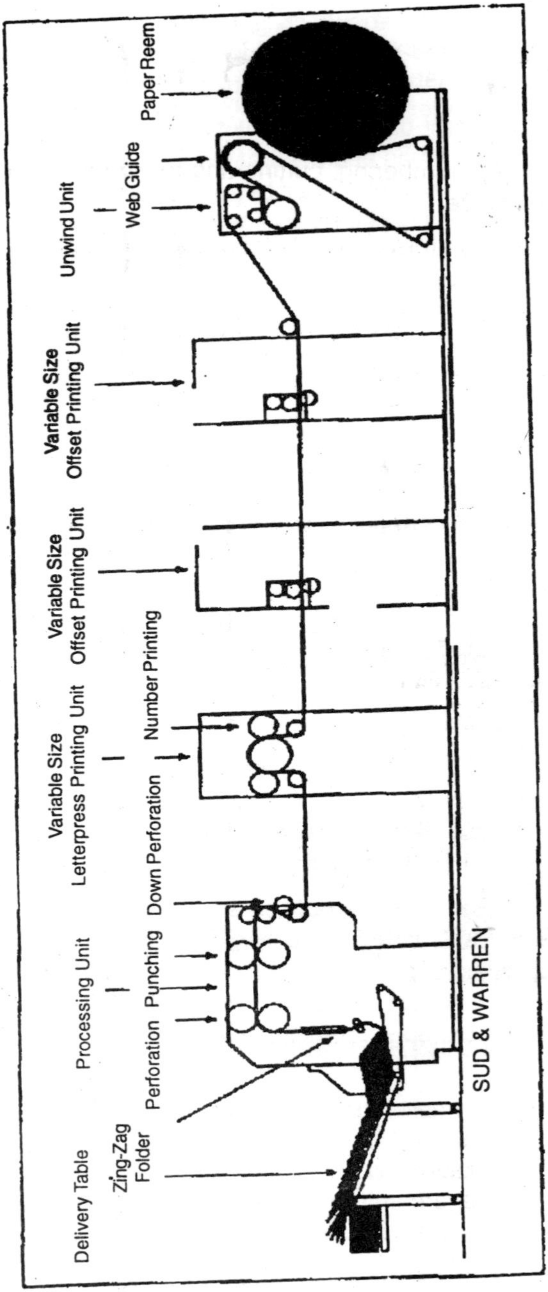

Figure : Business Form Press

This means in offset printing changing the plate and blanket cylinders only since the impression cylinder on a web press is generally a continuous drum. Some adjustment may be necessary to the inking and damping roller settings to compensate for the change in cylinder diameter. But for direct relief printing, only the plate cylinder needs to be changed. It may require with change-size presses to make similar adjustments to ancillary units such as perforating, punching, numbering, cutting and folding in order to accommodate the change of size.

Ancillary operation are very prolific in forms production. Hole punching is essential for form feeding through computers automated equipments are employed for filing and storage. Slitting is used to cut webs of paper to single- form width and to remove unwanted trim. Perforating is used for separating forms. Numbering is required on a high proportion of printed forms, stapling, crimping or line gumming will join sets of forms together. Spot carbon patches may be printed on the reverse side of the form where information is required to pass through to subsequent parts of the set only at discreet positions. Multitec, Bandhu Systematix Pvt.Ltd., Sud & Warren Pvt.Ltd & Convert Aid Enginneers are manufacturing Business form presses.

Figure : COSMOS 2614 PRINTER

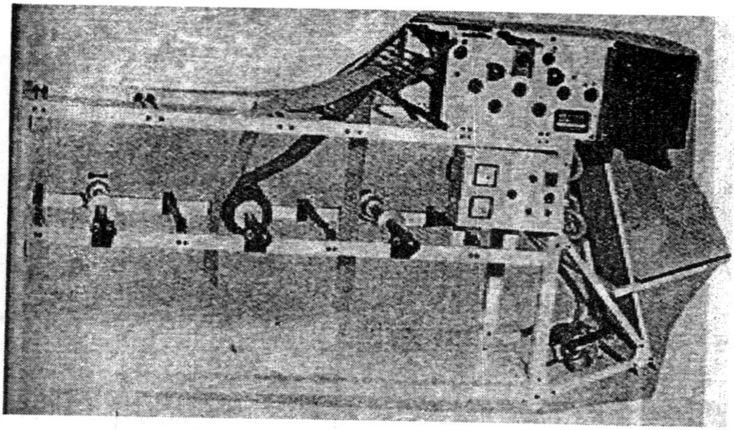

Figure : Galaxy 3614 Pack Collator

Figure : Orbit 1614 Converter

Multitec (Faridabad)

16

Specifications for Offset Publications

Quality means things to different people. Most printers want to produce quality work, and most customers except it. But how do you make sure that you consistently produce quality that's acceptable to your customers? A print buyer with a good original artwork should be able to rely on his printers to produce a satisfactory results.

Unfortunately, there are aspects of the printing process that can lead to problems. Proofing, platemaking and press-work occur in different departments and may be done in different companies. Nowadays it is common for positives and proofs to be purchased from one house, and printed reproduction from another house. Perhaps it is not too surprising that process houses often make proofs to a standard, that cannot be matched on a printing machine. Obviously, that gives the customer a misleading expectation of the final product. This can result in customer complaints, plate remakes, lost press time, general disappointment and perhaps lost orders in future. Similar problems occur when, as often happens, colour separations made to differing standards come together on the same printing forme. It is impossible to adjust the printing machine settings to optimize all the separations simultaneously.

When these problems occur, it is in the nature of things that the pressmen and the press-room supervisor will get the blame even though cannot control what happens at the proofing or separation stages. Production machines are inherently less flexible than proofing machines, and so it follows that the early stages in printing process must be organised, so that a good and predictable quality can be achieved on the typical production machine.

If all production machine could be made to print in a similar fashion on a given paper, and all proofs were made with the same characteristics as the production run, then all colour separations could be made to the same standard. Compromises on the printing machine would be avoided, proofs would be meaningful, customers should get a satisfactory result. All this can be achieved by working to a common specification throughout the process. This is the missing ingredient in the Indian printers quest for excellence.

Research Institute for Graphic Arts, RIGA, through a group of leading Graphic Art industry experts, has compiled a set of Standard Specifications for Offset Publications(SSOP), after intensive study of similar efforts made by a number of international research organisations, especially PIRA (UK), GATF (USA), FOGRA (Germany) and FIPP-International Federations of the Periodical Press (Europe). The pioneering work of these organisations has been widely acknowledged internationally and the specifications formally adopted as industry standards by association of advertisers, publishers, printers, process houses, ink- makers, plate manufacturers and paper mills. This has contributed greatly to enhancement of reproduction values and consistency of print quality in those countries.

STANDARD SPECIFICATIONS FOR OFFSET PUBLICATIONS

1. DENSITY AND DOT GAIN ON COATED PAPERS

Solid density and dot gain on coated papers should be within the limits set out below. The specification applies to dry ink film only. Density readings should be balanced across the full width of the control strip, within the tolerance of +/- 0.05.

Solid density on Coated Paper

	Solid density	*Peoofing tolerance*	*Production tolerance*
Yellow	0.95	+/-0.05	+/-0.07
Magenta	1.20	+/-0.05	+/-0.07
Cyan	1.35	+/-0.05	+/-0.07
Black	1.60	+/-0.05	+/-0.07

% Dot gain and tolerances on coated paper

	Gretag/FOGRA Strips		*Brunner Strip*	
	40%	*80%*	*50%*	*75%*
Sheetfed	18%+/-4%	11%+/- 2%	19%+/- 4%	14%+/-2%
Web Offset	25% +/- 4%	15% +/- 2%	26%+/- 4%	18%+/- 2%

2. DENSITY AND DOT GAIN ON GLAZED NEWSPRINT

Solid density and dot gain on Glazed Newsprint paper should be within these limits. The Specification applies to dry ink film only.

Solid density on Glazed Newsprint paper

	Soild density	*Proofing tolerance*	*Production tolerance*
Yellow	0.85	+/-0.05	+/- 0.07
Magenta	1.05	+/-0.05	+/- 0.07
Cyan	1.20	+/-0.05	+/- 0.07
Black	1.50	+/-0.05	+/- 0.07

% Dot gain and tolerancces on coated paper

	Gretag/FOGRA Strips		*Brunner Strip*	
	40%	*80%*	*50%*	*75%*
Sheetfed	28%+/-4%	16%+/-2%	28% +/-4%	19%+/-2%
Web Offset	35% +/- 4%	20% +/- 2%	35% +/-4%	23%+1/-2%

Note : These readings are based on Gretag D122, 142, 152 or 162 models densitometers accurately calibrated. To overcome the problem of variations in readings provided by densitometers of different manufacturers, pre-proofed colour reference bars will be provided by RIGA. The ink densities on these bars should be read on a densitometer used in the press and adopted as standards for proofing and printing.

3. SCREEN ANGLES

Preferred screen angles are :

Cyan	75	Degrees
Magenta	45	Degrees
Yellow	90	Degrees
Black	15	Degrees

4. COLOUR SEQUENCE

The preferred sequence, especially for proofs is : Cyan, Magenta, Yellow, Black. The three process colours should be in the above-stated sequence; Black may be allowed to float in any position.

5. INK QUALITY AND COLOUR

Inks should conform to BS 4666 which is the standard used in Europe and is now generally referred to as "European Shade Standards" conforming to CE113/67 ISO2846.

6. PROOFING SPECIFICATIONS

The five most important factors which cause mis match between the final print run and the supplied proofs are paper, set of positives, colour sequence, ink shades, ink density and dot gain. Following specifications should, therefore, be observed most carefully to ensure that printed results closely match the supplied proofs :

6.1 Proofing Paper :

Ideally, Proofs must be produced on the same variety of paper, that is to be used for final production run. Where this is not feasible, paper for proofing must have similar surface characteristics, gloss, absorbency and opacity as the productions paper.

6.2 Set of Positives :

It should be ensured that the proof is produced from the set of positives, which are to be used for final printing and not from a set of duplicates.

6.3 Colour Sequence & Ink Shades :

Must be the same as for final production run (as specified above under item nos. 4&5).

6.4 Proofing Densities & Dot Gains

It is technically possible to produce proofs on a flat-bed proofing press, using highly saturated inks, with higher ink densities and lower dot gain than is feasible on production sheetfed or web offset machines. This must be avoided. Proofs densities and dot gains must conform to the specifications stated above. If necessary, tack of the ink may be reduced or special inks obtained from ink manufacturers, to increase the dot gain on the proofing machine to bring it up to the standards specified above.

6.5 Progressive Proofs

Following sheets should be provided in a set of progressive proofs :

(1) Cyan

(2) Magenta

(3) Cyan and Magenta Mix

(4) Yellow

(5) Cyan, Magenta and Yellow mix

(6) Black

(7) Four colour mix

6.6 Control Strips :

Standard Control Strips must be incorporated across the sheets - in all progressive proof sheets - to facilitate densitometric evaluation of ink shade, density, dot gain, trapping, resolution and register. (SSOP recommends use of Gretag CMS, FOGRA PMS 80. GATF or Brunner control strips.)

6.7 Markings

The readings taken at the time of proofing must be recorded on the four-colour copy of the progressive.

7. VIEWING CONDITIONS

Correct viewing conditions are essential for effective appraisal and should conform to BS 950 Part II of ISO Standard 3664. The two standards are almost equivalent in effect; both relate to viewing colour transparencies and their reproductions. A colour temperature of 5000 K degrees is specified, together with light intensities for viewing transparencies and prints.

When viewing transparencies, the surrounding is important. The transparency on the viewer should be surrounded by an illuminated area at least 50 mm in width on at least three sides. Transparencies smaller than 70 x70 mm should have an illuminated surround, the area of which is not greater than four times that of the transparency. Illuminated area in excess of this should be covered with opaque mid-grey material.

8. PHOTO-MECHANICAL PRE-PRESS PROOFING

Dupont Eurostandard Cromalin and 3M Matchprint proofing systems are reasonable alternatives to a paper proof. However, utmost care is required while proofing, to ensure that the manufacturers control bars are incorporated on the proofs and the pigment densities and dot gain factors conform to specifications outlined here.

9. INK TRAPPING

The formula described in the explanatory notes in section three, should be used to evaluate effect of overprint colours. Trapping should be such, that the three-colour solid overprint of the primaries approach a saturated, neutral black. Control strips provide for a three-colour overprint step next to a solid black to facilitate this evaluation.

17

Proof and Proofing Techniques

A proof is an intermediate document produced before the actual printing operation. There are various types and qualities of proofs depending on the purpose for which they are intended. They may be used internally as part of the planning, processing they are intended. They may be used internally as part of the planning, processing and quality control aspects of productions, or they may be used externally for the information of the customer who may wish to have a preview of the finished job.

Simple Proofs :- Simple proofs are mainly monochrome or line work and could be the direct out put from a desk top systems a text setter, a camera or other pre-printing sources. Where the text is intended to be checked for only spelling, punctuation, etc. a simple photocopy may suffice. However, where the image quality is to be examined e.g., colour density or tonal effects etc. it may be necessary to produce an in-on-paper proof from printing plates.

Proof presses have a different basic design from lithographic printing presses. On most proof presses the plate is laid upon a flat bed instead of a cylinder, Having the plate on a flat bed gives easier inspection for the proofer and better facility for attaining and controlling variables as well as doing minor corrections.

It is important that the proofer is able to adjust and establish as many variable factors as possible, such as :

(a) Position(register) of the plate and the substrate to the press. Calibrated micrometer adjustment is essential for colour halftone work, combined with firm and positive location of the plate and substrate to the bed of the press.

(b) Pressure between the plate and blanket, and between the blanket and the substrate is usually controlled, by varying the height of the bed by means of a fire screw-threaded mechanism. Variation in pressure, in both instances, will affect the transfer of ink and the size of halftone dots.

(c) Inking and damping rollers accurate setting is essential. The balance between the applied films of damping solution and ink is a critical factor, in determining colour values and image strength.

CHINESE OFFSET PRINTING PRESS

COLOUR PROOFING :- For colour halftone work it is essential to have the press set to the finest precision standards. The slightest variations in dot size due to different pressures, ink variations etc. will misrepresent the tonal values of the image. The slightest false movement of the substrate, plate of blanket will give bad result in visible misregister of the superimposed colours.

COLOUR SEQUENCE :- The order in which the colours are superimposed will influence the appearance of the final result. There is no standard sequence for printing the four colours - yellow, cyan, magenta & black. It should be also remembered that there are jobs where more than four colours are required. It may be necessary for the proofer to try a variation of sequences in order to achieve the best results to match the artwork.

Control strips will usually be included in the margins to enable both proofer and printer to measure factors such as ink density, trapping, dot size, register, etc.

FILM LAMINATES : Film laminates, are based on sheets of light sensitive coloured films matched to the same standard as printing inks. Exposure to the appropriate colour separations and subsequent development, will produce four-coloured, transparent separations, which can be laminated in precise register for viewing either against a white substrate background or in a transparency viewing booth.

PIGMENT/TONER LAMINATES :- Pigment/toner, laminates buildup successive layers of coloured images on white base substrate. A transparent, light sensitive film is laminated to the base and exposed to the first colour separation. This leaves, the image area receptive to a coloured pigment powder, matched to the appropriate printing ink, which is dusted across the film to address to the image. Any excess dust is removed. A second film is then laminated over the first, exposed to the next colour separation left and so on until the complete proof is attained.

INK-ON-PAPER :- Ink on paper proofs can be made on eighter a proofs standard base of the actual job stock. The base is first coated with a light sensitive, water based ink of the desired colour. Exposure to the appropriate colour separation will fix the inked image areas to the substrate. The ink in the non-image areas can then be washed away, leaving the coloured image on the base. A second coating then follows and the procedure is repeated until the complete proof is obtained.

INK-JET- PROOFS :- Ink-jet-proofs are based on the principles of ink-jet printing, whereby fine droplets of ink are forced through apertures and deflected to the appropriate areas, on the substrate via microprocessor control. Any substrate can be used and computerised control makes this type of proofing system, suitable for use in conjuction with colour scanners. Ink droplets are just a few microns in diameter and they can be applied under high pressure, at the rate of many thousands of droplets per second, per colour. Resolutions in the range of 100 dots per continuator are currently possible.

VIDEO PROOFS *(soft Proofs) :-* This gives a preview of the job while it is in progress. A high resolution colour monitor is located at a suitable station in the system and programmed to display images in the graphic art terms with respect to dots, or lines per centimetre and proportions of colours in terms of yellow, magenta and cyan percentages. These may be linked to cable or network systems for viewing at more than one point and may also have provision for hard copy printouts.

ELECTROSTATIC :- Electrostatic colour proofing is based on as the same principle as monochrome electrostatic copiers. But the degree of control necessary to produce an acceptable colour proof is of a far higher standard than the normal office copier. Computerised, high resolution laser printers, are the basis of these devices, and this makes them particularly suitable for use in conjunction with colour scanners, where the digitized information in the scanner can be used to produced the colour proof.

PROOFING

Proofing is the very important aspect of the production process. Professional management of proofing tools enables to match the final job closely to the client's expectations at the lowest possible cost.

Thus it is important to consider both the role of the proof and the proofing technology itself.

The following are the main reasons for making proof :-

(1) To approve or if necessary correct a job before it is printed.

(2) To show how design decisions will be realized when the job is printed.

(3) To form a basis for agreement on what will be produced by the press and paid for by the customer.

The key point is that the proof should always match the characteristics of the printing process not the other way round. The role of the proof changes from a chent designer communication to a client-printer communication. Up to this time, most proofs will be generated as part of the design process and will include thumbnails and roughs to show the initial design concept followed by visuals and laser proofs of the layout and typography.

From the point when the job is sent to production and film and plates are output, the purpose of the proof is no longer to support the exploration of different design possibilities but to demonstrate that the production process has correctly implemented the specification and design. Both cases, the emphasis is on approval, and, in a well-managed job, approval of each stage in production is reached as quickly and efficiently as possible.

The main proofing stages are as follows :-

(1) Visual or rough—This allows approval of the designer's concept of the job before complete page layouts are produced.

(2) Design proof—the layout proof shows the finished design, with all the text and image elements in place. But it may not be possible to show exactly how illustrations will appear on the final job at this stage. It can also be used as a reading proof for the text if the copy has not already been finalised.

(3) Intermediate proof—Proofs made from the color separations before they are assembled with the other page elements usually called scatter proofs can be used to show how color separations have been carried out and how the images appear at the size and drop chosen and with the selected halftone parameters of screen ruling and dot shape. Fully composed pages incorporating text and graphics elements of the page in position can be proofed as page proofs.

(4) Final color proof or contract proof—This closely simulates the finished job and may be produced on separate sheets or made up as a book proof to look like the finished job.

Proofing Technologies—There are different proofing techniques available and each technique meets a specific needs. Proofing system technologies can be divided into electronic proofs, photomechanical prepress proofs and press proofs. Digital proofs are output from the electronic image-makeup data, while photomechanical prepress proofs are made directly from the final films and press proofs are printed with the printing plates.

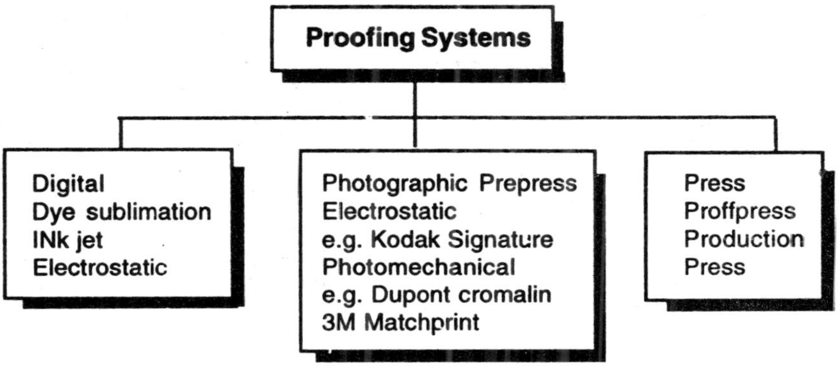

Digital Proofs :- Digital proofs, such as laser prints have been used by designers for many years as visuals and for approval of finished layouts and typography. The move toward desktop colour publishing has created a need for a digital proof, that shows how colour will reproduce and that can also be used as a final colour proof.

Electronic color proofing is a key enabler of digital color prepares, as it allows client approval of all page makeup and color separations and editing before the files are sent to the printer or to a bureau for output to film.

Electronic proofs are made directly from digital page and image data, at relatively early stage in production in comparison with other proofing methods. Most digital systems do not produce a conventional halftone dot, relying instead on dithering or continuous-tone printing to reproduce tonal values.

As with scanners, there are high-end, midrange and low-end proofing systems. Only the more costly high-end devices such as 3M Digital Matchprint, DuPont Digital Cromalin, and Kodak Approval will give a result in any way comparable to a conventional prepress proof.

The speed of digital proofing is dependent on the speed of data transfer and rasterization, as well as the speed of the imaging device. In some systems, the raster image is kept in the memory of the output device and multiple copies can then be made at a lower unit cost than the cost of a single proof.

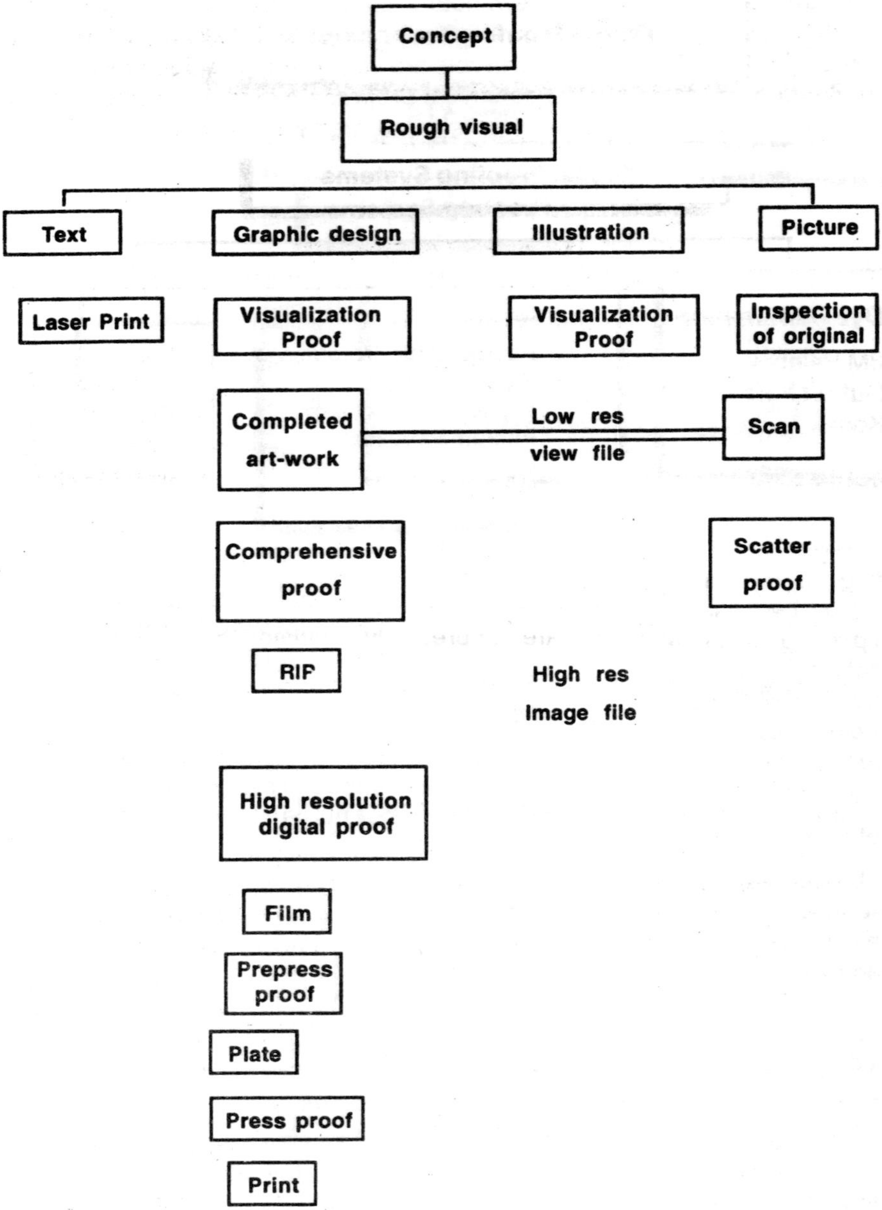
Concept
Rough visual
Text
Graphic design
Illustration
Picture
Laser Print
Visualization Proof
Visualization Proof
Inspection of original
Completed art-work
Low res view file
Scan
Comprehensive proof
Scatter proof
RIP
High res Image file
High resolution digital proof
Film
Prepress proof
Plate
Press proof
Print

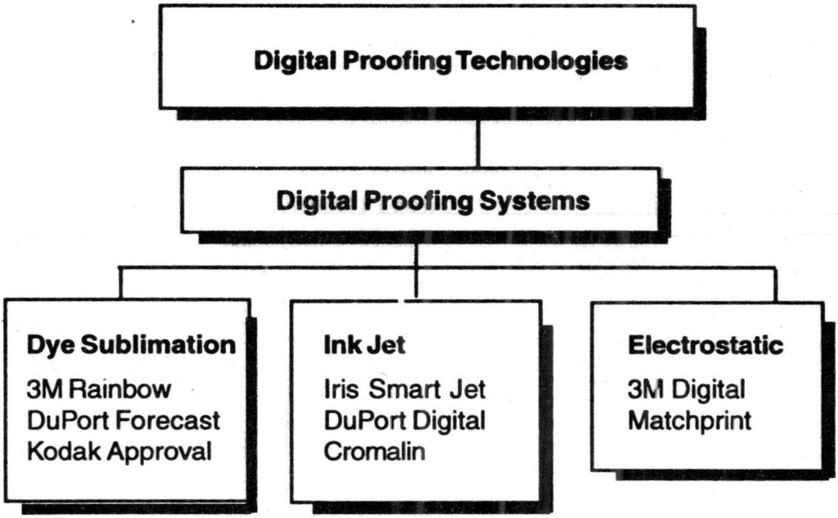

Digital proofing systems give the user a great deal of control over printing characteristics, with complete freedom to adjust tonal reproduction curves as necessary. However because the colorants are very different from printing inks, color values are not precisely simulated. Special colors are not available in electronic proofing systems and must be simulated with the closest CMYK combination. If the customer requires very highly accurate final color proof, digital systems can be used as a substitute for a photographic or press proof as on today.

The three types of digital proofing technologies used are electrostatic, ink jet and thermal systems.

Electrostatic Systems—Laser printers are regularly used for proofing elements of page design and typography. Color laser printers have proved to be an invaluable method of visualization proofing and when used in conjunction with an appropriate RIP and interface, provide quite reasonable results. One major advantage of the color laser printer is that once the date has been ripped, pages can be printed at the speed of the marking engine. The color laser printer is at present the only digital proofing systems that can produce multiple copies of proofs economically.

One high-end system uses liquid toner to produce high-resolution electrophotographic proofs. This has the advantage of working at a much higher resolution than a day toner system.

Ink Jet—Ink jets used for color proofing are different from low-end page printers. High-end ink jet proofing devices use Hertz technology to produce variable-size ink droplets, giving an approximate simulation of halftoning. Ink jet printers can use many different types of paper as long as they have the right level of absorbency.

Ink jet printers generate ink droplets, either by forcing a continuous stream of ink through a nozzle (with the unwanted droplets deflected electrostatically to a waste gutter) or by propelling droplets on demand according to the image requirements. Drop-on-demand desktop ink jet printers propel ink by thermal or piezoelectric methods. Thermal ink jets of bubble jets head a tiny amount of ink unit it vaporizes, its expansion forcing liquid ink through an aperture at a high speed, while piezoelectric system rely on the properties of quartz and ceramic materials to change shape, when a charge is applied, mechanically forcing ink out through an opening. Continuous ink jets are used on web offset process or in binding lives to add customer-of batch specific information. Most systems are effectively binary-droplets cannot vary in size-but some continuous ink jet printer have the ability to send several droplets to the same location as the paper, resulting in the droplets merging and spreading as they arrive at the paper surface. The number of droplets that can be sent to the same spot determines the number of gray levels that the process can achieve.

In a bubble jet printer, ink is vaporized by heat, forming a small bubble that forces an ink droplet out of nozzle. The bubble collapses when the heat is switched off, allowing more ink to be drawn into the nozzle. The cycle of heating and cooling takes place thousands of time each seconds.

Most desktop ink jet printers can produce resolutions comparable to low-end laser printers, at around 300 dpi. At this resolution, working at upto 20,000 cycles per second, a twelve jet array takes a minute or so to print a single page. The lower resolutions and faster droplet generation of press mounted continuous ink jet allow them to print at the full production speed of the press, which may be as much as 70,000 copies per hour.

Ink jet devices can print on most papers, although drying is slow on stocks with low absorbency, and wicking (ink spreading into the paper fibres) can occur on highly absorbent materials like newsprint.

Desktop ink jet printers produce colour documents and visualization proofs, but higher resolutions are needed for approval proofs, and even these cannot reproduce 'the halftone screen exactly as it will appear in the final printed job. With some color ink jet devices, the ink used is solid at room temperature and briefly liquefies as it is heated prior to printing. It then resoludifies immediately upon reaching the paper, providing a sharper image and stronger colors since the paper does not absorb this ink as it would any normal liquid ink.

Dye Sublimation—In this proofing device, tiny resistors are heated and vaporize a colored wax on to the proofing substrate. Like most digital proofing systems, thermal printers do not simulate the halftone screen, but by varying voltage that is applied to the resistors, the data controls how much colorant is vaporized and the resultant density of color on the proof. The result is similar to a continuous tone photographic print.

Substrates are usually restricted to papers with a special finish to accept the colorant. Consumable costs for day sublimation printers are higher than for other systems but the capital costs tend to be lower.

This technology has been very successful. Dye sublimation devices such as the 3M Rainbow are gaining acceptance as intermediate proofs and even as final color proofs.

Selecting a digital proofing systems—It depends on the following factors :-

(1) Output size-
(2) Substrates
(3) Data format
(4) Halftone capability

Advantages and uses of different proofing methods

Digital Proofs — advantages :-

(1) Eliminates expense of correcting final films
(2) Can be produced by designer or publisher
(3) Approval can be reached also at an earlier stage

use this method when :-

(1) Substantial alterations are likely
(2) Film is not being used (in direct-to-plate systems or in digital colour printing)
(3) Accuracy of simulation of final result is not critical

Soft Proofs — advantage :-

(1) Eliminates time and cost of hard-copy proofs

use this method when :

(1) Hard copy proof is not required
(2) Interactive correction is desirable

Photomechanical Proofs—advantages :—

(1) Accurately shows detail, such as halftone dot structure

(2) Eliminates the cost of correcting plates

(3) Cheaper than press proofs

(4) Allow approval at an earlier stage than press proofs.

Use this method when :-

(1) A single copy of the proof is enough

(2) Simulation of ghosting, printing inks & etc. is not critical

Press Proofs — advantages :-

(1) Economically many copies can be produced

(2) They can be printed on both sides and made up as book proofs

(3) Press proofs can be printed as the job paper

(4) They use printing inks with the same hues and overprinting characteristics, as those used on the production press.

Use this method when :-

(1) Multiple copies of the proof are required.

(2) A very accurate simulation of the final job is needed

(3) Clients have difficulty in interpreting other types of proofs.

Proof Correction Symbols

(1) Reduce dot size

(2) Increase dot size

(3) Reduce ink density

(4) Increase Ink Density

(5) Decrease Contrast

(6) Increase contrast

(7) Improve detail/modeling

(8) Too hard, soften

(9) Too soft, sharper

(10) Uneven tint

(11) Make good, broken line

(12) Improve register

(13) Correct slur

(14) Passed for press

(15) Reproof

However, electronic prepress could play a part in bringing about a change to more quality-conscious methods of working. Digital proofing tools have the potential to be used to produce a full visualization of a job at the design stage and support interactive alteration, until it can be finally approved. If the job could then be passed into a well-controlled production process working to known standards, no further proofing would be necessary.

18

Modern Printing Process Ink- Jet

In the past, the quality of ink- jet printing has not been good enough to prevent the recipients of direct mail products from throwing them unopened into the dust bin. If the mailshots are accompanied by the wrong name, house number, or post code, one can hardly expect a positive response to the advertising messages contained there in. No wonder then, that the main expert (technologists) in ink-jet systems for graphics arts applications, including Domino, Image, Scitex Digital Printing and Videojet, are constantly striving to improve speed. quality & accuracy.

Now-a-days, the majority of personalised direct mail ink-let printing, has been in mono, with the occasional use of spot colour. However Scitex Digital Printing, is emerging as the pacesetter for developments in high-speed, high quality, four-colour ink-jet technology. The other manufacturers are showing evidence of lower-cost developments.

Scitex Digital printing demonstrated a prototype full-colour (CYMK) digital press, based on continuous-flow ink-jet imaging. This technology will result in full-color 100% variable printing systems. that will provide fast set-up, quick turnaround, letter quality, direct-to-paper digital printing with variable output.

Colour images are produced by printing droplets of Cyan, Magenta. Yellow and Black on top of each other. This technology prints with multiple dots, at each pixel location, giving exact control of colour and density. Colour depth and saturation are also said to be enhanced by this process. A potential single row of jets is provided to enhance edge definition, registration and positioning.

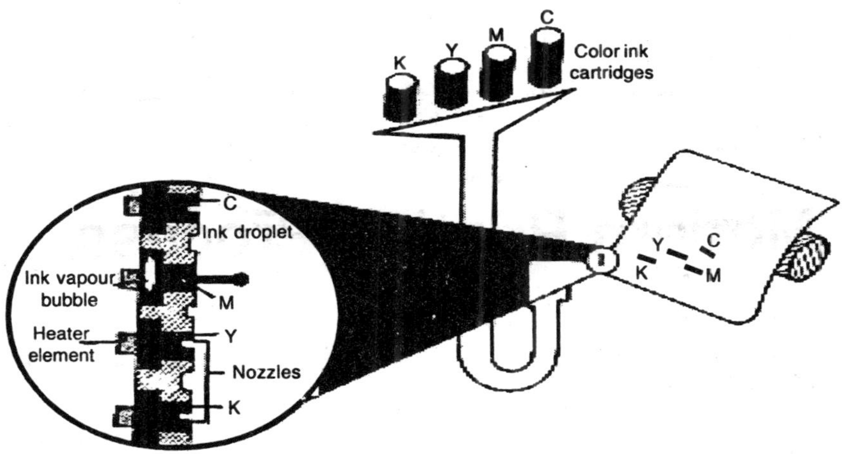

Figure : Ink - Jet

Suppliers have already promised for a device with low operational costs and speeds, that will considerably exceed the 200 ft./min. demonstrated. Development is in the early stages.

Benefits include complete variability of layout, contents and graphics, compatibility with industry standards for page layout and data transfer, the flexibility to set up and print jobs, ranging from short runs to high volumes on the same equipment. In addition, it gives the ability to personalise every document and to offer high-volume production at a low cost per page.

However, not every one will have a requirement for four-colour ink-jet print, nor be able to afford the press, which is expected to costly. Of greater concern to the other ink-jet suppliers is improving the speed, resolution and flexibility of existing technology.

Until the new press is available, ink-jet equipment prints only black-and-white or spot colour. It serves the commercial printing, business forms, direct mail, promotional graphics, catalogue magazine, lottery and gaming industries, as well as inplant printing and mailing operations. This is taking white paper at one end of the production line and transforming it into a full-colour, personalised and innovative leaflet or mail shot. Because this takes place in one continuous operation, optimum cost-effectiveness is achievable.

The ink-jet print head have been added to Zirkon eight-page four & five-colour mini-web presses. Digital printing offering customers the benefit of two komori 16-pages web presses able to personalise products, as well as an 8-pages Heidelbery-Harris. It is able to print, personalise and in-line finish in excess of 45000 products an hour, all in one pass. The three systems can all personalise in red & blue as well as black on both sides of the web. Most continuous stationery producers tend to personalise on just one side, due to cost implications and finish as a separate operations.

Ink-jet had come a long way from its original application for straight forward product identification. Ink-jet is no longer confined to Coca-Cola bottles. The technology has left this market and is offering versatility and creativity to markets previously not involved in ink-jet. With the capacity to print upto 40 inch width from a combination of one and 4.25 inch heads at speeds of 1000 ft. a minute, the quality and flexibility can now offer is outstanding.

Magazine publishers are now able to personalise their products on the outer wrapper and inside and have a full-selective capability. Postal codes and addresses are printed on the wrapper, while two personally-addressed coupons are added to the back cover containing e.g., a 10% discount on any tickets and a special offer for a Sunny's cricket bat.

Such capabilities could also be of benefit for subscription renewals or to encourage a higher reader response to advertised products, by supplying a ready-printed card containing readers' details so that they don't need to spend time filling it in themselves.

This level of sophistication is only possible through powerful control systems, which operates ink-jet printers to provide inside/outside addressing, personalisation, numbering and postnet bar coding.

Suppliers also offer the smaller editor 2 multi-position ink-jet controller for up to 24 print-heads, providing upto 48 lines of print. An extensive range of fonts and graphics is available, enabling numerous messages to be printed. One of its advantages is the ability to control multiple print heads located in different areas of the production line.

Ink jet print-head can be used to reproduce a variety of bar codes, fonts, user-generated graphics, logos and photographs.

Image also acknowledges that the way forward for addressing or presonalisation using ink-jet, is to achieve higher resolutions and multiple fonts readable in postal equipment, which are becoming essential with the higher speeds of mail handling systems.

As ink-jet improves in accuracy, resolution and speed, the only limitation to its wider use in commercial printing applications is the imagination of the end-user.

Ink-Jet

Image sprayed on to the paper with water based inks.

Many different ink formulations including inks with etching or acid bases for printing on coated stock

Problems: since they are water based, the images will smear if they were to get wet.

Since ink is water based, sometimes it is absorbed into the paper, resulting in a weak image.

Since ink jet produces bar codes through a sprayed image, bars formed at higher densities (closer together) become more difficult to read correctly. Up to 700 sfm.

Laser

Light impacts on photo-sensitive, hollow cylinder or drum. Laser beam hits and comes off a 6 or 12 sided prism and rotating mirror.

The beam of light rotates to the drum, swinging from the operators side to the gear side for one full line of image, usually 240 dots per inch in both height and width.

The image, exposed to light, is on the drum. Drum rotates to a hopper containing toner with developer, then rotates to the paper, and, usually, a corona wire (on the reverse side of the paper) pulls the image off the drum onto the paper.

Image is then treated by heat (either a pad from the back side or infrared light above the printed image).

Images are usually more on the gray side. Image sits on top of the paper and does not impregnate the paper.

To protect image, coating the bar code is recommended.

The printed image is light black or gray. The larger the image, the grayer it will be because of the inherent nature of the photo sensitive drum.

Average speeds from 90 to 130 pages per minute.

Ion Deposition

Image cylinder holds the image to the surface with an electrical charge.

The image head is a hollow tube or chamber, with a small hole in the bottom. The air in the Chamber is electronically charged separating plus and minus Ion's. The minus (-) or negative Ion's leave the chamber through the hole in the bottom.

The (-) negative ion's, place strokes or images on the image cylinder to create a negative image.

The image cylinder rotates to a hopper containing toner. The negatively charged ion's attract the toner to form the image.

The image cylinder rotates and transfers the image to the paper as it passes by.

* Under pressure: 200 lbs per linear inch.

* Shears: There is 1 degree angle between the impression cylinder and the image cylinder, so that the character image is sliced off on to the paper.

*As in the laser process, the image sits up on the paper and should be coated for protection.

Speeds run from 90 to 130 pages per minute and much slower.

Manography

Honeywell bull engine is the hardware. Print head (Coil) generates a square spot on a drum which rotates to a chopper containing toner.

Toner is then magnetized in the hopper and applied to the image on the drum.

The drum then rotates to the paper, at which point a corona wire with an opposite charge will pull the image off the drum on to the paper. It is then heat set.

Problems: When the toner is fused, it will fuse to itself.

The image sits up on the paper and will not impregnate the paper itself.

Image can chip or break off.

Sometimes there is teflon in the toner to help in the fusing process. This prolongs the life of the print head, but the chemical reaction harms the corona wire.

Finish the form with a protective coating.

90 pages per minute (approx. 82.5)

Led

Led-Light emitting diodes

Similar to the laser. Has a photo sensitive drum. Image head is similar to a bar print head with pins or grids. The diodes light the readout in a sequence as the drum passes by.

Image cylinder sees the image and takes it off the drum. Voltage is either increased or decreased. The cylinder rotates to a hopper with an electrically sensitive toner, which is attracted to the image charge.

It transfers the image to the paper using a corona wire with the opposite charge. It is then heat set.

Very slow. Only one company (Japanese) selling in the U.S. Breaks and chips. Only 5 to 7 pages per minute.

Thermal Direct

Special heat sensitive paper. Print heads are a series of pins that heat up and once heated, the pins impact on the paper burning a black dot where they touch the paper.

* Customer/Lose time heating the pins in the print head.

* Only a narrow image area (Approx. 4").

Thermal Transfer

Uses a wax or poly-based ribbon. Normal papers can be used. The image head is heated and image released to the paper from the ribbon.

Advantage over thermal direct is the lower cost of paper and its shelf life. Very, Very Slow; Much slower than the laser.

Electronis Art & Assembly Network

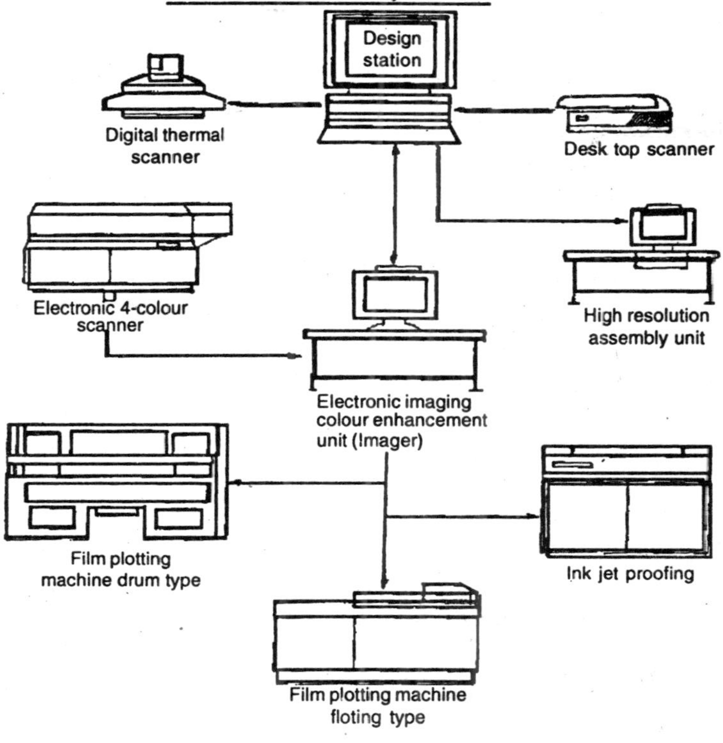

Film Plotting Machine

This is the final station of the process. Once the design is approved for printing, plate film must be created. This machine uses a small laser beam to expose film a resolution of 2,540 lines per inch. It produces negatives upto 34" × 48". Digitized information from the electronic imaging unit is networked to the film plotter, where it exposes each negative to produce each colour of the design. This is the first point at which film has been created throughout the entire process.

Basically there are two types of plotters : (1) flatbed plotters, where film is laid flat and then electronically laser exposed and (2) drum type platters, where film is hung and then wrapped around a large drum that spins and is laser exposed while turning.

Each of these electronic stations, when net worked, form a complete design-to-plate-ready film system.

Suppose a customer has an idea for a design. The customer can draw a rough sketch. If the customer is in Calcutta or Varanasi and the engraver is in New Delhi, the customer can fax the sketch to the engraver's electronic art station.

Fax in hand, the design station operator can scan the black and white fax into the electronic art system of the desk top scanner and send it to the design station. The image is now on the screen in a matter of minutes. The operator now can smooth rough edges, set type and position the type in the right places. Type can be curved, wrapped around irregular shapes or set flush left or right.

Once all the copy is in position on the screen, the operator assigns colours to the elements. The screen on the design station has all colours available to it. A colour wheel can be displayed on the screen, allowing the operator to mix and match his own colour.

Once colour are assigned, the design is ready to be proofed. The digitized information is sent from the design station to the digital proofer, where a proof is generated depending on the complexity of the design, this can be done in less than an hour.

The customer has a colour proof very quickly, with a minimum of time and money invested because no film, stripping or separation have been made.

Another benefit of the electronic art system, is the ability to design on 3,1/2" floppy disks. If a design requires promotion or revision, the previous version can be called upon the system, revision made and either proofs or plate-ready negatives created.

Once the proof is approved, we can begin to create a high resolution file on the computer. Remember that what we scanned in on the desktop scanner was low resolution and only for design.

Now we can scan in transparencies or reflective art on our high-resolution, four colour, scanner. The scanner then networks the colour separation to the electronic imaging unit, where it's colour enhanced. The separating is then sent to assembly unit, where it is electronically stripped together with the type and art created on the design station.

Black & White mechanical elements that were low resolution scanned for design on the desktop scanner e.g. company logos, must now be scanned in on the, high-resolution scanner and net worked too the assembly unit for positioning in the design.

The design is now ready to be digitally networked too the ink jet proofer. The important thing is that now we have made a proof of four-color process and line cop, without generating any negatives. If changes are required, we simply manipulate computer information and can then generate another proof.

Once the proof is approved, the design is ready to be made into film negatives. The design is now networked to the high resolution computer film plotter.

Some customers are setting up low resolution design stations in their own plants and through a telephone and modern are looking into the trade shop's high resolution electronic imaging system. This technology allows the printers to do their own design in the shop, by using the inexpensive desktop scanner and design station.

The best thing about the electronic system, is the ability to hook up low resolution, front and equipment that is cheaper to run rather than the very expensive, high resolution computers and film plotters.

This systems takes a lot of the graphic burden off the high-resolution equipment and puts it on the front-end, low-resolution equipment, where designing is cost efficient.

COMPUTER TO PLATE TECHNOLOGY (CTP)

Today big printers are investing in computer-to-plate (CTP) systems that by pass film entirely. With electronic imposition of complete flats, it makes sense to go a stage further and expose plates instead of film, eliminating several prepress production stages.

Imaging directly to plate also does away with the distortions introduced in film-based plate making. It is possible to resolve fine image detail more consistently, which is a special benefit with frequency—modulated screening where very small spots are being imaged.

Offset plates are quite thin. If the base material is flexible enough, they can be loaded in an imagesetter in roll form and exposed. Alternatively, dedicated plate exposure units commonly known as platesetter can be used for metal plates. These systems can in corporate automatic register punching.

Sometimes CTP loses out to conventional film-based platemaking due to the speed and low capital costs of conventional platemaking equipment. e.g. When duplicate plates or reprints are needed, a CTP system has to repeat the imaging process, while a conventional platemaking system has only to re-expose the film. Film is a convenient storage medium that is cheaper than archiving rasterized data on magnetic media. Proofing can also be a problem unless digital proofs are accepted as contract proofs.

It is not possible for every element in a job to be supplied a digital form, and then it is necessary to combine film produced conventionally with the digital data.

The platesetter receives the imposition instructions and exposes the digital information, masking the areas that are missing. The device then uses the register marks as a guide to position the films & expose them conventionally.

Computer-to-plate technology (figure Y) eliminates the film output and stripping stages (Figure X) in the production workflow.

Alternatively, film or ârtwork can be scanned and converted into digital form. Very high resolution scanners are used to avoid image quality loss on line artwork, such as type with fine strokes and to avoid the need to descreen halftones, which risks changing color values One type of scanner designed for this purpose converts the scanned data directly in postscript code, which can then be inserted into a postscript CTP workflow.

Film is a convenient storage medium that is cheaper than archiving rasterized data on tapes or optical discs.

When alternations are required the new matter is either stripped in manually or the films are re-output. It is more or less impossible to edit rasterized CPT data. Changes must be made to the customer files and new plates produced whenever alterations are required.

Platesetters are usually very large as they must be able to handle the full plate size for a given job (plates cannot be assembled from strips like films). Some models incorporate automatic plate handling mechanism and can accept plates upto 48 × 66 inch i.e. 1200 × 1650 mm. They may incorporate work flow management, including queing and preparing files while the previous job is being exposed, automatic compensation for shingling and placement of color bars, register marks and fold and trim marks. These systems at the very end of the prepress production workflow, are installed at printers or occasionally at larger trade shops.

In contrast to conventional photographic methods in which the whole plate is exposed from film simultaneously, in direct to-plate systems, only the area under the laser is exposed at any one time. Consequently the plate coatings for direct-to-plate systems must be more sensitive. Such coatings are too sensitive to handle in daylight and unlike conventional plates, they must always be handled and loaded in the dark.

By using high-sensitivity plates and multiple blue laser exposure heads, exposure at 4000 dpi can be similar to those for conventional diazo and photopolymer plates.

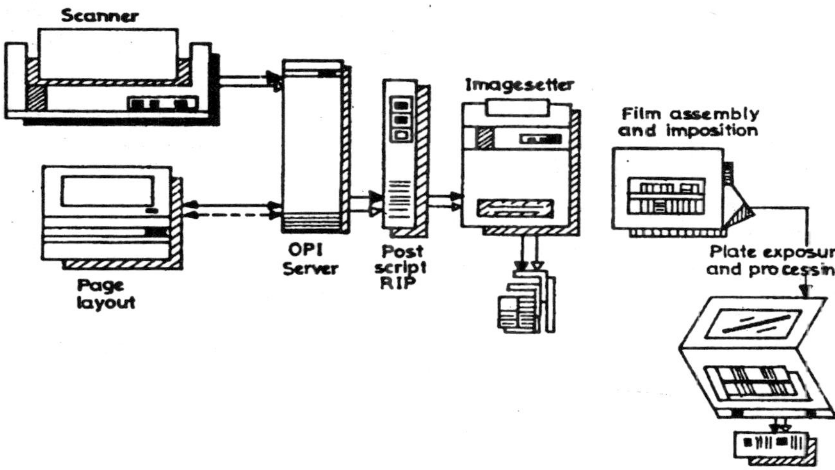

Figure : "X"

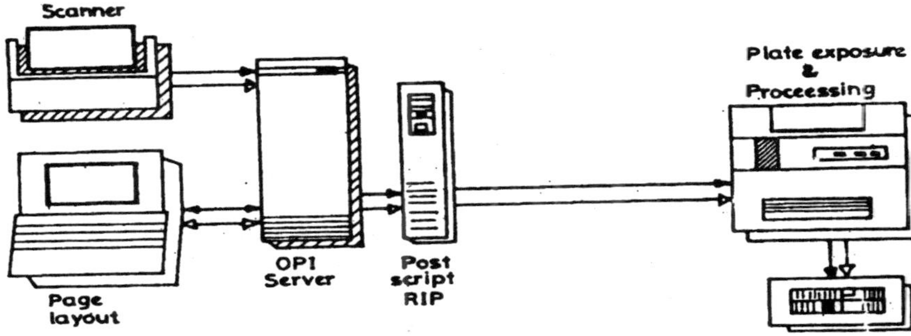

Figure : "Y"

Computer To Plate

It is not all that long ago since publishers supplied manuscript or typescript for the typesetter to re-key. Indeed some still do — but with the advent of computers and universal languages like Postscript an increasing number of authors and publishers do their own key-stroking and page make-up and supply the printer with a disk for outputting on to bromide or film.

A number of prepress houses are now able to offer a CTF (computer to film) facility which involves outputting eight pages of A5 on one piece of film already imposed with, correct heads, backs and gutters. Four of these are then exposed on a Stoesser print down frame for platemaking in very quick time. Some imagesetters are able to go from disk direct to 70mm film for automatic imposition. One of the problems are the moment is that generally speaking image setting film is much more expensive than negative film for platemaking. However, that situation is unlikely to continue. The aim has got to be to avoid the use of film altogether.

This brings us to a consideration of CTP (computer to plate) and as the technology evolves so some platemakers are struggling to keep up. CTP involves the customer supplying an electronic disk in Postscript or similar so that the prepress supplier can use this to go direct to plate thus avoiding the use of bromide and film altogether.

Table I Krause Laserstar — technical data

Model	LX110	LX170
Maximum Plate Size :	850 ×1100 mm (33.5" × 43.3")	1400 × 1700 mm (55.1" × 66.9")
Maximum Image size :	780 × 1100 mm (30.7" × 43.3")	1250 × 1700 mm (49.2" × 66.9")
Resolution (Variable) :	400 —1000 lines/cm (1016 — 2540 lpi)	
Exposure Speed @ 400 lines/cm: or @ 1016 dpi)	286mm/min (11.2"/min)	174mm/min (6.8"/min)
Exposure Speed @ 500 lines/cm or @ 1270dpi):	182mm/min (7.2"/min)	174mm/min (4.4"/min)
Exposure Speed @ 1000 lines/cm or @ 2540 dpi):	45.6mm/min (1.8"/min)	27.6mm/min (1.1"/min)
Data Transfer Speed :	6 MBit/sec	

Table 2. Krause Laserstar — plate exposure times

Plate Size mm	Image Size mm	Image length mm	Model	Exposure Times @ 400 l/cm (1016 dpi) Minutes	@ 500 l/cm (1270 dpi) Minutes
OB605 x 730	510 x 720	720	LX110	2.5	4.0
2C740 x 925	630 x 915	915	LX110	3.2	5.0
3B795 x i030	710 x 1020	1020	LX1I0	3.6	5.6
51035 x 1270	880 x 1260	1260	LX170	7.2	11.3
61140 x 1410	990 x 1400	1400	LX 170	8.0	12.5
7B1325 x 1610	1190 x 1600	1600	LX 170	9.2	14.3

There are at least two solutions available on the market that address large format press sizes and these are from Krause and Misomex. Interestingly they approach the problems in quite different ways. The essence of such solutions is that the disks supplied should be as Postscript files created using QuarkXpress, PageMaker, Ventura and other software packages. When disks are supplied they should be accompanied by laser proofs to ensure there are no conflicts between systems and no surprises when the job is printed. No doubt, as confidence in CTP grows laser proofs will be redundant and publishers will use unaccompanied data transmission "over the wires".

The Krause solution involves RIPing individual pages and storing the subsequent bitmap on the PSU (page storage unit). At that point the information is read onto exabyte tapes each usually containing one complete book. The tape is then kept until the plates are required at which point it is read onto the IPU (imposition processing unit). Once the imposition parameters have been installed the plates are loaded onto the platemaker and the plates exposed and processed ready for use on the press. The exabyte tape can be stored in a tape library for future use as required for reprints.

Nervous publishers will be able to see a proof before plates are made but obviously this will slow down the whole process and the use of a large plotter is necessary to produce the proofs.

The approach taken by different manufacturers in exposing the electronic bitmapped data direct to the printing plate is interesting. The Krause solutions involves a large and heavy platesetter with about 5 tons of granite in the framework to give stability to the laser. The plate, up to SP65 size, is loaded on the concave bed and slowly moves underneath the laser which exposes the image through a prism onto the plate. The technical data and plate exposure times for the Krause Laserstar are presented in Tables 1 and 2 respectively.

The Misomex solution is based on their step and repeat system. The laser head is clamped by vacuum on to the plate and four A5 pages can be exposed at one time. The head then moves (rather than the plate) and exposes the next four pages. There are of course pros and cons for each solution but at this stage Misomex claims that their system is almost twice the speed of the Krause system. That of course is not the whole story. Misomex are working with Cascade in developing pre-platemaking protocols and hope to have an actual system to show shortly, while of course Krause already has systems in the USA and Europe working in printing plants.

Savings are achievable in time and cost due to fewer people being needed and, of course, without the intermediate film, higher quality can be achieved as well. With help from Misomex, Create Publishing Services and Redwood Books (both BPMG companies) cooperated to produce their first CTP book in the summer of 1994. This was a book of old photographs and text and the quality achieved was much higher than similar books produced conventionally.

A continuing and unresolved problem is the limited suppliers of plates and the price being asked. The main supplier to users of existing Krause systems is Kalle, but they propose to market their plates in the UK at about 50 per cent more than the price of conventional plates. The same plates are available from the same supplier however at very much cheaper prices in Ireland and even more so in Germany and Sweden.

However, other suppliers are working hard to break this present monopoly and these include Polychrome, Du Pont, Horsell, Fuji, Agfa and others. However because of plate technology some will work better than others on the two systems. For example the laser on the Krause platesetter is strong enough to be able to penetrate and expose onto the N90 plate but could be too strong for other plates. No doubt as systems are installed increased demand will drive down prices as will increased competition and both Misomemx and Krause will develop their systems to work with the various plates.

This then bodes well for book manufacturers and for book publishers who will need to work closely together to extract the major advantages this front end revolution has to offer.

CD ROM

A short while ago CD ROM was unheard of but now everyone has the word on their lips. Purchasers of home computers insist that the facility is available on their hardware and publishers *of* books are now among those

GATF
NovaDom

SOLD

rushing to fill the newly created opportunity. A number of large UK publishers including Dorling Kindersley and O.U.P. are now producing titles on CD ROM. DK published five in 1994 and a number are already on stream for next year. Part of the excitement will be, for example, to see blood flowing through veins in the diagrams of the human body.

On the other hand having seen this sort of thing once who will want to see it again? Overall the CD ROM facility will best be used for journals and for dictionaries, directories and other works of reference. This is servicing a new and growing market and should not impinge too strongly on the traditional book market. It is estimated that strong growth over the next three years will mean that many books will be marketed through this medium.

Printing

Digital printing for monochrome and colour is now offered as a solution to very short run printing requirements both for loose leaf and conventionally bound products. Suppliers such as Indigo, Heidelbery, Exikon and Agfa are now offering presses which are able to accept digital data. The first Indigos for bookwork in the UK are at Eastern Press and other companies are watching this initiative with interest.

The Rank Xerox Docutech is now well established as a means of producing short run books. A number *of* these are in operation at book printers engaged in printing small numbers of publishers out of print titles, which are finding new loeases of life in the market place.

The more conventional means of book production — on large perfecters and web offset presses seems to be growing too. While no longer available from Crabtree the perfecters are nevertheless still being built — both the Sovereign and the Print X, and finding purchasers too. Second hand presses are still in demand.

Short run colour, around 4,000-10,000 copies, is an area yet to be fully exploited. Jarrold are investing in this market and will shortly be joined by one or two other companies aiming at producing short runs of the very highest quality print. The advent of stochastic screening and waterless printing will continue to drive quality standards even higher. Webs for monochrome and two colour books are now also well established. The front runner is of course the Timson T32 press and developments are awaited from Strachan and Henshaw who have not really been able to convince the market that the Variquik press is the way forward.

CTP is not a separate , stand-alone technology. It is a part of an entire electronic operation. CTP can simply be thought of as an advanced output application. As a result, printers with largely conventional prepress operations will experience frustration and massive work-follow distribution, trying to step into CTP in a long or high jump.

To ensure a smooth transition and high probability of success, printers with conventional prepress operations would be well-advised to implement a gradual change over process. There are three stages in this process-two page film imagesetting (composed single pages & spreads), imposed full-size signature (four or eight page) film image -setting, and plate setting.

Each stage requires a certain technology, and the technological infrastructure is cumulative-the technology of each stage builds on that of the previous step.

The first step towards CTP is proficiency with basic colour electronic prepress, including output of composed single page and two page spreads Requirement for this are as follows:-

Colour image input. Bringing a colour image into the system as an image file requires a digital scanner of some sort of a digital camera.

Simple networking. Image setters and their RIPS require a networking environment in order to exchange files with one another. Either is the most common protocol for small networks such as the basic electronic prepress operation. It provides theoretical speeds of up 10 million bits of information per second, though typical real world through-puts are much slower.

Trapping. Traps can be applied by desktop trapping packages, RIP-based trapping modules or high-end work stations.

Post script capability. Virtually all major desk top programmes and output devices employ the postscript page description language.

Digital page assembly. This is where layout takes place, since the purpose of all these stages is the output of complete pages, digital page assembly is essential.

Digital colour correction. Simple image retouching is the most common form of colour correction. It is a prerequisite for any type of sale able commercial colour.

Next is the output of imposed film files, which requires the addition of a large-format image setter to the system. In terms of learning and capabilities. This is probably largest single step. For many printers, this stage will actually comprise two stages-four-page film setting. However, printers who plan to use only smaller format presses like 20 × 28" sheetfed or eight page webs, will stop at the four-page format.

Printers who plan to use only 40" sheetfed and or 16 page web will probably want to go directly from 2 page image setting to 8 page imposed film.

To be effective, the large-format image setter must be supported by several new capabilities.

Open prepress interface (OPT) refers to the ability to use a low-resolution version of the scanned colour image for page make-up and image-placement, replacing it with the full output- resolution scan at the moment of output. This is essential for efficient network operation is a largely Macintosh or PC environment. High resolution colour files are too large for Macs & *PCs to* handle comfortably and will show a typical ethernet-based LAN (local area network) to an unacceptable level.

Imposition capability allows the correct placement of individual pages within a master file. Many prepress operations find themselves handling both desk top pages and pages prepared by a high-end system. It is, therefore, desirable that an imposition programme be able to handle both types of pages in a single imposition.

Up to this point, colour proofing needs can be adequately met with a conventional film-based (analogue) colour proofing system, possibly bolstered by a desktop digital colour proof. Before taking the final step into direct-to-plate production, however, Dupont believes it will be necessary for the printer to install contract digital colour proofing and educate customers in the use of these proofs.

The most courpicuous difference between the imposed film stage and full CTP is totally reliant on a digital proof.

At present, some printers are dealing with this by running parallel analogue and digital proofs of the same job and showing them both to customers. This assumes that the digital and analogue proofs are tightly calibrated, which is absolutely essential. They must both also give a close match to the final press sheet. Of the pioneering printers using CTP methods, many claim that moving to computerized platesetting, while a major step, is not as critical as the imposition and digital proofing stages. It does, however require the addition of its own enabling technologies:

A platesetter is essentially an imagesetter capable of handling metal plates.

Advanced networking. As platesetters are integrated into the production network, the volume of data increases dramatically. Standard Ethernet networks are inadequate for high-volume operations, which should use higher performance networks such as fast ethernet, ATMs and FDDI.

Electronic press control bars are an important requirement of the CTP work flow. In a CTP environment, they are generated electronically and must be exposed in the proper position an each plate. In addition to their traditional functions as devices to monitor print conditions during a press run, these colour bars will also be used for calibrating the platesetter to ensure consistency between plates.

Undoubtedly, CTP is the wave of the future. To use its maximum benefits printers need only take care to approach computer-to- plate methodically and with full awareness of the steps involved.

CTP or not?

The number of computer-to-plate systems on offer is growing almost daily. Brand name identification in some cases is becoming confusing. Even competitors on the supplier side are entering OEM agreements. Gerber uses pressetech for the latest unit which, like systems from Greo (Trendsetter) and optronics (Aurora) are bringing prices for CTP systems, albeit primarily in the smaller A3 format, down to under USD 100000.

Du pont study in North America has found that more than 60 per cent of printers there have no plans at present to invest in CTP. The companies switching over are mainly large groups like Donnelley, Quebecor, publishers, press and specialist book printers and producers of manuals. Black-and-white users for the time predominate.

PDI announced a CT system for bimetal plates.

Polyesters plates have gained in plate life, quality of images and "should be considered for CTP applications" says Denis Mason, a US print consultant. In his view these plates, together with the new budget price systems, could appreciably speed the spread of CTP use.

Misomex 5040 SR, as well as systems from Krause and purup are specially designed for CTP applications in package and folding box printing.

Creo and Mitsubishi printing presses have linked up to offer complete colour profiling and management from digital platemaking via ink setting and press control to even integrated finishing operations, Heidelberg Data control is also -like MAN Roland pecom-covering the whole gamut of controls from pre-press to finishing.

Telecom giants like AT&T and sprint offer communications systems for CTP user so that from one digital entry point remote platemaking at print centres worldwide become instant. Donnelley have developed an interface program for such applications.

Plates & films

Specially formulated to meet the requirements of stochastic screening are Anitec Craftsman Ultra offset plates. Western Lithotech, now part of Mitsubishi Chemical, has a new range of Diamond two-side coated plates

for "almost waterless" printing and LA-1 and LY-1 laser/exposure digital plates for use in imagesetters and direct platemaking.

Polychrome has teamed with Toray in Japan to offer CTX thermal plates, made by the polychrome plant in Columbus GA. The waterless plates are available for both conventional and laser exposure.

Brazil's IBF plate and film manufacturing company, celebrating its 25th birthday, has developed good markets especially for plates in North and South America, as well as Britain and continental Europe.

Gerber, Creo and optronics offer CTP systems for thermal plates for these systems for and those Kodak and (soon) 3M, too.

Dry film processing, pioneered by xerox verde and now polaroid is on the March. Linotype-Hell introduced the Drysetter for the latter system. This manufacturer also presented a 16-page CTP system utilizing Delta technology including post script Level 2 interpretation and screening. The system is available initially for Herkules, Quasar, Drysetter and Chromagraph R3060. The German manufacturer also announced that over 1500 Topaz colour scanners have been sold so far. The new Topaz Robot flatbed transparency scanner allows batch scanning.

ECRM introduced the knockout 4550 laser imagesetter for colour work, apparently based on a Konica engine Monotype/Scangraphic Express Master that enables complete colour newspaper pages to be produced within 40-60 seconds.

Dalim, who have gained Dr Lutz Thiele, who was formerly on the board of LInotype-Hell, as CEO, offer systems facilitating ideas, such as Twist, which permit different prepress systems to be run compatibly. Escher-Grand, of Canada, who have just opened a European branch in Belgium also offer systems solutions. Pantone Hexachrome, the six-colour system, cam now be proofed via Afga.

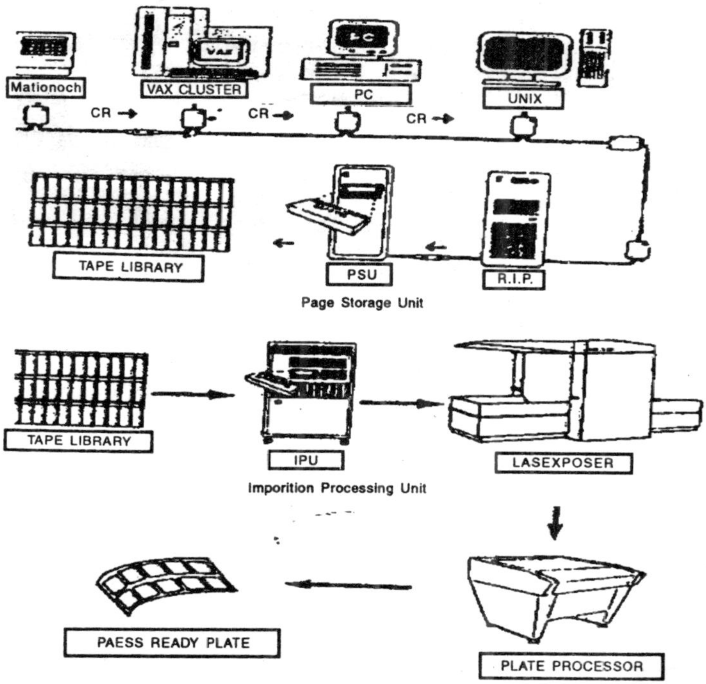

SCANNERS

Select Scan

Select Scan's high density range of 3.6 D and maximum resolution of 4000 ppi enables it to produce scans comparable in quality to that of top-of-the- line drum or CCD scanners. Its 6000-element, linear array CCD registers 16 bits per color data capture, providing superb detail even in the darkest of shadow areas. Selectescan's convenient flatbed design and off-line mounting capabilities allow it to meet the high productivity requirements of commercial printers, service bureaus, color trade shops, and others, Incorporating the most advanced CCD technology and specially designed optics and lenses, Agfa Select Scan delivers superior image sharpness and color fidelity with every scan, making it the perfect scanner for producing high-quality advertisements, magazines, and catalogues.

Features

Select Scan provides two levels of scanner driver software applications:

Fotolook for simple yet effective scanning operations, and Select Scan 1.0 for sophisticated scanning controls, like selective color corrections unsharp masking and smoothing, etc.

Selective color corrections during scanning to minimize the need for subsequent image manipulation.

* 16 bits per pixel for deep shadow detail.
* State-of-the-art aspheric illumination lens for scratch and dust minimization.
* Unique custom lens system to scan images at any resolution up to 4000 pp1.
* Single or batch image scanning with simultaneous off-line mounting using additional slide holders.
* Slide holders that handle various film formats.
* Extensive documentation including Agfa's acclaimed An introduction to Digital Scanning guide.

Horizon plus

Horizon plus is the impressive successor of Agfa's popular Horizon scanner, which is in worldwide user by thousands of prepress professionals. It has been developed using the latest electronic and optical technology.

The Horizon plus is a versatile flatbed CCD scanner, designed both of reflective originals up to A3 format and transparencies as large as 24 cm x 34 cm. The ability to "batch" scan multiple color positive and negative films, color, grey scale and line-art reflectives, and even screened originals, provides outstanding flexibility and productivity.

Horizon plus scans color originals even faster than its predecessor. Set to "speed" mode, scanning is up to 200% faster than the normal mode.

With a resolution of up to 1800 ppl in color, significant enlargement of originals is possible without noticeable quality loss. Thanks to its 36 bit color depth and 3.2 D sensitivity range, Horizon plus captures even the smallest tonal variations in any original. *When* color images need to be printed in black-and-white, a neutral density filter provide perfectly balanced results.

The redesigned focusing mechanism and improved optical components transfer maximum image detail into digital format. Where sharpness is lacking in original images, variable unsharp masking (USM) is adapted to scanning resolution, contrast levels and image density. Details are sharpened without emphasising grain or running shadow areas. The Agfa Foto Look scanners interface provides automatic or manual contracts in aimplaiogical menus to achieve the best possible results.

Horizon plus is delivered with a frame with rotating holders for 35 mm, 6 Cm x 6 cm and 4" x 5" films. Optional holders are available for batch scanning, uncut 35 mm strips and framed 35 mm slides.

Technological development is only of interest when the final product remains affordable. Horizon plus adds features across the board. Top performance in terms of cost, productivity, quality and flexibility make Horizon plus an ideal scanner for the price-conscious profession.

Arcus II

Arcus II dominates the A4+ desktop color scanner market. Successor to the award-winning Arcus family, it features 36-bit pixel depth, a true density range of 3.O D, resolution up to 2400 ppi for continuous tone pictures and an astounding 3600 ppi for line art. Clearly this is the desktop professional.

A transparency scanning unit is built into every Arcus II, taking full advantage of the impressive CCD technology, which is capable of registering 68 billion colors; No details are able to hide in even the darkest shadow areas!

Such an impressive range of leatures and additional bundled software might appear to be outside your budget. You will be surprised to find that the cost of professional -quality scanner is really " desktop".

Duo Scan New

- Flatbed and slide tray-twice the functions. Not the cost.
- Twine Plate Scanning Technology
- One -pass 36 bit color scanner
- 2000 x 1000 dpi-optical resolution
- Mac/PC compatible.

A New Large Format Imagesetter From Scangraphic

The most obvious effects of the continuous change in the printing industry are shown in the pre-press area. Digital work procedures become increasingly accepted and demand a consistent realisation of the companies regarding the input and output of digital jobs. Highest productivity and high throughout with simultaneous rationalisation of complex and manual work procedures determine the work process of modern and future-onented pre press companies. However, print house which in the past years were often only specialised in montage, copy and print, now also require an increased supply of modern imaging systems to achieve a more efficient work flow. Because instead of film, digital data is increasingly supplied also to print houses, which therefore begin to invest in pre-press equipment again.

High demands on imaging systems

High performance, throughput, quality, precision and flexibility are the increasing demands placed on today's imaging systems. These are the criteria which determine the economic viability of such systems.

The scangraphic family of drum imagesetters, first introduced to the market in 1986, has always possessed these characteristics, thus ensuring productivity and quality for typesetting and reproduction tasks. The new Scantext Apollo laser imagesetter is also based on the tried and trusted internal-drum technology onto stationary material, a technology which scangraphic established for the Post script output, thus setting a standard which has become world-wide for colour imagesetters.

The principle of exposing images onto stationary photographic material within a drum guarantees exceptional precision, register conformity and reproducibility, an extremely rapid and thus optimal dot structuring and overall conformity of dot density.

The geniality of this principle is as simple as it is effective. To deflect the helium-neon laser beam onto the photographic material in the drum, a rotating mirror is used to deflect the beam to every pcint at the same distance and at the same angle. During this process, the material is not transported nor is the laser beam modulated, two factors which could lead to errors in the dimensional stability. Thus, it is the tried and trusted Scangraphic indrurn principle which fulfils the high demands of a colour imagesetter. Then geometrical accuracy and high reproducibility are the indispensable characteristics of the exposure of colour separation.

Continuous Progress in quality....

On the grounds of its comprehensive experience in the manufacture of drum imagesetters and vivid exchange with its customers, scangraphic has always continued to improve performance and quality of its imagesetters. The scantext Apollo thus now offers six different exposure resolutions with which the spot size can be varied between 8 and 22 my. These resolutions of 1016,1270,1905,2540,3810 and 5080 dpi and the variable spot size are the guarantee for optimum recording quality and the very best reproduction of the entire tonal value scale in black and white, colour and the frequency modulated scantext High Fidelity screening process. This high quality, however, also profits fonts, including those with the finest serifs, graphic elements with hairlines as well as negative text and graphic elements.

During the construction of the new imagesetter, the main goal was to establish a balance between recorder speed and performance and to also ensure the highest throughput from the creation of a complete sheet to the final exposure. The overall concept of Scantext workstation and imaging system takes into account that the capability of one component may not be blocked by another.

The main objective here was to reduce the total output time for a typical A1 color sheet (eight A4 pages) to approximately four minutes, which corresponds to a maximum exposure time of less than one minute per color separation, and all this at a resolution of 1000 lines/cm or 2540 dpi. The typical recorder speed is 145 seconds at 1270 dpi for the full imaging format of 1072 × 770 mm. At the highest resolution, this results in an exposure performance of 56 seconds per A4 page.

On the one hand this does, of course, require enormous RIP performance such as that offered by scangraphic's current Raster image processor, the scantext Presto-a software RIP developed on the principle of the Adobe CPSI and furnished with Pixel Burst hardware accelerator.

On the other hand, however, an output engine is required which can transpose the high RIP performance of the Presto into exposed film.

The scantext Apollo drum imagesetter, built to leading edge design specifications, fulfils these demands with the highest precision and at the highest speed. In combination with the new scantext combo RIP concept this has been achieved in a completely new way. All data which is made available for the generation of fully imposed sheet exposure is first transferred to the scantext combo workstation, which is a combination of server, imposition station and software RIP on one platform. The data is imposed and then calculated on the combo hardware. The special feature of the scantext combo work flow is the ability to generate the high-resolution bitmaps required for the exposure in the background, and to thus minimize the time previously caused by the RIP process. Scantext Apollo exposes the complete sheet at its writing speed, as no further calculation times are needed for the postscript generation.

To cater for the increasing demand for imposed exposure, scangraphic has constructed a new drum designer the Scantext Apollo which features an extended circumference. This enables the Aplio to process jobs with a maximum hight of 770 mm. The 770 mm corresponds to an effective drum area of 85 degrees.

The exposable format was conceived on the basis of the facility to expose eight impsoed A4 page including all additional elements such as register marks and cut marks. The exposure format of 1072 × 770 mm fulfils these requirements and covers the A1 oversize format. This means that an A1 poster in oversize format, four A3 jobs, eight A4 pages or 16 A5 pages can all be exposed on a single film.

The scantext Apollo is capable of processing different types of material (film or polyester plate) which are stored in exchangeable supply cassettes and allow the imagesetter to be rapidly adopted to the various formats without any complicated mechanical adjustments being made.

Photographic material can be selected in any intervals and in width between 460 and 1082 mm. According to these material widths, scantext Apollo allows the application of various imaging formats and the use of polyester plates that conform to most popular printing plates.

Punch systems compatible to printing presses up to A1 oversize

The scantext Apollo is equipped with a punch system that supports the Bacher 2000 or stoesser film register. The new feature of centered material feed allows centered register punching of film or polyester plates for all printing formats within the range of 460 up to 1082 mm punch side width.

By using the integral punch system, the user can copy directly form the exposed film to the printing plate without having to carry out time consuming work of remounting the exposed film for the subsequent copy process to the plate, or repunching.

The laser system of the scantext Apollo is based on the tried and trusted helium-neon laser which is already in use on other scantext laser imagesetters. The use of a helium-neon allows the application of cost - effective, standard film material. In comparison with laser diodes, helium-neon gas lasers are characterized by a better consistency of intensity and considerably enhanced dot sharpness. The latter is becoming increasingly important for output quality in connection with frequency modulated screening techniques, and in particular for the application of scantext High Fidelity screening.

The high rotational speed of the mirror drives required for high exposure speeds placed extreme demands on the mirrors themselves. They must be light, stable, and must retain their optical characteristics even at high speed. With an eye on future developments, scangraphic has departed from the conventional all-metal mirrors as these are simply too heavy. The use of beryllium, on the other hand, is extremely hazardous due to the high toxicity (beryllium oxide). Instead, a solid glass model of scangraphic's own design was used with the mirror surface composed of vacuum metalized aluminium with a special optical coating and protective covering.

RIP and RIP Interface

The performance of the scantext Apollo imagesetter is considerably influenced by the high mechanical speeds, the resolutions and the raster image processor, for these reasons, Scangraphic offers exclusively external software RIPs for the Apollo. The most powerful software RIP, scantext presto, is based on the adobe CPSI concept which enables interpretation of post script data on a software basis and uses a computer running on a standard platform (Sparc technology). A special ASIC, the pixel Burst accelerator board, is utilized to accelerate the process of pixel calculation.

The scantext presto covers the entire spectrum of applications used in modern publishing systems. It can process all Level 1 and Level 2 Post script jobs and in addition to the original type 1 fonts contained in the scantext Digital type collection, can also process all original Type 1 fonts irrespective of the manufacturer coupled with the high performance of the scantext Apollo, this facility to utilize toutilize proprietary DTP programs and fonts gives the user decisive advantages in speed and system independence.

The scantext Apollo is equipped with two RIP interfaces as standard the Li2/Li5 interface and the FPI fast Pixel Interface, the scangraphic standard for control of scantext imagesetters. The Li2/Li5 interface enables the connection of the scantext presto and Scantext Rip Express. For connection of the Rip Express, the Laser Bus interface is required as an option. All of the recent RIPs from scangraphic support the fast FPI which allows a pixel clock rate of upto 125 MHz.

Technology aspects

For the electronic control elements of the Apollo imagesetter, scangraphic has departed from the now outdated VME bus, for reasons of better serviceability, flexibility and the promotion of EMV, scangraphic has implemented the principle of distributed intelligence on the the basis of CAN bus technology (Controller Area Network). The individual components can be tested independently of each other and the main system, thus simplifying service aspects. 01 the nine components contained in the electronic systems of Apollo, five are absolutely identical, achieving rationalisation of spare part management, in addition, this concept enables remote diagnostics of the imagesettter, which involves considerable saving in time and costs.

Every scantext Apollo is equipped with an internal data bank containing parameters relevant to the materials and which can be expanded and modified at any time. If a new material for printing plates is utilized which, due to its mechanical stability requires other speeds, for the transportation of the materials to and from the drum, the parameters for this material can then be defined by the user and entered into the data bank.

The scantext Apollo embodies scangraphic's experience in the development of high-tech products for the graphics industry. In combination with the scantext combo workstation, the output on film or polyester plate meets the demands of a modern and efficient work flow. Scantext Apollo combine the tried and tested scantext Apollo combines the tried and tested Scantext in-drum principle with state-of-the art mechanical and electronic technologies.

The Scantext Apollo is thus the ideal output medium for fully imposed sheet assemblies, and offers completely new perspectives for the digital work flow in the pre-press industry.

Scantext combo: optimum job Management

Prior to the introduction of the Scantext Largo imagesetter, Scangraphic had defined the basic principles for a new RIP concept. This concept is based on the understanding that recorder speed and RIP have to he adjusted such that neither component in the production chain has to wait for the other . As the slowest element of the chain determines the throughput, the speed of both RIP and imagesetter had to be adjusted with each other. This concept is currently successfully applied with the scantext Presto software

RIP in combination with the scantext Apollo and Scantext Largo imagesetters.

Combo combines several work procedures

Scangraphic has now added a unique and productive feature to this RIP philosophy -simultaneous imposition of complete sheet assemblies and ripping on one work-station prior to the imaging process. This extended concept, called scantext combo, runs on a Unix workstation with sparc technology, and combines server functions, imposition software and software RIP in an extremely economical way-all on one plateform.

Data output is rationalised considerably, and complex calculation procedures are reduced to a minimum bits is an important requirement for efficient and highly productive application of large format imagesetters.

The combo Philosophy

Scantext combo may be explained very easily by a practical example during the production of printed matter such as magazines and periodicals, the work flow may be interrupted due to missing pages or because of last minute correction of a single page. Why should the production have to wait until the missing pages are available?

The philosophy behind scantext combo is to take into account these missing pages in the layout sheet of the imposition software , but they are not relevant for the continuing production process. All available pages of the job can be imposed and ripped for the final exposure. The RIP renders the data as bitmaps in the final imagesetter resolution and saves them temporarily on the system. When the missing pages are available, they will be ripped and included into the complete job.

The whole job, stored as bitmaps on the system, is then sent to the imagesetter. There is a considerable saving in time, because the imagesetter only outputs the data as precalculated bitmaps. The RIP process during exposure is therefore unnecessary.

Imposing with scantext

Production in the pre-press area demands imposed sheet assemblies including all control elements such as cutmarks and color control strips. The last step prior to the plate copy, the generation of the montage films, is not to be made manually at the light table but electronically at the monitor using an imposition program. PostScript data form applications such as page Maker, Quark X-Press and Freehand are entered into the program, analysed and automatically checked for completeness by the program. The previews of the pages are included into the imposition layout, and the complete imposed sheet is stored for the exposure. For the assembly display on the monitor, new resolution previews with a resolution of 72 dpi are generated by the scantext RIP.

Scantext combo uses the advantage of the imposition software and includes them in the combined concept of imposing and ripping. Combo's special feature is the ability to generate the high resolution bitmaps exposure in the back ground, and to thus minimise the time by the RIP process.

Preparation for the Exposure with scantext combo

The sheet assembly is carried out at the imposition terminal by using the lowers file generated by the RIP. An automated work step of scantext combo-program, the bitmap control, checks whether the high-resolution bitmaps needed for the exposure are already available or whether these data were still have to be calculated by the RIP. Once all data (bitmaps of the page) is submitted, scantext combo starts the actual output process. The complete sheet can now be output at the writing speec or the imagesetter, without additional RIP times.

Imagesetters are for imaging

With this technique, scangraphic has optimized the RIP process to such a degree that the customer is unaware of the lengthy processing time of several colour pages. The RIP process is carried out as a background operation so that the imagesetters only has to image.

This procedure is especially suited for large formes or multiple copy exposures output on scantext Othello imagesetters (eight A4 pages each) for output of fully imposed sheet assemblies.

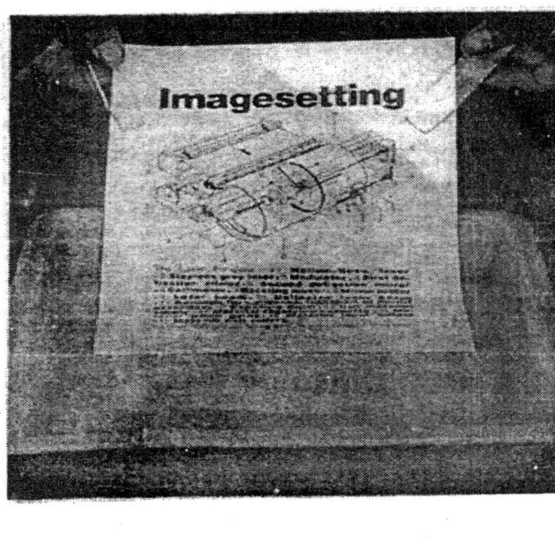

DIGITAL COLOUR PRESSES & COMPUTER TO PRESS

There is no universally accepted definition of the term "digital printing". Probably this is why it is a subject of discussion. In fact there are two basically different points of view are those who in the computer Industry and another the printing industry.

For the computer Industry, every printout of data on paper is an example of digital printing, whether it is in black and white or colour. As result the digital printout machine is known as "printer" which to the printing industry means the technician who runs a press.

Plateless

The printing industry speaks of digital printing when the computer is connected to the printing press and the digital data produce. The printing forme directly in the press. This means plateless information transfer or non impact printing.

How did it Start— Heidelberg unveiled the world's first digital printing machine at Print 91 exhibition in Chicago in the form of the GTO-DI, thereby initiating a new direction to the development of printing machines and the printing industry.

At Ipex 93 in U.K. two innovative machines were introduced but only could print up to A3 size and did not become available until 1994. They are the E-print 1000 from the Israeli firm Indigo and the Chromapress from the German-Belgian firm Afgo-Gevaert. The latter machine has been developed by Xeikon, a firm established by Agfa-gevaert and became available through Xeikon's own channels of distribution under the name of DCP-1 (Digital Colour Press).

The computer industry has an urgent need of digital printing for the continuation of its growth. To enhance the value of the computer and promote its sales it must be possible to print out the colour screen display, including pictures, cheaply & rapidly on paper.

At the 1992 Seybold congress in San Francisco, the then president of Apple Computer, John Sculley pointed out that conventional offset printing was unacceptably costly affair for the short runs required by the Computer Industry. He almost challenged the entire suppliers of printing machinery to develop further a means of printing off colour pages at much lower cost, other wise the computer industry would have to develop such devices for itself.

Digital Printing for the Printing Industry

As a matter of fact, the whole subject of digital printing runs counter to the interest of the progressional printer. The question is why should user spend more money to buy an additional machine simply to sell shorter runs? Should he/she split up his/her long runs into a number of short ones? His/her customers would gain the advantages of reduced stock and rapid updating. But again would they pay extra for these ?

Would customer be prepared to pay an express charge for same day delivery of a few hundred sheets, as it is acceptable in the photo business?

Naturally there are some good reason for adding a digital printing service to conventional printing.

Digital-printed jobs can be delivered more quickly, short runs can be produces more cheaply.

Digital Printing for pre-press companies

Pre-press services have far more reasons than printing companies to take up digital printing. They control the production of data and the process. They already have data. With digital printing, they can extend their services and increase their turnover. Production can be ultra-rapid if one makes use of the synergies of the chain of digital data.

A page can be created, imaged and printed in fifteen minutes from the digital camera and text entry to the finished print.

These possibilities are, of course, dependent on the purchase of a digital printing machine plus finishing equipment, which represents a high investment.

Digital Print

To day printing technology has become an industrial process. With standardization and automation, prices have come under pressure. Anyone who wants to make money out of print must lower his costs, both on press and pre-press stage. From the cost point of view it makes little difference if savings are obtained in the manufacturing process or new markets opened up by digital printing.

Several other possibilities are open to the printer such as

(1) automatic imposition

(2) digital platemaking

(3) automatic platechange

But in any discussion of digital printing it should be made clear within whether one in speaking of small A3, medium A2 to A1 or large formats, because the price/performance ratio is different in each case. It is known hat digital printing in A3 size from single copy to long run, is already available n the International market with a variety of machines.

Gravure periodical printers, who engrave cylinders directly from digital data, can witness that the disproportional organizational effort involved, could hardly be justified in offset, with its comparatively single platemaking procedures. Rapid automatic plate change, on the other hand, will certainly reduce offset production costs.

For medium size, whether suitable digital printing machines will be an offer, or whether the combination of computer-to-plate and quick, automatic plate change will bring the required cost reduction for cheap, short colour runs.

If we look at from the offset printer's point of view, a reasonable requirement would be the hybrid digital printing machine, capable not only of imaging plates produced off-line, whether by conventional or digital means.

19

Newspaper Printing: Letterpress, Offset, Flexo And Anilox

The rapid rate of progress in newspaper printing during the past few decades will continue into the future.

When Export -Polygraph first appeared in the spring of 1953, the world of newspaper printing was relatively routine, with newspapers almost everywhere printed in the classic letterpress process with lead stereotypes. Then, in the sixties, sweeping changes in prepress triggered off a period of turbulent development, as photosetting rapidly drove stereotypes from the market and paved the way for offset to challenge the letterpress supremacy.

Revolution in the press hall: offset storms the letterpress stronghold

Although invented at the end of the 19th century and successfully marketed in its country of origin— the USA— by Goss, the leading supplier of that time and now a member of the Rockwell group, it was not until 1964 that offset spread to Europe, in the form of a double-width newspaper web press for a customer in East Berlin. Having secured this foothold in the European market, however, offset soon gained wide-spread recognition, and many sales were recorded in subsequent years by press manufacturers such as Koenig & Bauer, MAN Roland, Wifag in Switzerland and Plamag in East Germany.

Offset printing plates were less expensive than lead stereotype plates and could be produced faster. Parallel to the introduction of dedicated offset equipment many newly delivered letterpress machines were converted for photosetting by retrofitting saddle plates onto the cylinder to take BASF nyloprint or NAPP photopolymer letterpress plates.

While other manufacturers in the early seventies were abandoning the construction of letterpress machines, one company which decided to retain this sector was Koenig & Bauer. In addition to expanding its offset activities, it developed high-speed web presses with wrap-around plates and an output up to 40,000 iph. The nyloprint or Grace plates were clamped magnetically directly onto the plate cylinder.

A spectacular machine which is still fresh in the memories of many newspaper printers today was the Koebau-Jaumbo-Courier delivered to the Gazet van Antwerpen in 1974.

The press had three folder formers side by side and web width of 2.52 meters(8ft 3 in), making it the widest newspaper web press ever built.

Tailor-made technology is the trend of today

Today between 80 and 100 per cent of all newspaper web presses delivered by leading press manufactures--MAN Roland/MAN Plamag, Koenig &Bauer (KBA), and Wifag in Europe, Rockwell Graphic systems (Goss) in the USA, and Mitsubishi, TKS and Ikegai-Goes in Japan--are offset presses with conventional film inking units and brush-type, turbo, or spray dampening units. Different performance classes, from less than 30,000 to more than 40,000 iph, and a variety of printing -unit configurations are available. There is growing demand for customized solutions to individual and colour flexibility, number of pages, size of circulation and automation, Leading press manufactures satisfy this demand with modular built machinery offering a wide choice of options.

Rapid progress has been made in the development of inking and dampening units drive systems, reelstands , folders and-above all-automation equipment. The benefits are reduced makeready and waste, increased output, improved quality in multicolour printing, and easier operation. Whereas the fundamental laws of mechanical engineering (highest manufacturing quality of the cylinder, gears bearings etc.) still hold true for top-quality printing, the electrical systems in the press hall and mailroom have undergone just as revolutionary a transformation over the past 15 years as other industrial areas, and have been outpaced only by the speed of technological change in prepress.

Cumbersome mechanical controls for protective devices, vast cabinets packed with relays and switchgear and often as long as the press itself, huge operating panels with a phalanx of push-button and display instruments—these have all long since vanished from the press hall. Whereas in 1971 -measuring values for ink keys were still being transferred from drum or flatbed film scanners into punched tape and conveyed to the control desk together with a punched job card, by 1973 press manufacturers were already installing contactless semi-conductor controls. These were followed in 1975 by programmable logic controllers (PLCs) and, from around 1977 onwards, a move towards decentralized press controls.

Advanced console technology is now standard equipment

Modern console technology for press presetting, remote control and computerized monitoring of the printing process was introduced in the early eighties and is now a standard feature on high-performance presses. Today many control functions are integrated into the printing units, reelstands, and folders. Data buses link the decentralized process stations with the operating and systems stations. Thanks to the electronics revolution the entire control system can now consist of a single computer with a range of peripherals

Standard interfaces provide an EDP link between press and related areas, logistics, and accounting. Production data, malfunctions, waste-copy totals and much, much more can all be logged by computer and printed out as hard copy.

Electronics can tap enormous productivity reserves

While promoting fierce competition to newspapers from the audiovisual media, the boom in electronics— frequently described as the third industrial revolution-has also mobilized previously unimagined productivity and quality reserves in press manufacture through the invention of increasingly powerful electronic components . Electronic equipment now accounts for 20 per cent or more of the production value of a newspaper press-confirmation that challenges also create opportunities. The resulting improvements in print quality, colour and immediacy of news coverage (through later editorial deadlines) have helped newspapers compete more effectively against satellite TV, videotext , the home video and other new media.

Developments since the mid—seventies have led to ever faster machinery in a wide variety of satellite configurations, with reversing gear and numerous web-lead variations for added flexibility in colour and production. Although almost anything demanded by the newspaper industry is now technically possible, economic considerations have resulted in a growing trend towards less complex technology and greater standardization in industrial newspaper printing.

Counter-trend towards less complex technology

Manufacturers have responded to this trend, by developing less complex press configurations. These include four-high towers with H-or arch-type printing couples for four-colour perfecting: stacked Y-type units for printing two colours recto and one verso or four colours recto and two verso: or stacked satellites without reversing gear for four-colour perfecting. Anilox keyless inking technology, i.e. anilox letterpress and offset with oil based inks or flexography with water-based inks, are further examples, In comparison to the familiar and proven satellite units, the four-high tower configurations has the advantage of dispensing with impression cylinders and reversing gear while still being able to print four colour on both sides of the web, four-high towers are available both in the classic configuration, with four identical arch-type units and four operating levels and in the Hversion where the second and fourth units are up-ended and there are only two operating levels.

The benefits to the printer are considerable: simple webbing-up, straightforward web leads (with no web-guide rollers between printing couples), easy access to the units, ergonomic operation and full flexibility in colour leads for each individual page. Additional advantages are the shorter makeready and wash up times and longer service life, while the less complex drive technology reduces initial investment costs.

In the US newspaper industry, flexographic printing with water-based inks has gained a significant market share of more than 20 per cent with over 1300 printing couples sold. At Associated Newspapers in London, the world's largest flexo newspaper installation, comprising eight KBA flexo-courier presses with a total of 144 printing couples , prints more than two million newspapers with a high colour content every day. Flexo's big advantages lie in the rub-resistance of its inks, the absence of show-through on thin stock and its extreme brilliance in four colour printing. In Italy over 180 Cerutti flexo printing couples are currently in operation , most of them on single width presses. Elsewhere in Europe, however, flexo has still to make the breakthrough.

The last bastion to fall to keyless inking was multicolour newspaper offset printing, with its inherent risk of ink emulsification due to the need for dampening. In may 1989, after several years of testing and optimizing an experimental printing unit, Koenig & Bauer's subsidiary Albert-Frankenthal demonstrated multicolour newspaper printing for the first time on an anilox offset press, with a four-high tower configuration. There are now around 1500 anilox offset printing couples in operation worldwide, some 800 of them in Japan. Japanese keyless inking units, a number of which are also in use in the USA , are not , properly speaking, anilox units since they do not incorporate the basic essential-an anilox roller— and have so far been limited exclusively to black and white printing.

Technological and market leader in multicolour anilox offset newspaper printing today is the KBA group, which to date has sold twelve anilox offset presses (with a total of 294 printing couples), including a giant installation for a newspaper publisher in Indiana, USA. Nine of the twelve presses are already in operation, printing some, one million high quality four colour newspapers every day. MAN Roland is due to deliver a large anilox offset press to Finland in the autumn of 1994. Other European and American manufacturers are still at the development stage.

Anilox offset has made tremendous progress and has matured into a viable and promising alternative process for economical, high-quality multicolour newspaper printing.

The newspaper technology of tomorrow

The rapid rate of progress in newspaper printing during the past few decades will continue into the future, with press manufacturers fast becoming suppliers of "complete" systems. Automatic roll stripping and splice preparation stations are now available from most leading press manufacturers. Within the web press itself, rationalization is in full swing, as is the replacement of manual activities by robots and other automation modules. Automatic plate change, computer to plate (i.e. plate exposure directly in the press), remote conversion of folders to different types of production, customer targeting of newspapers through new technologies in the folder and mailroom—all these have become reality, and many other developments are in the pipeline. Only time will show which are worthwhile. Technical progress, while continuing to present a tough challenge, will also help strengthen the position of the newspaper as one of the leading vehicles of advertising and information.

Anilox - Offset

The unmistakable trend now-a-days towards more colour is news paper printing has led to the use of printing towers with up to 8 inking units. This means that as a conventional (imported) 80-page newspaper press with five towers and e.g. 40 inking units the operation team must pre-set upto 1600 ink keys, may then have to adjust them after proof-printing and must also keep a check as them during running -on.

In order to present and control the inkkeys comprehensive electronics and a link to the pre-press stage via plate-or film scanner are necessary. Higher investment costs and more expensive maintenance and service are the inevitable result.

In contrast to conventional offset inking systems, where extensive automation is the rule, the anilox inking system uses a technologically less complex concept which dispenses with the need for a comprehensive ink-roller frame and ink keys: a constantly even film of ink on the printing cylinder plates is ensured by means of a doctor system together with an anilox roller and forme roller. Thus to a great extent the printer is relieved of the need to check and regulate the ink unit and can concentrate as the dampening unit.

Except for the inking system an Anilox offset printing unit consists of the same basic components of the same basic components as a printing unit for conventional offset.

Layout of an anilox offset arch-type printing unit

1. Ink throuth with pump
2. Chamber doctor
3. Anilox screen roller
4. Ink forme roller
5. Ink rollers
6. Plate cylinder
7. Blanket cylinder
8. Dampening duct roller
9. Brush roller
10. Dampening distributor
11. Spiral roller
12. Dampening forme roller
13. Paper web

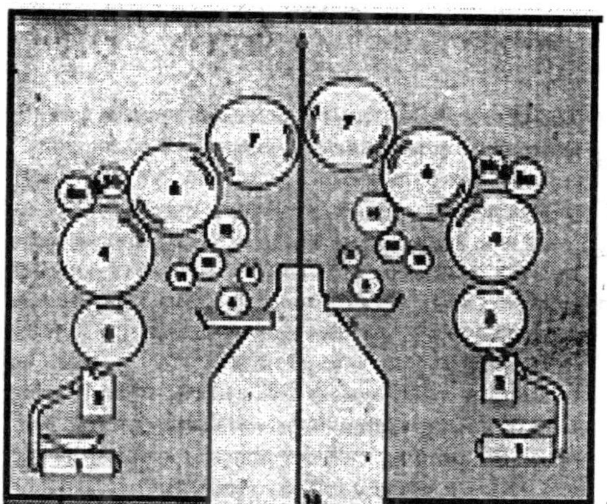

Anilox inking unit on KBA newspaper offset presses

Layout of an anilox offset arch-type printing unit

(1) Ink trough with pump
(2) Chamber doctor
(3) Anilox screen Roller
(4) Ink forme roller
(5) Ink roller
(6) Plate cylinder
(7) Blanket cylinder
(8) Dampening duct roller
(9) Brush roller
(10) Dampening distributor
(11) Spiral roller
(12) Dampening forme roller
(13) Paper web

At the heart of the anilox inking system are the anilox roller and the doctor system.

The doctor system, has three fundamental tasks:

(a) To fill all the anilox roller cells evenly with ink

(b) To wipe excessive ink from the anilox roller surface

(c) To wipe smooth the profile created on the surface of the anilox roller through contact with the forme roller and the printing form, and to direct the ink back into circulation.

Now-a-days doctor blades have a life-expectancy appropriate to the demands of everyday practice.

Half width and quarter-width doctor systems can be used in the inking units. This is to satisfy the demand for increased colour in newspaper. Because of quick-lock mechanisms, valve mechanisms, system orientated auxiliary devices and an accessible layout at all operating levels, replacement of the doctors and inking systems anilox and forme rollers is now quite a simple matter.

A slight surplus of printing ink is directed continually into the doctor blade chamber by ink pumps; excess ink and that wiped from the anilox roller is returned into circulation The individual ink troughs are filled by means of ink-feed systems. This can be done manually on small presses and fully automatically on large ones.

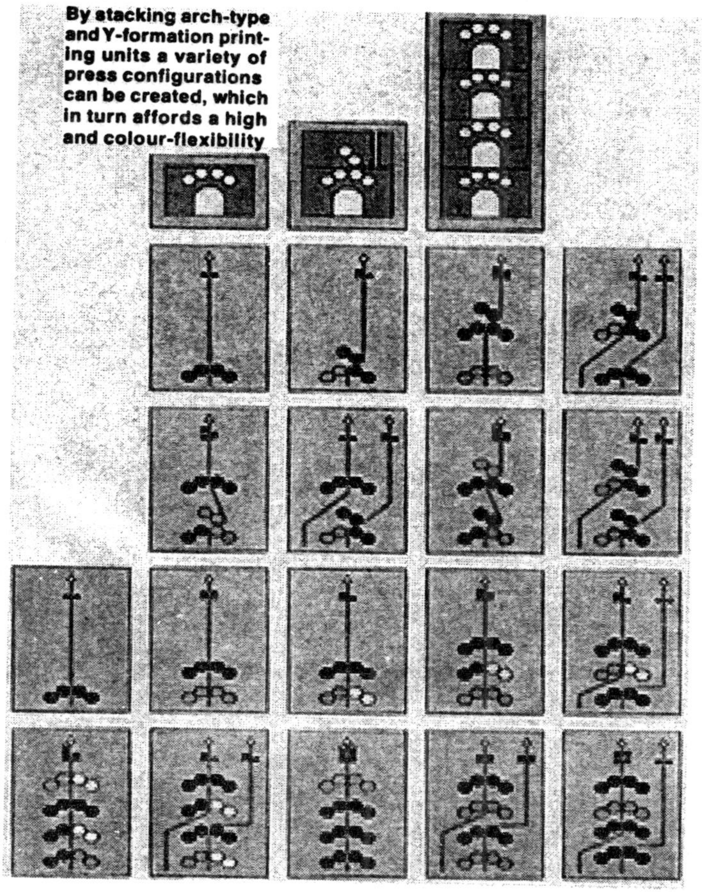

Because of the hydrophilic quality of chrome only ceramic coated, laser-engraved anilox rollers are used in ANILOX OFFSET. These have an additional advantage over chromium -plated rollers in that their life-expectancy is very much longer. Thus the ceramic rollers used show no recognisable signs of abrasion even after a year in daily production. In order to achieve a good ink transfer the cells on the ceramic roller are coated with a this layer of copper.

The depth, angling and geometry of the cells, the cell/cell wall ratio the number of cells per line, the surface temperature of the anilox roller during running -on the viscosity and pigmentation of the ink-these and many other factors all determine the ink laydown behavious on the anilox roller.

DAMPENING UNIT

Because of the shortness of the inking device in ANILOX OFFSET the range between tonning and emulsifying is narrower than in the conventional inking unit . A precise dampening unit is therefore a top priority for an optional ink /water balance.

An instant response to alterations in the dampening procedure and a rapid run clean of the plates at the start of production are guaranteed with this contact free system. The dampening-duct roller drive enables finely controlled metering at all machine speeds.

In addition a page-wide adjustment of the dampening application is also possible according to subject and dampening liquid requirements.

This shutters are available in a variety of levels ranging from manual setting, to mechanical setting on the outer press surface, to motor-operated setting controlled from the console and complete with display. The double-walled dampening water trough insulated with polyurethane foam ensures an even dampening -water temperature from production start-up right through to the final printout.

Due to this contact-free system of operation contamination of the dampening liquid, recycling of paper dust and ink back into the dampening-water preparation and circulation system and the resultant alteration in the chemical composition of the dampening liquid (e.g. PH valve) are almost completely eliminated.

PRESS CONFIGURATIONS

Compact press configurations with high production and colour-flexibility are in ever greater demand in the newspaper industry. The ergonomics of the press itself, i.e. ease of access to the inking and dampening units as well as to the plate and blanket cylinders, also plays an important role.

All mechanical, pneumatic and electrical operating elements should be arranged in such a way as to enable work to be carried out on the printing unit safely, flawlessly, rapidly and precisely.

Against this back ground there is now also an increasing demand in Europe for the 5 to 8 couple tower configurations which originated in the USA for the blanket to blanket printing process. These offer a technically less complex and more printer friendly alternative to the satellite units which have so far dominated. One of the merits of the tower concept, quite apart from its economic efficiency and ergonomics, is the simplified web-lead.

The basic configuration is the so-called archtype printing unit for 1:1 printing, which can be supplemented with a colour hump to form a Y-formation for verso printing (2:1 printing) . Thus by stacking arch-type or Y-units it is relatively simple matter to extend the unit to form anything upto and including an 8-couple tower (4:4 printing) in order to suit specific production needs. The range of possible configurations is shown in the accompanying illustration.

Now-a-days all the newspaper offset presses manufactured by KBA can be equipped with the anilox inking system.

With regard to initial colour-register waste appreciable reductions in make ready times and start-up waste are achieved using the anilox inking system. This ensures good, consistent quality across the whole width of the press and throughout the whole course of production from start-up to print-out.

KBA can supply two or three armed reel stands, both equipped for fully-automatic pasting once the given reel end diameter has been reached and designed for web speeds up to 12.5 m/sec. The reel diameter can measure up to 1270 mm. This is designed for automatic or manual paper-reel feeding and loading.

KBA folders in the 2:3:2 system, 2:3:3 system and 2:5:5 system are available in the following versions:-

(1) With adjustment of the over-fold or underfold during operation by means of an external hand wheel and calibrated display.

(2) With press-button conversion from single to double production.

(3) With automatic conversion from single to double production.

(4) With collecting-and jaw-cylinder diameters adjustable during operation.

For large numbers of pages gear folders are also available. There are following advantages of the KBA offset technology:

(1) less start-up waste,

(2) even ink transfer during production,

(3) even ink thickness across the whole width of the machine,

(4) negligible ink-misting,

(5) simple operation,

(6) optional accessibility due to ergonomically constructed, identical printing units,

(7) reduction in maintenance costs,

(8) production with customary ink adapted for anilox inking units,

(9) production with customary colour separations,

(10) production with customary paper types.

The printing quality is no different from that of good conventional offset. Dot gain and printing contrast are also comparable.

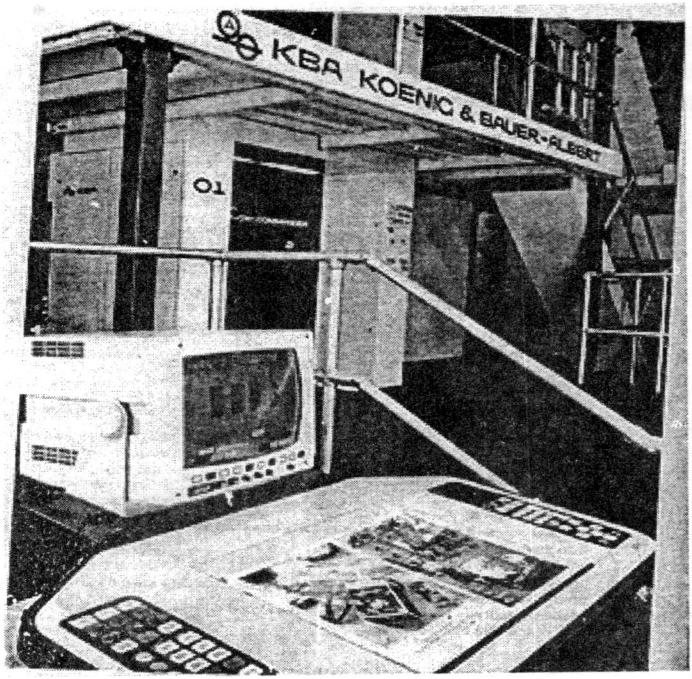

WATERLESS OFFSET PRINTING FOR SHEET FED & WEB FED

The OFFSET process still contains the potential of increasing quality and technological improvement; the composition of the dampening fluid is one of the items which has found particular attention. Reducing the amount of alcohol in the dampening fluid has become an important goal, desirable because it affects the protection of workers and the environment, and it also influences costs. In this context, waterless offset is fast emerging as an alternative. The process had been known for almost as long as wet offset, but up to around 10 years ago, there were no practical printing plates. Waterless offset is, however, accepted by an increasing number of users since Marks-3zet introduced Toray industries waterless plate from Japan to Germany in 1982.

Sillicone Instead Of Aluminium

In the waterless process, a sillicone coating replaces aluminium as the hydrophilic surface to keep non-image areas free of ink. Special inks are required to make possible, in cooperation with the sillicone coating, the separation of printing (text and pictures) and non-printing areas.

For many years, MAN Roland has been gaining extensive experience with waterless offset. This enabled the company to put Toray plates on sheetfed presses and recently also webfed, presses since have been constantly improved, including their use for waterless offset printing.

Best Results With Temperature Control

Practice has shown that constant temperature is required to stabilise the waterless offset process. The Sillicone coating of the Toray plate starts to pick up ink, that is, to scum, when the temperature of the plate surface exceeds a certain limit. Too low temperatures have a detrimental effect, particularly in solids. This is due to a developing picking tendency. Equipping the press with effective temperature control of the inking units keeps the process within the required temperature range.

The users may switch to wet offset without problems even when the presses are equipped for water less offset.

An efficient temperature control for plate cylinders has been developed. The unit blows cooled air from the outside onto the plate surface . Fens suck off the surrounding air and conduct it through heat exchangers. These adjust the air temperature to the value which is ideal for waterless offset and the energy used is optimised by circulating the cooling air through the cooling beam. The fan speed can be controlled individually if desired. This permits to compensate very quickly for temperature variations over the plate width. A combined cooling unit with two separate water circulating systems supplies, as required, either cooled dampening fluid for wet offset or cooled water for the heat exchanger in the cooling beam.

Temperature Control Of The Inking Units

The temperature of Roland 700 units is controlled via the inking units, a plant-proven method. Temperature-controlled water is conducted through the ductor and two distributors which are located in the immediate vicinity of the four plate inking rollers. An infraed sensor constantly monitors the temperature of the plate surface; water is automatically heated or cooled if the temperature deviates from the pre-set value.

There are two variants of temperature control of the inking units;

* Uniform temperature control for all printing units; All the units are supplied with water of identical temperature this method is sufficient for most applications.

* Individual temperature control of each printing unit; This variety is recommended where special inks, or inks with differing temperature stabilities, are used. Separate water circulations make it possible to eliminate eventual scumming in a printing unit by feeding only the affected unit with water of proportionally lower temperature.

In addition, two options are offered for temperature-controlled distributor roller surfaces; Rislsan or copper coating. Copper has a more favorable effect than Rilsan because of its superior temperature conductivity. It is therefore recommended where mainly waterless offset is to be used.

Another option recommended for waterless offset is a sucking system in the sheet feeder area which helps to prevent the formation of hickeys.

Addition allym, dirt particles on the printing plate can be removed during the run with hickey catchers, yet another option.

Start-up Waste In Web Offset Reduced

An additional advantage of waterless offset is the fact that the process quickly reaches a stable system condition a marked reduction of start-up waste is possible on web presses because there is no need to reach an ink-water balance that is a major economical plus point.

Using Synergism

The success already achieved with waterless offset on sheetfed presses, encouraged the new technology for webfed presses, too. An important aspect is the fact that Toray, meanwhile, made available printing plates for longer runs which, in addition, are less sensitive with regard to handling than the plates of earlier generations. These new plates, therefore, can be used for longer web offset runs.

Waterless Technology For Heat-set Presses

All modern web presses are equipped with temperature control via the inking units. The temperatures achieved in the inking units are, however, too high for waterless offset. This is due to the milling work of the rollers and to mechanical heat increase.

Systematic development has shown that, generally, waterless offset will work only if appropriate measures are taken in the printing unit to reach constant surface temperatures on plates through-out the run.

The dampening fluid which is transferred to the plate when printing wet offset and the evaporation of isopropanol alcohol provide for constant cooling of the plate surface. This cooling effect does not, of course, exist with waterless offset. Therefore, manufacturers developed a cooling air fan system which is mounted in place of the dampening units, these are replaced by inserting a cooling air beam the press is equipped with an operating console for air cooling and temperature monitoring. Additionally, a cooling system of sufficient capacity is required.

Active Air Cooling

Active air cooling for blanket and plate cylinders is used in addition to temperature control via the inking units to keep the surface temperatures of plates and blankets constant within a closely defined range. This system consists mainly of several radial ventilators arranged in a row and a water-cooled heat exchanger. It enables the operator to adjust the amount of cooling air and therefore, temperature across the width of the plate. An infra-red sensor constantly monitors the plate temperature, another sensor picks up the surface temperature of a distributor roller. The water temperature for cooling the inking unit is calculated by a comparison between the distributor and plate temperatures. Inks are comparatively viscous. It is therefore important to create constant temperature conditions in the ink duct/duct roller area. The water-cooled duct rollers are supplied by a separate water circulating system. Temperature control valves take care of keeping the duct roller temperature at a constant value of around 25 degree C.

Making plates for the waterless process does not require any major changes. The plates are exposed in conventional vacuum frames and the use of control elements makes it possible to establish the correct exposure time with a stepped exposure. the use of diffused light has no determental effects.

The plates are processed in an automatic processor, using appropriate chemistry. The equipment does, of course, require the investments of capital. The chemicals are not subject to labelling requirements. They may be disposed of in the normal way, and this applies to the Toray plates as well, plates may be corrected after development. Once ready for printing, the plates require slightly more careful handling because they are more prone to scratches than normal offset plates-care, therefore, has to be taken when bending the plates and mounting them on cylinders.

THE SECRET OF THE WATERLESS PROCESS

The special feature of the Toray plate-the secret of the waterless process-is the silicone coating it carries. It repels the ink, thus replacing the dampening fluid. The coating lies on top of the light-sensitive photopolymer layer which is coated onto an aluminum plate serving as carrier. The sillicone coating is protected by a transparent plastic film. The plate is exposed with ultraviolet lamps through the protective film. The light hardens the polymer coating, creating a firm bond with the silicone rubber coating. The protective film is removed after the exposure.

Subsequently, the developer (a swelling fluid) is applied. The swollen areas are rubbed off while the photopolymer coating remains as the ink carrier in the image areas. The areas exposed to light develop a hardened sillicone coating which serves to repel the ink in non-image areas.

In the waterless offset process, the halftone dots are on a slightly lower level than the surrounding sillicone coating. This permits the transfer of large amounts of ink, increasing the print density. The smooth plate surface makes it possible to transfer the halftone dots on film to the plate, perfectly and with sharp edges. Waterless offset, therefore, is particularly suited for fine halftone work and for stochastic screens.

Printing plates for waterless offset cost about twice as much as conventional plates. Otherwise, the process offers many advantages. The print quality increases, the waste rate decreases, and the process is a perfect answer to environmental concerns.

The processing of the Toray plate is rather similar to the processing of plates for wet offset but the quantity of chemicals required is much lower. Waterless offset permits lengths of runs of around 300,000 sheets. The further extension of this length is currently under development.

Cylinder Machines: Hand and auto-fed cylinders are both made by several reputed manufacturers. While users are generally satisfied with both the price level and the quality of the machines, there are reservations about the lower output of the indigenous machines as compared to the older models of imported machines. Users also complain about the obsolete design and the engineering aspects of the machines which, in certain cases, have resulted in an unacceptable level of down time. There is therefore, an immediate need for Research & Development and quality control machines in order to upgrade the quality of these machines. Manufacturers of autofed platens is being phased out owing to falling demand.

20

Newspaper Production Technology

Software to paginate a broadsheet newspaper page works in conjunction with the Editorial System Software. Pagination software, works on a standard platform IMB-PC compatible system. Both the Editorial and Pagination software operate simultaneously on the same system.

The body copy of the story can be generated in the Editorial system, either by key-entry-or import of the story from a wire service or the word processor. The composition of the body copy is done automatically by the system, without the need to enter any command. The status column of the Editor gives information on the depths of the story in newspaper lines and in other units of measurement. The status is also given on the active font, point size, leading, X and Y position of the cursor on the printed page, line length occupied and remaining. This information can be used to determine the size of the story-hole on the page grid.

Page Forms

As a one time effort, the basic constant elements on a page such as masthead, page slugs, etc., are created as page from files. These page form files get utilised by the system automatically while preview or print of any particular page.

Page Grid

Page grid is a screen representation of the newspaper page. It represents graphically on the screen the number of columns and lines. Every vertical line denotes a column and a horizontal line repeated at a frequency of 5 newspaper lines. The accurate position of the cursor is depicted in the right hand of the screen, which shows the column and the line on which the cursor is currently located. The screen also gives the default typesetting parameters of line length, gutter space and leading. A menu of available functions/ commands also appears on the right hand bottom of the screen.

Creating a story-hole on the page grid

Story Hole is a rectangular space reserved on the page grid, to accommodate the story which is active on the terminal. The story being edited on the terminal is the active story. Page grid can be accessed by pressing "E" in the Command mode.

Auto Registration

If the body copy and basic elements (Headline etc.) are ready, the story can be registered automatically on the page grid by a single key command, "A". This command will calculate the space required by the story and generate an appropriate story hole. This story hole can be placed at a required position on the grid.

To commence 'Registration' of the storyhole, a user has to press "R" on the page grid, after moving the cursor to the top-left corner of the story hole. Using normal cursor keys, the "Story-hole" is drawn. Registration is completed by pressing "R" again. During the process of drawing the 'story-hole', system gives the user information on the number of columns and newspaper lines spanned by the story hole on the page grid. Columns with unequal depths can be defined visually on the page grid, after the story-hole is generated.

The geometry of the story hole, including number of columns and newspaper lines is transferred to the active story in the Pagination System and the story is now automatically composed with this geometry.

After the story elements like Headline, Intro, Slugline etc., are composed, the system will automatically feedback the user on the over/ under setting of the complete story.

Composing Story-Elements

All the story-elements can be composed from the menu in an easy way. The 'Menu' is accessed by pressing 'Shift+?' from the Command Mode.

To compose Headline, for example, press 'H' from menu. The screen is horizontally divided in two parts. The upper half denotes the default typesetting parameters which will be used by the 'Headline'. Any of these default parameters can be altered.

The Headline matter is keyed-in in the lower half. The status column on the right, besides giving information on typesetting parameters is also used to determine the headline-fit, since information is available on the space already occupied by the text and the space balance.

For any story element, it is possible to define the font, point-size, leading, justification and over/under line. It can also be specified to span a given number of columns, for a specified number of lines, to begin from a given column and also the ability to place it vertically at Top, Middle or Bottom, or at a specified line number within the story.

Composing the Complete Story

With body copy and all story elements in place, if the story is composed, it will automatically acquire the layout from the geometry of the 'Story-hole' to fit the story. The story can also be previewed on the screen, with facility for zooming.

Boxed Story

A story can be automatically encapsulated in a box. The rule thickness of the box is derived from the default settings. The story will be boxed, if the option of Boxed text in the 'compose Setup Menu' or in Story Elements Menu' is turned ON. The system will automatically alter the column width of the body copy to accommodate the box.

Rules between stories and Over/Underline for Story Elements

Rules of defined weight will be drawn automatically between individual stories. These can be both horizontal and vertical. Also, for story elements like by-line, the rules above & below are drawn automatically. Horizontal and Vertical space between rules and the text is also handled automatically.

Meta Editing

This is very useful facility available with the software for last minute changes on a page. The page as a whole can be accessed without opening any particular story. This is simply done by asking for a specific page grid identified by publication date and page number. The full page grid is displayed on the screen. Any story in the page can be selected for functions like sizing, edition, deregistering or processing in any other permissible way.

Advertisements

The Classified Advertisement Management & Processing System (CAMPS) allows you to book an advertisement, modify, inquire and cancel an ad, schedule the advertisements, generate page layout and screen or printer output as required. Apart from this, a history of transaction data is maintained and updated at specific intervals, CAMPS also provides several vital reports, which can greatly influence marketing decisions. Since there is an exact calculation of the number of word, line and size of the ad, the charges calculated by the system will be exact reducing the loss of revenue caused due to miscalculation. Moreover, the time spent in manual sorting of ads and page layout will be reduced due to automatic scheduling and layout, this greatly improves efficiency, saving a lot of money and time.

The uniqueness of the system lies in the fact that a user can configure the entire system as per his specific requirements. The system provides tools for designing pages and the elements of the pages. The user can configure the details of each publication, category, sub-category, appearance of different types of ads, rates and the discounts.

Yellow page software

Abacus Computer Limited with their expertise in the pre-press technology, have developed just the right solution for the Yellow pages directory publishers. The software, a combination of commercial data entry/ analysis module and an automated typesetting module has proven to be a very successful Yellow Page pagination software.

The concept applied to develop this product is Database Publishing.

The front-end module of Yellow Page software is dedicated to data capture, validation, archival and sorting. Various information analysis can be performed using this module, such as salesman wise statistics, business category wise statistics, etc. The data thus captured is stored in a Data base format. Data entry is an integral part of the Yellow Page Software. Depending on the type of listing/advertisement, the data entry formats are provided.

For boxed type ads (NB ads), the text is taken at the time of data and is formatted automatically.

The back-end module essentially is an interface to the typesetting software, Abacus SCRIPT. This module will take in the validated and sorted data from the frontend module and embed pre specified page parameters on the entire job and will also attach typesetting parameters to each of the database fields, depending upon the definition of the type of the entry (fee listing/bold listing/boxed listing, etc.) The system will also cater to the display advertisements by providing space asked for on the page. Dynamically, the depth remaining on the page will be calculated and page will be planned accordingly, automatically.

Features

The software is totally menu driven. Described below are the menu options, which will give an insight to the software capabilities and operations:

Parameter File Maintenance

Here, you can define Page Parameters. For example, as Paper Length, Paper Width, Number of Columns, Column width and Gutter Space according to your specifications. In the same way, index page parameters like number of columns, column width, gutter space, can be defined. Identification code, product code, reference code, page number, heading text, point size, line space and font to be used, can be defined.

Text parameter particulars like page header and product category point size and line space can be set. The input product may or may not be placed in a box. For different types of ads like free listing, normal listing, bold listing and NB ads typesetting parameters like point size, font type can be defined individually.

The software also allocates free space for dummy information. For example, on which page the dummy information is to be placed, number of columns and size of the advertisement. The chooser has three options (as example, the dummy ad should be on every 2nd page, or every 6th page, or every 10th page).

Advertisement Size File Maintenance

The option allows to enter the type of advertisement (normal or display), advertisement type in codes, number of columns and size of the advertisement. The advertisement types can be modified/deleted and a list of records present can be printed and displayed.

Area Master File Maintenance

Records can be added by giving an area code and area name. The enties can be modified, deleted, displayed and printed.

Category Master File Maintenance

Records can be added by giving category code and the category name. Also, a across reference code can be given. An indication can be set to show, whether an area wise sorting is required or not for this product category. The entry in the file can be modified, deleted or printed.

Salesman Master File Maintenance

This file maintains information regarding sales persons involved in sales of advertisement space. This information is further used to process the database towards analytical reports.

All file maintenance options are design definitions for the Yellow Page directory and as such, are one-time efforts.

After the data is entered, the page formating option flows the data into pages and generates SCRIPT files using the page parameters already specified. Similarly, the index (with cross reference) will be generated with headings, automatically.

Abacus Computers Limited has its origin in the fascinating world of computers in India in September 1983, with intention to impart technical consultancy services to the industry.

From the formative years, Abacus has shown abilities to be innovative. Abacus pioneered the Desk Top Publishing revolution in India. It has to its credit more than 150 newspaper installations and thousands and odd DTP customers in India. The impressive client list is exhaustive and evergrowing. The pioneer (Delhi and Lucknow), The Independent (Mimbai), Economics times (Ahmedabad), Economic Times (Kolkata), Navbharat Times (Mumbai and Lucknow), The Afternoon (Mumbai), offer eloquent testimonies to the cost-effective and most productive techniques developed indigenously by Abacus.

21

Flexography & Packaging Printing Plates

Earlier rubber plates had to be used as individual units and positioned on a cylinder using a mounter-proofer device. This was very time consuming. Often plates had to be repositioned on the press because of inaccuracies in the mounting equipment.

To-day, technology has been adopted from the offset and gravure industries. Images are stepped & repeated into multiple images on one plate, which is positioned mechanically onto a printing cylinder by a pin registration system.

Photopoloymer-printing plates have seen changes too. Today they are of predictable size and with no distortion across the cylinder, last longer and can be pin registered (not practical with rubber). They have also better register. Water-base ink technology has improved and has helped to promote photopolymer.

Rubber plates still rule a large market regment and will be around for a long time. They are economical to manufacture (about half the material cost of photopolymer). These can be made in minutes and can be ready for instant use. Photopolymers must have all solvents dried out before use so the plate will have uniform thickness.

Modern suppliers now make process printing plates using stastical process control. Art work for flexography now can be computer generated and all lines, rules, circles, ovals & masks can be cut on computer plotter with more accuracy. Designing is done on computerized design machine that quick & capable. Logos and other repeated materials can be stored electronically and called up to any size when wanted.

Moulded Rubber Plates

Rubber plates are flexible resilient and have the printing image in relief. They are duplicated from a mould, or matrix, made from an original pattern plate. Any number of printing plates may be made from this mould. Because it is made from a flexible material of either rubber or a combination rubber/plastics, the plate itself is flexible. These materials have excellent affinity for a wide variety of inks and the ability to release ink onto many different substrates.

Basis of Rubber Platemaking

Rubber plates are produced from moulds or patterns generated from the artwork through a photographic and etching process.

Photopolymer Plates

Photopolymer Plates are light sensitive materials. They take many different forms and are used in printing to prepare letterpress plates, offset plates, printing resists proofing films and flexographic plates. Photopolymers used for flexographic printing plates are similar to rubber plates in that they are flexible, resilient and have the printing image in relief.

Basics of Photopolymer Platemaking

There are many systems available for producing photopolymer flexo plates. The materials are either viscous liquids or solid sheets of appropriate thickness. Photopolymer materials, whether sheet or liquid are converted to flexographic plates in a direct production from a photographic negative of the artwork to be produced. Though the process may differ slightly from system to system, they all require the following :-

(1) Back exposure of base to UV light to harden (cure) floor and establish relief depth.

(2) Face exposure of surface to UV light through the negative to harden (cure) printing images.

(3) Washont in appropriate solvent to remove unexposed polymer and leave printing images in relief.

(4) Dry to remove absorbed solvent and restore gauge thickness.

(5) Post expose to final cure of floor and character shoulders.

(6) Finish plates with chemicals or UV light to remove residual tackness.

Preparation of the negatives is one of the most crucial operations in the manufacture of photoengravings. The finished black & white (mechanical art) art work is photographed using the equraving camera. Negatives then are made. Special camera leunes are used to pick up the finest detail, so any imperfections in the artwork are transferred to the negative.

Before artwork reaches the camera, it should be carefully inspected to make sure that elements are clean and have sharp line definition.

After photography, the film is developed to achieve maximum density and is inspected for defects. Defects in the negatives are corrected by painting over them with a commercial opaquing solution.

In single color jobs, only one negative is shot, usually from artwork prepared on a single piece of illustration board. For multicolor jobs, a negative for each color must be made.

If preseparated art is used each overlay is shot and negative made. But with composite or keyline art work, it is necessary to make composite negative for each color to be printed.

The color separation artist paints out all the colors but one on each negative leaving all details of that one colour on its own film.

Great care is needed to ensure that all copy is clean and sharp. In addition, imperfections such as pinholes, broken letters or other flaws should be carefully retouched. Center lines and registration marks should appear on the negatives and be recorded on the printing plates.

The next step is photoengraving — a process of etching a design into a metal plate. The image on the negative is first transferred photographically to a photosensitive coating on the face of a sheet of metal. Then the plate is etched in an acid bath.

In this way, the original piece of art that was prepared on paper or other medium now becomes a bas-relief image on metal. The engraved metal becomes the master pattern from which the rubber plates is made.

Today, metal is precoated with a light photosensitive material, ready for printing down an image. The emulsion side of the image-carrying negative is placed in direct contact with photosensitive coated surface. The two are locked tightly together in a vacuum frame and exposed to fight source.

When printing down, it is recommended that step gauge be used to ensure sufficient light exposure the negative should be wrong reading, with the emulsion down for surface—printing plates. The negative would be right-reading emulsion down for reverse-printing flexo plates.

Powderless Etching of Metal—About forty years back, a chemical process called "powderless etching" was developed. The process involved a special filming agent mixed in a nitric acid bath that acted to protect the sidewall or shoulder formation of the image from undercutting during the etching cycle. It is the well accepted etching method used today.

The printing metal is placed in a stainless steel etching machine where it is placed with a mixture of nitric acid, oil, etching additives and water initial the proper etching depth is reached. The hardened coating resists the acid's action while the unprotected areas are etched away. In this way, the acid bath creates an image in the metal's surface against a background that's been etched away in relief.

The entire etching procedure requires very precise control of the chemical solution, machine speed, bath temperature and timing. The powderless etching method it universally used and produces high-quality flexo engravings.

Finishing—the engraving is cut from the large metal flat after etching and carefully finished to remove imperfection the photoresist is removed from the face of the engraving and cleaned for subsequent reproductions of the mould. If all the photoresist is not removed, it can cause unevenplates or flaws in the finished plate.

Centerlines are cleanly scribed into the background of the engravings during finishing. These are accurately drawn to connect center marks that were left an all colors of the original art work.

It is important that scribe lines be cut deep, sharp and clean. They also must be accurately aligned with the center of the registration points.

Extra centerlines are sometimes provided for convenience in locating shiq sections such as price changes, weight, any nutritional clauses etc. to any part of the plate.

After the finished engraving in proofed and checked for quality, size and accuracy of color separation, it is ready to serve as a master pattern for moulding the matrix.

Qualities of a good metal engraving—The finished engraving should be checked for proper depth. It should be roughly 0.035" in fully etched open areas.

In fine type and within groups of letters or numerals where full depth cannot be achieved, a proportionate depth should be obtained, so that each element is clearly defined (as in the figure—A)

Ideally, the supporting sides or shoulder formation should be almost straight down, with a uniform smooth tper. Avoid excessive shoulders. A broad, stepped shoulder would tend to catch and luid up ink, causing a smeared and dirty print (as in the figure-B)

At the other extreme, avoid a badly indirect condition too, this would tend to lock the engraving into the mold and make separation of the two impossible without irreversible damage to the mold and possibly to the engraving (as in the figure-C)

Pinholes in the photographic negative, impurities in the metal or a lack of proper control of the etching both operation often cause pimples or tick marks to be left. If there are too many pimples or ticks, remake the engraving. Pinplex & ticks should be eliminated otherwise they can show up in the finished print as extra spots or splatter (as in the figure-D).

Figure-A

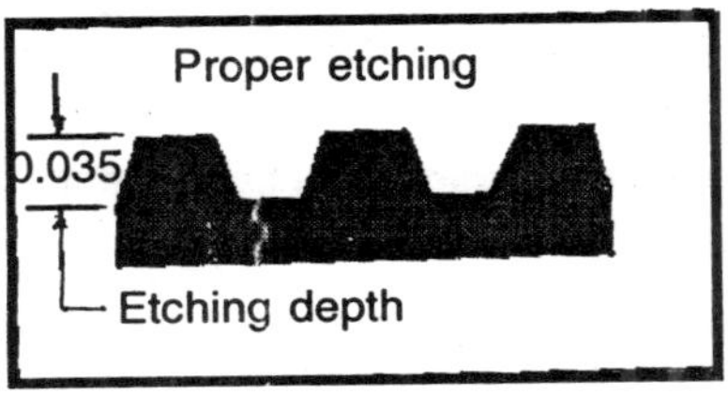

Figure-B

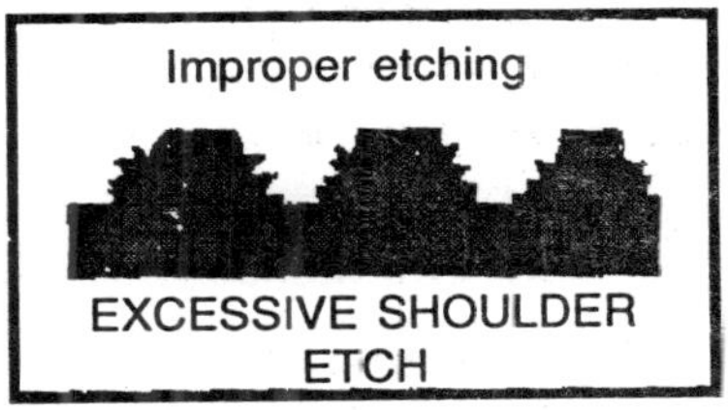

Figure-C

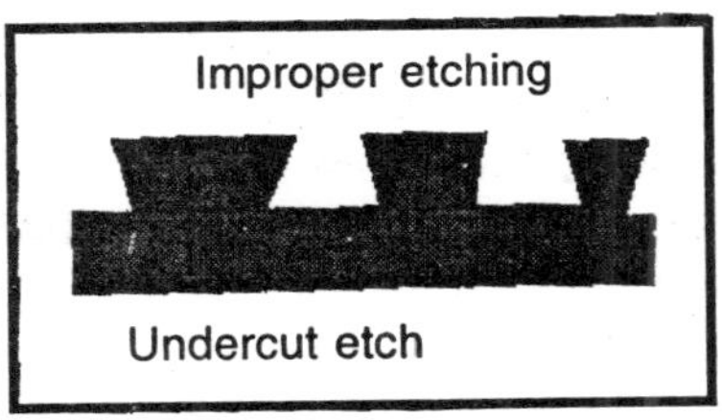

Figure-

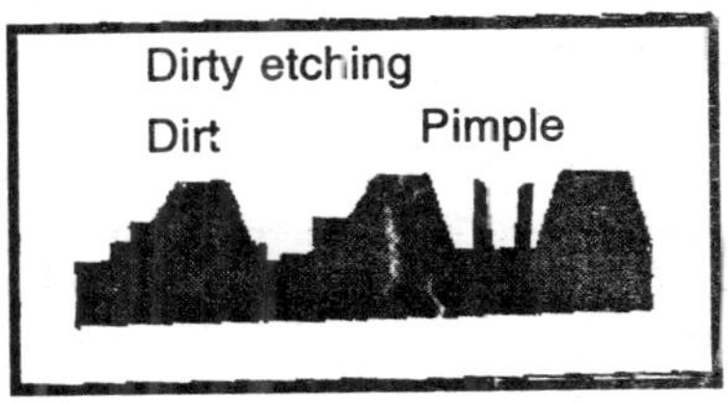

Types of Engraving

(1) Line engravings are made from original copy in which the subject matter is rendered in solid color or black such as lines, types, illustrations etc.

(2) Tone If the design contains gradations of tone, such as in photographs or wash drawings, need a halftone engraving.

Types of Metal – Photoengravings can be made from copper or magnesium. Copper is used mainly for fine screen reproduction and whenever fine tone or process color jobs are involved. Magnesium is an excellent engraving metal for producing high quality line originals. Sometimes it is used for coarse to medium upto 100 line-screen) process color work.

Hard-Durometer Photopolymer Master

Introduced in 1984 and gaining acceptance are rigid, photosensitive nylon photopolymer master. These plates come in a variety of thicknesses and with different backings, usually either stainless steel or aluminium.

The plates are produced in the same way as regular photopolymers. Using a negative, the plate relief areas are exposed to UV light and hardened. The unexposed areas are washed away with plain water. The plate then goes through a drying cycle to remove excess moisture, and post-exposure stage that further hardens the plate, before it is used in the molding operation. Once the master has been made, it is handled the same as a metal master for the molding procedure.

Nylon master materials are capable of molding very fine screen counts, upto 150 lines for process color printing.

Water-Washable Photopolymer Master for Molds—This type of **water**- washable photopolymer plate is harder and more heat resistant, than **the** nylon pattern plate.

There are some important features of this materials like etching washout with tap water, improved shelf life, fine pattern and halftone reproduction.

Molding Press Equipment is used to make the mold, and in turn, the printing plate. Because these are heatset and vulcanizing processes respectively, the press needs heat from steam, electricity or hot oil. Molding pressure is generated by hydraulic power to the bottom table or plate which travels up or down as required. The top platen is stationery in the press. The mold and printing plates are produced between these two accurately ground surfaces. Precise temperature control is necessary to produce flat models and even rubber plates.

Laser-Engraved Design Rolls— Laser-engraved design rolls can be created on nearly any base on which printing plates might be mounted : integral shaft cylinders, metal demounable cylinders or rigid sleeves. They also can be manufactured for most cylinder sizes, from the typical "norrow web" to "wide-web" types.

Rubber selection — The engraver has natural and synthetic polymers, which are not always available for moulded plates.

Polymer applications — The laser engraver usually will utilize thin, latex-like sheets of the chosen polymer to wrap the cylinder or sleeves, having first coated it with heat-activated adhesive to assure bonding during vulcanizing. The sheet rubber will be wrapped under tension and pressure, so care should be taken not to trap air between the successive layers. Excess rubber will be applied to allow for shrinkage to provide for later grinding to size.

The wrapped rubber will be tightly wound, inturn, with wet-shrink tape, and end plates put on the roll to prevent the rubber from escaping during vulcanizing.

Vulcanizing — The wrapped roll will be placed in an autodave where, under temperature & pressure, it will be "cooked" until the polymer is fused into a solid jacket, firmly bonded to the base. Cooking time will vary according to rubber thickness and roll size.

Griding and Polishing — The aim is to produce a stress free roll that is level. conentric, smooth and precise in dimension within 0.0001".

Design Factors — Because the cylinder surface is stable and seamless, stretching and shrinkage, as with rubber plates do not have to be considered in providing artwork for laser engraving. The engraver can then photograph, modify, step-repeat or otherwise lay out the images to meet the flexo printer's requirements for repeat (cylinder circumference) and print width.

Engraving the cylinder — There are minimum two different laser engraving technologies in use with reference to the way the laser beam is guided, both achieve the desired result by using the concentrated, high energy of the laser, to remove the polymer from the nonprinting areas. The rubber is atomized, leaving a clearly defined image. Depending as the technology used and the requirements of the specific application, engraving depth can be varied from cylinder to cylinder and the image profile can be vertical, sloped or stepped.

Proofing and Inspection —The laser engraved cylinder can be proofed an any one of several available proofing machines to see, whether it will print uniformly with minimum pressure. Proofs can also be made to assist in mounting plates on other rolls to be used with the lasered one.

The engraver inspects the cylinder surface, as well as its mechanical components, using special lighting and magnification.

USE & CARE — Laser-engraved cylinders, with proper use and care, are designed for long runs. As they are solid, integral units, they usually cannot be repaired if damaged.

On press, cylinders should be exposed to the minimum press pressure consistent with quality printing. As cylinders warm up on the press, pressure can often be further reduced. As soon as a run is completed, the cylinders should immediately be removed and cleaned, using any one of the available cleaning agents made for the purpose.

By using fluid & a soft brush, a cylinder can be cleaned quickly and without damage.

If the cylinder is to be used again, it should be stored suspended by its journals or by a rod through the bore, in a cool area, and be loosely wrapped to allow the solvents to evaporate, while protecting it from direct fluorescent light or sunlight.

Cylinders also should not be stored near electric motors, as the ozone they produce will attack most polymers.

Advantage — Laser cylinders are often used for multicolored fine-image designs requiring tightness of registration, hard to achieve in plate mounting for jobs involving the precise location of eye spots or for design being matted to cutting dies or patterned embossing rolls.

Photopolymer Plate making

The negative is the single most important element in photopolymer plate preparation. It is a light stencil that controls image formation during exposure of the photopolymer.

Types —There are a number of photopolymer types and constructions from several manufacturers to meet the requirements of each flexo market segment.

ALCOHOL & WATER-BASE INKS— Most photopolymer materials available to day work best in these ink systems, used most often in flexographic printing. Most sheet photopolymer compounds will utilize polymer compounds that resemble natural rubber in their construction & solvent compatibility.

Backing Materials — Normally photopolymers are supplied or produced with transparent and dimensionally stable polyester backings. Some offer a metal backing for use on magnetic or mechanical lockup printing cylinders. Because the backs of these materials is not transparent to the UV light, they come with a preset relief depth.

Compressible construction — A compressible photopolymer plate construction is available in both sheet and liquid. These plates have a compressible foam layer within the plate between the photopolymer image layer and the dimensionally stable polyester backing.

Durometer— Usually photopolymer plate materials exhibit a cured plate durometer hardness of 45 to 50, Shore A, which developed as the right hardness for general film and paper converting with rubber plates.

To day there are a variety of materials with differing durometer hardness available from photopolymer manufacturers worldwide.

Platemaking from sheet photopolymer

Preparation :- Provide 4.0 density negative (matte finish) and plate material of proper gauge. Cut plate to size of negative (SO) Allow 1/2"to3/4" clamp strip on two edges.

Back Exposure :- Turn on exposure unit. Allow 5 minutes of warm up. Place plate materials face down on the vacuum frame. Set timer. Back expose to determine relief depth.

Position Negative :- Turn plate over, wipe with antistatic cloth or brush. Strip off cover sheet. Clean negative, place it on plate emulsion side down.

Face Exposure :- Cover negative with edge strips. Roll out cover sheet. Pull vaccum. Inspect negative and plate for perfect contact set timer Face expose.

Washout preparation :- Turn off vacuum. Strip off negative. Inspect plate. Turn on washout unit. Allow solvent to reach operating level. Set brush pressure.

Washout Step one :- Clamp plate, avoiding copy and illustration areas. Lower bush. Set timer for brushout time, Brush plate.

Drying :- Start dryer 30 minutes before use. Remove plate from washout unit and place it flat on dryer shell. Set timer dry until swelling is gone. Check by measuring dimensions and making for return of caliper.

Platemaking from sheet photopolymer

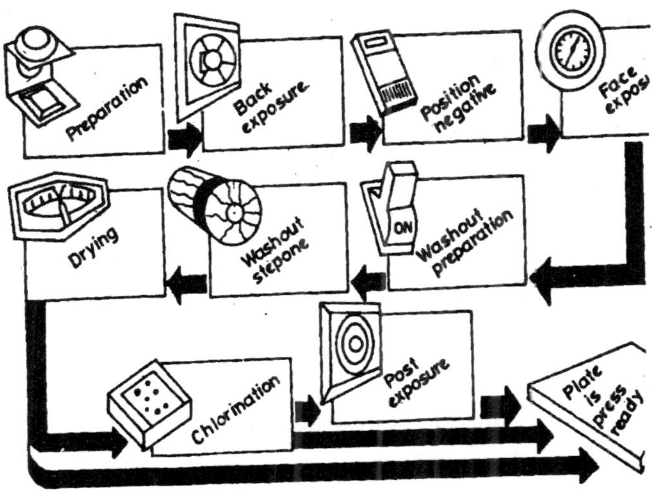

Chlorination :- Immerse plate for recommended time. Remove. Rinse immediately with water. Blot excess. Put plate in dryer 5 minutes.

Post Exposure :- Simply place face up in the vacuum frame and expose it with no covering.

ADVANTAGES WITH PHOTOPOLYMER PLATES

(1) Better Print quality : Sharper line images, better ink transfer, predictable halftone results

(2) Longer plate life :- About twice that of rubber plate

(3) Plates are more reusable :- Less distortion when mounting and remounting

(4) Step and repeat now possible :- The plate can carry multiple images.

(5) Predictable plate gain :- Unlike rubber plate gain, it can accurately be determined and compensated for in art.

(6) Better registration because of stable backing :- 90% coverage plate will now register with a 10% coverage plate.

(7) Prepress proofing :- Cromalin, Match Print and color keys are all accurate reproductions of film used for plates.

(8) Pin registration :- Now possible and more accurate than plates mounted on the mounter-proofer.

(9) Less skill required :- Less skill is needed to mount plates. With pin registration, any one can do it.

(10) Less mounting time :- A six-color, 4-up job can be mounted with pin registered photopolymer plates in about 30 minutes as opposed to 8 hours or more for rubber

(11) Eliminates mounter-proofer operation

(12) Less press downtime to register plates. Less plate cleanup is needed. When rubber plates are mounted on a monter-proofer they still do not register on press.

(13) Less file space :- Less file space is required for film than for metal and molds. Negatives will not deteriorate like metal or molds and can be easily duplicated.

(14) No engraving or mold cost

(15) Faster plate production than for metal molds and rubber.

(16) Lower labour costs :- More square inches of photopolymer can be made per man-hour and relatively unskilled labour is needed.

(17) Larger plates :- Sizes up to 80" × 48" are possible

(18) Lower costs :- Process work will cost only about half as much as for initial printing plates to press as copper, proofs, models and rubber. Proofing is also more accurate with photopolymer prepress proofs.

(19) Less time :- Process work can be done in half the time (or less) that is takes to make copper engraving, models and rubber and can be stepped and repeated in multiple images so there are fewer plates to register.

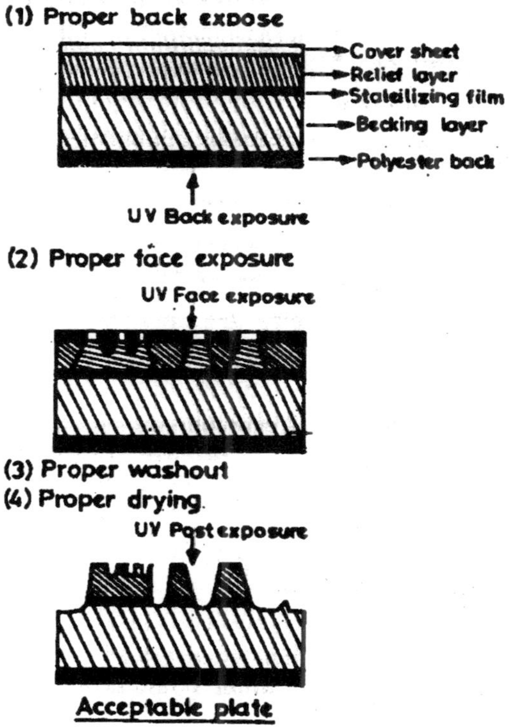

STEPS IN MAKING LIQUID PHOTOPOLYMER PLATES

ENGRAVING AND PRINTING PLATES

PLACING NEGATIVE AND COVER FILM

COVER FILM

NEGATIVE TRANSPARENCY

LOWER OPTICAL GLASS PLATE

Steps In Making Liquid Photopolymer Plates

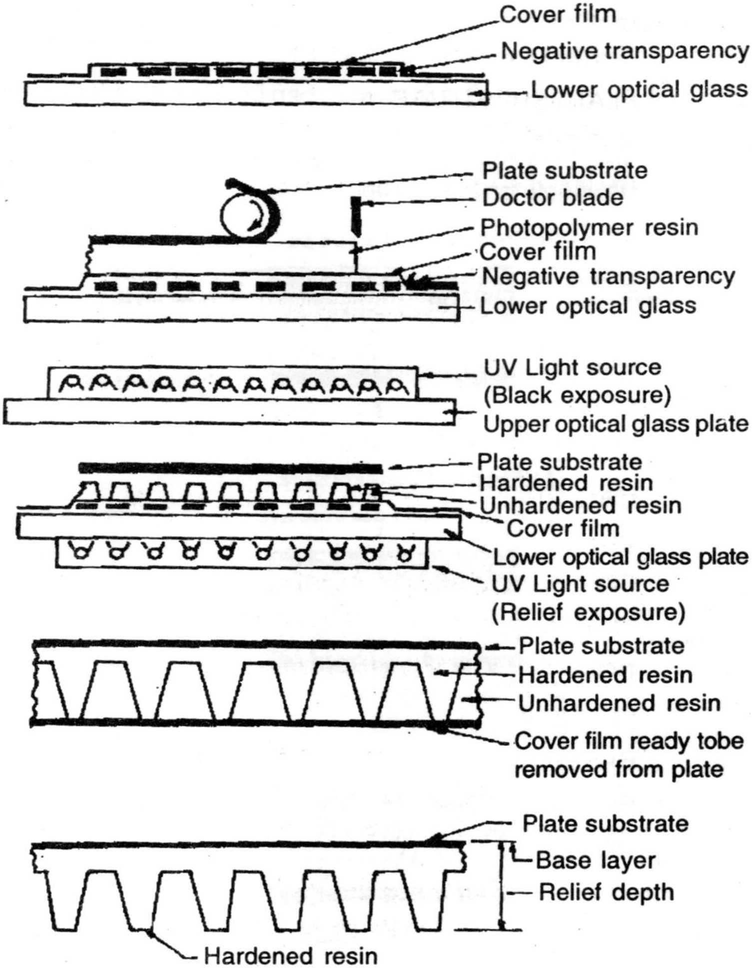

MAKING THE COATED PLATE

PLATE SUBSTRATE
DOCTOR BLADE
PHOTOPOLYMER RESIN
COVER FILM NEGATIVE TRANSPARENCY
LOWER OPTICAL GLASS PLATE

EXPOSING THE PLATE

UV LIGHT SOURCE

(BLACK EXPOSURE)

UPPER OPTICAL GLASS PLATE

PLATE SUBSTRATE

HARDENED RESIN

UN HARDENED RESIN

COVER FILM

LOWER OPTICAL GLASS PLATE

UV LIGHT SOURCE

(RELIEF EXPOSURE)

EXPOSED UNWASHED PLATE

PLATE SUSTRATE

HARDENED RESIN

UNHARDENED RESIN

COVER FILM READY

TO REMOVE FROM PLATE

COMPLETE DEVELOPED PLATE

PLACE SUBSTRATE

BASE LAYER

RELIEF DEPTH

HARDENED RESIN

Plates for corrugated Printing :-

Plates used to print corrugated board are quite similar to flexographic plates and as in flexography, can be molded rubber, hand cut engraved rubber or photopolymer. The big differences is the overall height of the plate. Presses that convert corrugated board are made to use a plate and carrier (mounting material) that have an overall thickness of 0.280". In this case, the plate is normally 0.250" and the carrier 0.030". Today some converters have been using a 0.155" plate and a 0.125 sponge carrier.

In making the plate, there are steps similar to those used in flexography. If the product is molded rubber plate, some kind of a master pattern must by generated. If it is a photopolymer plate, final negatives must be made. With hand-engraved rubber plates, a mechanical layout must be created.

PRINCIPLES OF FLEXOGRAPHY

The flexographic printing systems consists of following four basic parts.:

1. Fountain roll
2. Anilox (ink-metering roll)
3. Plate cylinder
4. Impression cylinder

Fountain Roll – It is generally covered with natural or synthetic rubber. It is positioned to rotate in a reservoir of flexo inks. Its purpose is to pick up and deliver a relatively heavy flow from the reservoir or fountain to the metering roll. Because the fountain roll is usually driven slower than the metering roll, it has the effect of wiping the latter.

Anilox Roll – It is usually chrome plated and is covered with tiny, engraved cells anywhere from 80 to 500 per linear inch. The roll's job is to supply a fine film of ink to the printing plates, mounted to the next roll in the sequence, the plate cylinder.

Plate cylinder – The plate cylinder is usually steel and is installed between the anilox roll and the impression cylinder. The plate's raised impression picks up ink from the anilox roll & transfers it to the substrate.

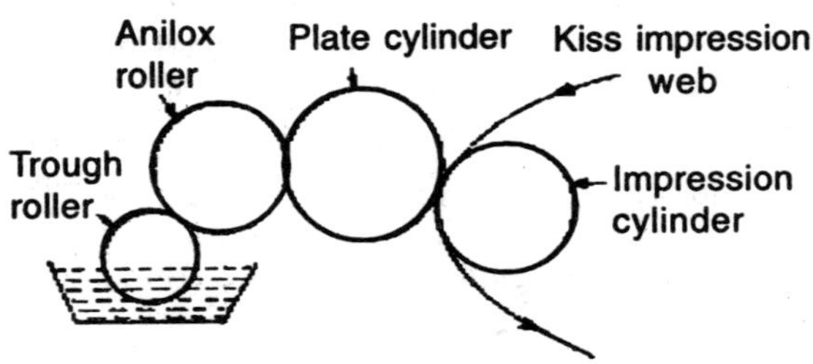

Flexography

Impression cylinder – This is smooth, highly polished and supports the substrate when it contacts the printing plate. The speed of the impression cylinder must match the plate cylinder's, the anilox roll's and the substrate's.

The flexographic process prints on the rotary principle. It has a plate cylinder and an impression cylinder .

It utilises flexible rubber plates either as separate stuck onto a carrier sheet and wrapped around cylinder, metal backed rubber clipped on to special cylinders.

It points from a reel and the web run through the units can either be re-winded on subjected to either in-line separation as desired with the necessary equipment.

It utilize fluid volatile inks-solvent based, which dry in seconds. Driers are provided to aid evaporation of the solvent after each printing unit. In addition, a large oven drier is provided for overall drying of the printing unit. In addition, a large oven drier is provided for overall drying of the printing before rewinding etc.

Print Pack (Faridabad)

The flexographic process is well suited to printing on metallic foils and plastic films because-

1. The web tension and control design and rollers arrangement is such that very good control of this difficult stock is possible.
2. The ink is solvent based which also frees it from the limitation of normal letterpress inks in for keying the ink to the foil, plastic etc.
3. The ink can be made transparent (using dyes) or opaque (pigments) for special effects on these materials.
4. It gives wider range and brilliance of colours.
5. It takes split of second to drying of ink.

Flexo's basic advantages of low all round cost coupled with simplicity of operation enabled it to move into other markets such as books, textiles, wallpapers and children comics and it has more recently established itself as a viable process for newspaper printing.

Hand -cut plates or cylinders may still be obtained from specialist houses, where simple and straight forward designs are required, but most flexo surfaces are now either duplicate rubber plates, photopolymers or lases engraved.

Duplicate rubber plates generally produced via a negative, which is exposed to a light-sensitive coating on a zinc plate. Light passing through the negative hardens the coating in the image area, and subsequent development washes away the coating from the non-image area, keeping the hardened coating as an acid resist. The plate is then etched in an acid bath to produce a rigid flat original plate within appropriate etch depth for duplicate platemaking. The image will be wrong-reading as in all direct relief printing processes.

A mould or matrix is then made from the zinc by placing it in contact with a sheet of matrix material such as a Therm setting plastic and subjecting it to heat and pressure in a moulding press. Heat will soften the matrix material and pressure will force it into the recesses of the relief plate. This is a critical operation as temperature, time, pressure and the dimensions of the materials must be correctly measured and controlled to obtain desirable results.

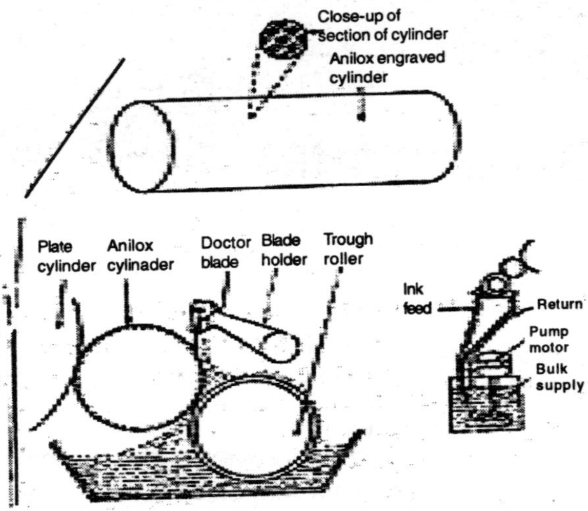

Figure : Principal of Flexography (a) The anilox system, (b) Reverse angle doctoring, (c) Principal of Circulating ink system

To produce the final printing plate the matrix is now placed in contact with the plate material and again subjected to heat and pressure in the moulding-press . Any number of plates within reason may be taken from the matrix and matrices may be saved for future reprints.

The plate material is usually a compound of natural rubbers and synthetic materials such as copolymers plus additives like sulphurs, carbons, plasticizers and resins each with a definite purpose, either to aid the moulding process or to give particular required properties to the final plate.

Plate hardness (durometer) will very from 30° s (soft) to 65° s (hard) depending on the purpose. Soft plates make for easier and quicker setting up, although they are more likely to distort under printing pressure, which will affect the reproduction of fine lines and dots. For this reason plates intended for accurate reproduction of detail work, such as colour halftone printing will tend to be a harder durometer.

Care must be taken at every stage of production to ensure that the plates are accurate to a caliper of plus or minus a few hundredths of a millimeter and within the tolerances for the printing image dimension. Most rubber compounds will shrink in the moulding process, and where this is known to occur an allowance can be made at the original stage. It should also be noted that rubber plates are made flat and will distort when curved and mounted on to the printing cylinder. Where final print length is critical this also can be compensated for at the orientation stages, provided that the degree of distortion is known in advance. As we know distortion is a function of plate caliper and cylinder radius (photopolymer relief plates)

Flexographic photopolymer plates- Flexographic photopolymer plates, need in the first instance to be resistant to the solvents commonly used in flexo inks, and this will affect the choice of polymer. They are usually thinner than moulded rubber plates, which may be several millimeters in caliper, whereas photopolymer plates rarely exceed 2.5 mm and plates of 1mm or less are obtainable. To avoid distortion of the relief image under printing pressure the plate may be composed of layers of material of varying hardness- the printing surface layer being of a comparatively hard durometer to resist distortion, while the base layer may be of a much softer durometer to absorb the stress of printing pressure. It is also possible to coat photopolymers material onto a cylinder or sleeve for the purpose of printing continuous designs for packaging, wallpapers, textiles etc..

Earlier, photopolymers wee criticized owing to their poor ink receptivity when compared with rubber, but modern compounds have improved this situation and photopolymers are steadily replacing moulded rubber plates, due to their simpler techniques, lower overhead, faster production and the fact that there is no need to produce an original zinc plate or thermosetting matrix. Where print length is critical, it is quite reasonable to expose and process photopolymer plates on the curve and thus avoid the distortion problems which occur with rubber duplicates.

Laser engraved plates and cylinders are produced via computerized scanning equipment **in which** a helium-neon laser beam examines the artwork, as it revolves below the beal on a copy cylinder. The scanner relays digitized information to the central computer which analyzes the information and initiates the cutting of the image by means of a co laser beam located above an engraving cylinder, rotating in synchronization with the copy cylinder.

This is the fastest ways of producing flexo printing surfaces, as it is a direct copy-to-cylinder system without the intermediate stages required in other systems. Though the initial capital expenses is higher than for moulded rubber plates or photopolymers, but as well as speed, computer control allows a great deal of flexibility of image reproduction. Receptive designs or images can be stepped and repeated from a single master, and continuous designs can be seamlessly joined.

Rubber is the preferred image medium as it is easily vapoursied by the co laser and will receive and release ink readily. For economy, plate patches may be mounted upon a carrier sheet so that only those parts of the cylinder, which contain a printing image will be engraved.

Flexo plates are usually proofed on especially designed proof presses, which will take the actual cylinders or sleeves which will be used on the printing press. Printing cylinders are easily removable on mestflexo presses to allow for changes in print length and for pre-proofing. It is also possible to store complete sets of cylinders with their plates intact for reprints.

Most proof presses print one colour at a time using modified inks and inking systems to allow for their slower rate of production and the time interval between colours.

Individual plates are usually fixed to the cylinder by means of a double-sided adhesive. The first set of plates is placed in position by reference to a key layout, which shows the location of each element in the design. A proof is then taken and used as a basis for the location of subsequent sets of plates by cross referencing to the key layout. As each set of plates is mounted, further proofs are taken and super imposed on the previous images until the work is complete.

Complete cylinders as in the case of continuous photo-polymer designs or laser engraved cylinder are quicker to proof as there is no need to mount individual plates for each element of each colour on each cylinder.

Most mounter proofer devices are intended primarily to ensure accurate registration and image size of the plates on their various cylinders . Other values such as colour and image qualities relating to such factors as dot gain, ink film density and ink trapping will be only approximate.

Designs - Flexo printing units comprise a plate cylinder, an impression cylinder and an inking system of two rollers only plus an ink trough.

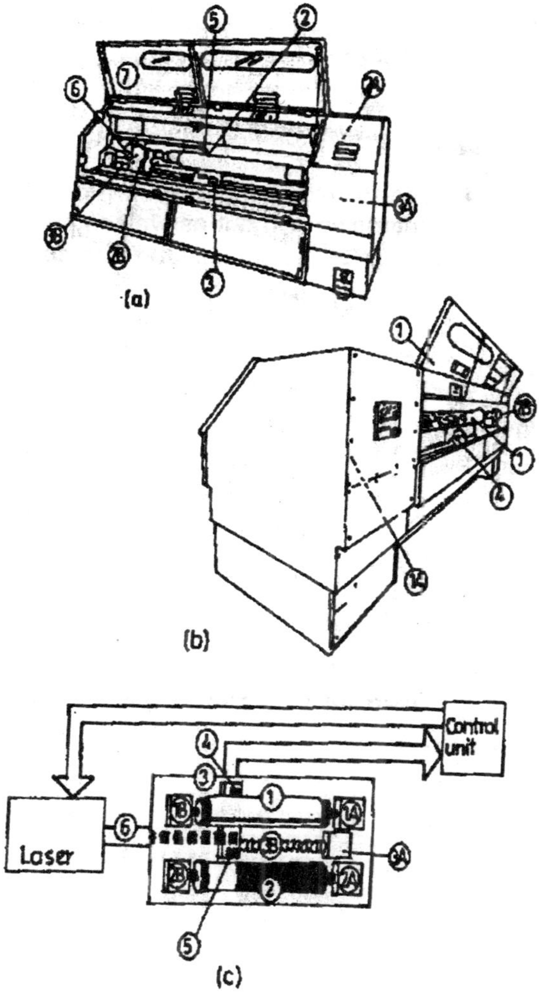

Fig. Laser engraving of flexographic plates and cylinders. (a) Engraving side. (b) Scanning side.

1	Artwork cylinder (unjointed copy)	3	Cross-side assembly
1A	Headstock drive and encoder	3A	Lead screw
1B	Tailstock	4	Scanning head
2	Engraving cylinder (seamless)	5	Engraving head
2A	Headstock, drive and encoder	6	Laser guard
2B	Tailstock	7	Guards

Plate cylinders - The plate cylinders are solid steel or alloy cylinders upon which the plates are mounted with a double sided adhesive. The cylinders are removable to allow for different circumstances of cylinder to be employed to cater for different print repeats and pre- mounting and proofing prior to printing. This is in contrast to the pattern, with web offset and photopolymer relief presses, which have fixed cylinders of a standard size.

Continuous design cylinders, will also have a metal base, but the photopolymer or rubber will form a complete overall layer in which the image is contained, as distinct from the conventional multiplate system.

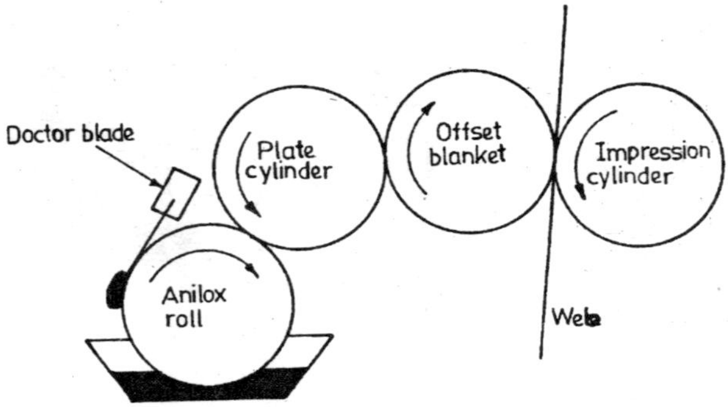

Flexo offset Printing Station

Sleeves - are a simple and inexpensive alternative to solid cylinders. These are fibre glass tubes, upon which the printing image is held. The sleeve may be simply slipped over a body cylinder or mandrel on the production press or proof press and locked into place. Sleeve have the advantages of light weight, quick mounting, low cost and easier handling when compared with solid metal cylinder, provided that the press has been specially designed or adapted to accommodate them.

Impression cylinders - are generally plain steel cylinders geared into the press drive. Their circumference is not necessarily the same as that of the plate cylinders, as is in the web offset and photopolymer relief presses and do not have blankets. Their main function is to act as a support for the web and as an impression resist for the plate cylinders as well as providing a drive to the other cylinders in the unit.

Inking system and ink trough - Inking system on flexo presses comprise only two rollers. A rubber covered trough or fountain roller sits in the ink tray, containing the fluid ink and feeds the second roller, which in turn applies ink to tie printing cylinder. There are no ducts of keys as with other systems .

Ink is metered first by adjustment of the gap between the two rollers and then by a cell pattern is referred to as the banlilox' screen and will determine the ink film thickness which is laid upon the printing surface. Different shapes, sizes and depths of cell patterns may be used according to the type of work and the substrate in use. For solids and line work on absorbent paper stocks a course, deep pattern of 25-30 cells per centimeter may be used, whereas colour halftone work on non-absorbent plastic film may cell for anilox rollers with more than 150 cells per centimeter .

The two roller inking system on its own give satisfactory results in many instances, but for the finest ink control a 'doctor' blade will be used. This is a fibre or metal blade, which rides on the surface of the anilox roller against its direction of travel. The function of the blades is to shedder the surplus ink cleanly from the surface of the roller, leaving a precise quantity of ink in each cell; to be deposited on the web.

Generally ink is pumped to the trough from a bulk supply and constantly re-circulated to keep it clean and of consistent quality . An independent variable-speed motor drive is often fitted to thinking rollers to control the initial ink flow, which may vary with press speed, ink viscosity etc. The motor drive to the inking, system may also be used to keep the rollers' idling' during press stops and to avoid ink drying on the rollers.

There are three main ingredients-colouring matter, resin and certain additives for specific purpose.

Colour - Is generally provided by means of dyes or pigments, or a combinations of these. Dyes are usually transparent and dissolve completely in the solvents, whereas pigments may be opaque, have a particle size and form a suspension in the solvent, in combination with the resin to give the ink its body.

Resins are usually synthetic materials used to give a substance to the ink and the final printed image. It is the resins which will key the ink to the substrate and the choice of resin will depend largely on this factor. For absorbent stocks, the resin will need to give the ink sufficient body to prevent it penetrating through the other side, causing the fault of strike through. For plastic films the resin must have a chemical affinity for the substrate, in order to ensure good adhesion.

Solvents are used to dilute the ink to a consistency which will enable it to flow through the inking system, fill the anilox cells and transfer easily to the web. The flow property or viscosity of the ink. This can be measured in the laboratory by special purpose viscometers, or more simply at the press side by measuring the time, a given quantity of ink takes to flow through a known aperture in a flow cup.

Solvents are chosen for their ability to dissolve the resins and carry the pigments and for their compatibility with the rubber or photopolymers printing surface and the substrate. For these reasons the number of solvents which can be employed in flexo inks is quite limited.

Most are classified as alcohol, ranging from the traditional methyl and ethyl spirits to propyl and butyl alcohol. Various other solvents may be added to the ink to give it or the final printed film, particular properties.

The solvent may comprise more than go percent of the volume of an ink but once it has fulfilled its function of carrying the pigments and resins to the substrate, it is no longer required and may be allowed to evaporate or be forcibly evaporated, by means of accelerated drying devices on the press.

There are two safety factors which must be taken in to account, when using evaporative solvents.

1. Most of the solvents referred to are flammable and a concentration of fumes can ignite under certain conditions.
2. Random evaporation of solvents will pollute the atmosphere and could contravene health & safety requirements.

Evaporation must therefore be contained and controlled.

Printing units - Flexography printing units are initially arranged in three ways-

1. Stacked
2. In-Line
3. Central impression cylinder.

Combinations of these arrangements can be used for particular purpose and specially designed units are obtainable for publication work.

Stack presses-stack presses have their printing units positioned around a central frame, a web being threaded from one unit to the next. These are very common type of flexo press, as they are compact, of moderate cost and versatile in use. Stack presses are normally used for single-side printing but, perfecting can be arranged by altering the gearing on one or more of the units and re-threading the web. Up to six units is quite common and more than this can be provided if required.

In- line printing units - In-line printing units are a simple and logical way of arranging the press elements. All units are accessible and at convenient working height. Any number of units can be included for front and back printing, and other in-line operations, built in as and where required.

Though in-line system may occupy more floor space than stack presses, but this can be preferable for extra large presses printing on stiffer materials and where multiple in-line operations are undertaken.

Central Impression - Central impression presses have several plate cylinders, grouped around a single impression cylinder. This design is suitable for closer register work as the web is held firmly on one impression cylinder during the whole printing cycle. This is mainly advantageous when printing on extensible plastic films where stretch units. Central impression extensible plastic films where stretch or shippage could occur when printing on separate units. Central impression printing units print on one side of the web only, therefore reverse-side printing will require an additional unit or units depending on the need and nature of work.

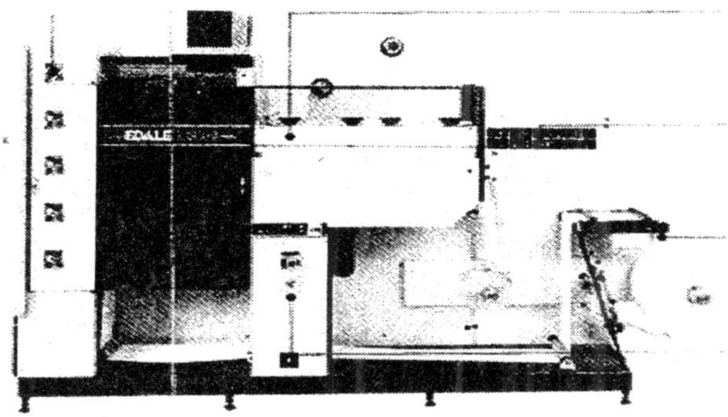

REPRODUCTION PHOTOGRAPHY AND IMAGE ASSEMBLY

Flexography and photographic imaging in various forms like all other printing processes. On packaging and imprinted product design, it's necessary to illustrate products & their uses.

Objectives - The objective of reproduction is to prepare film intermediates so that good" ink -on-substrate" reproduction can be obtained with flexography. For black & White originals, the criteria for good reproduction are image sharpness, resolution and tone reproduction.

Line work - With line work originals such as text display type & graphic careful focusing of the camera, correct exposure and film processing can assure a sharp image.

Halftone - When photographs or artwork are made into halftone film intermediates; good tone reproduction comes from placement of the highlight dots, shadow dots and midtone dots in relation to the tones of the original. The highlight dots must be the smallest, the flexo press can consistently print & must be placed in the lightest part of the original image. Similarly the shadow dots are the largest dot area, the tlexo press can consistently print before the tones change to solid print. These dots should be placed in the darkest sections of the original placement of the midtone (50% dot area) in relation to the highlight dots of the original determines what the reproductions will look like. The closer the 50% dots are in relation to the highlight tones of the original, the more image contrast the reproduction will have.

The further the 50% dot is from the highlight tones of the original, the less highlight-to-midtone contrast, the lighter the reproduction will be and the greater the amount of shadow detail it will have.

Tone compression - To got a pleasing reproduction, careful placement of the 50% dot is necessary. Because the tonal range of the original photograph or artwork is usually greater than the potential tonal range of the printing process. Control of where the compression is taken with in the tonal areas of the original will decide whether the reproduction succeeds or fails. Compression is controlled by properly exposing the halftone negative.

Process colour printing (process printing) - It describes the technique of full-colour reproduction of a subject, rendering the original in full, continuous tone colour as in paintings, colour transparencies and full colour photographic prints.

This is achieved in printing by converting the continuous tone copy to halftone, separating the colour into Yellow, Magenta & Cyan process colour and printing these colour sequentially in register with each other. This three colour process is frequently converted to a four-colour process "by using black to improve the contrast and /or tone balance of the reproduction.

This process is the same for all printing methods.

Light -Sensitive Materials - To capture an image, photography needs a material with a light-sensitive surface. "Film" is the most familiar term. Photographic film consists of a film base or support, with a light sensitive emulsion layer on one side (the emulsion side) and a backing, or pilloid, layer on the other.

Film also implies a flexible, transparent material that can be bent without damage to itself or to an image recorded on it; this means a material that is easy to handle in subsequent prepress image-assembly and plate making steps.

Reproduction films There are many kinds of films for graphic arts reproduction photography.

Halftone photography depends on this imaging, characteristic. Because of this trait, time work and type reproduce with sharp, clean edges Now lith films are used primarily in camera copy image- capture applications, both time and halftone work and on same colour scanners that use a halftone contact screen to generate halftone separation films.

Today non lith or rapid access films are more popular. They produce their high- contrast, high-opacity or high density images without the infectious development effect. These films are usually restricted to image-assembly operation where their rapid processing cycle of 90 seconds dry to dry are most attractive . This new generation products can be used in image capture applications of camera and colour scanner and in image assembly.

Photographic papers - Since paper products have fewer prepress applications choice in photographic paper are much simpler than for films Normally, paper images are used in image assembly when an opaque paper print is required, either as a proof or as part of a mechanical assembly's makeup. This requires a paper product that has similar exposure and processing characteristics to films used for image assembly.

A major exception is in photo typesetting operations where headlines and text galleys are produced as photo typesetting papers. This is highly specialized image capture application that combines high speed imaging features of photo typesetting equipment with the high-contrast imaging of phototypesetting papers.

Diffusion transfer materials - Other way to use photographic papers is indiffusion transfer imaging. This technology differs from conventional prepress photographic applications, in which an image is captured by camera exposure, producing a film negative. The negative, with its image reversed in polarity (whites are black and black areas are clear) is used to expose another piece of photographic paper or film to return the image to its original polarity.

Conventional photography is a two-step process involving two separate exposure -processing cycles, the first creates a negative image, the second uses the negative to create the original, positive image.

Diffusion transfer combines these two steps, the image is captured using a camera exposure as a sheet of diffusion -transfer negative material. This exposed material is processed through a simple, single solution processor in which a sheet of diffusion transfer receiver material is sandwiched to the solution soaked negative material as it leaves the processor. After a short transfer time, the two sheets are pulled apart to deliver. a positive print, that is ready for use in mechanical assembly.

Photographic characteristics - Two basic photographic traits have to be weighed when choosing the right product for either image capture (including camera, scanner and phototypesetter) or an image-assembly application that involves contact exposures 1. Light sensitivity or photographic speed, the amount of light energy needed by product to reproduce an image and 2. Spectral or color sensitivity, the range of colt the photographic material can see .

Photographic speed - Prepress operations requires a wide range of photographic speeds. A separate product is needed for each speed. The range includes products with very high sensitivity that need only a microsecond (one-millionth of a second) of exposure to products that are so low in sensitivity that they can safely be handled in normal room light.

Phototypesetting equipment that operates at speeds up to 3000 characters per second, needs photographic materials that can record high-quality images, with exposures of only 5 to 10 microseconds.

At the other end of the sensitivity -speed scale, room light-handling materials may require up to 20 seconds of exposure to a high intensity light source to record the image in a contact image-assembly application.

Colour or spectral sensitivity - Colour-spectral sensitively, of photographic materials has been broadly broken down into following three categories:-

1. Blue sensitivity : insensitive to red and green
2. Orthochromatic: insensitive to red, green & blue.

Colour in the original must be recorded on the photographic material, if the image is to be reproduced satisfactorily. e.g. a colour picture or art work may contain reds, greens & blues. If all these colours are to be recorded in their proper relationship a panchromatic material must be used. However, if the original copy is black & White line art work or black & white photograph, an orthocromatic material can be used. It a contact exposure is to be made through a film negative, blue sensitive material is acceptable.

If a panchromatic material is used, it must be handled in total darkness.

Orthochromatic material can be handled under red" safelight conditions that allow the operator good working visibility.

The Emulsion : The emulsion layer is a gelatin coating that contains silver halide crystals in suspension, they are the light-sensitive elements that produce the image after development. The emulsion is coated as one side of the base material.

The film backing : To counteract the "pull" of the emulsion side as it swells and shrinks during the wet-to-dry processing cycle, another gelating coating is applied to the opposite side of the film base. This coating contains special colour dyes to absorb light that passes through the emulsion layer during exposure. This absorption characteristic is important to control internal light scattering and achieve good sharp image resolution. This layer is known as antihalation packing or simply the " back " side of the film.

Colour differences between emulsion & backing : The emulsion side is generally the lighter side. The difference allows the operator to position the film correctly for the exposure.

Surface characteristics : Very small matt particles are embedded on the emulsion side of films and papers and on the pilloid (base) side of films. The number and size of these particles vary from one product to another and are based as the intended use. This feature is of major importance to materials used in image assembly and contacting. Contact exposures require large sheets of film or papers sandwich to the image-carrying film, with their surfaces in intimate contact . A vacuum contact frame is used to achieve this. Any air trapped between the sheets would of course, degrade the finished image. Male particles on these surfaces provided a pathway that allows any trapped air to bleed away uniformly.

Photopolymer plate requirements- Smooth surfaced photopolymer platemaking usually demands a pretty high number of matte particles on the emulsion side of films used to expose the plates. Special matte films are generally used as final films.

The process camera A process camera is usually used for prepress operation. There are two basic types of process camera—

1. Vertical - in this form, the movement of copy holder, lens and film holder is vertical
2. Horizontal- With this one the movement of these part is horizontal.

A combination horizontal/ vertical also is available in this, the lens-to-coup holder image path is redirected through a mirror arrangement at a 90 degree angle allowing that part of the camera to function as a vertical camera, while the lens-to- filmholder section operates as a horizontal camera.

Mechanics of Design Preparation of Flexo Press

1. The maximum number of colors needed to reproduce the design is dictated by available print stations on the flexo press.
2. The effect of special printing procedures such a web -reversal limits, printing to three-colour face and three- colours back on a six colour printing press.
3. Remember color sequence, especially when the usual light-to-dark color progression is changed for some reason.
4. Hold to register tolerances that suit the type of press to be used.
5. Avoid placing large solid areas and fine details such as small type, tints, fine filigree or halftones in the same color whenever possible.

6. Consider color trap tolerances to minimize color-to-color registration problems.
7. If correct registration is not avoidable, it should be changed to a limited print area whenever possible.
8 Be aware the ink fill-in and distribution problem inherent in reverse printing.
9 There are shig change problem inherent in reverse printing.
10. Avoid placing straight head-edged solids, horizontally across the web in favor of diagonals curves, weavy and irregular leading edges to minimize press vibration and bounce.

Product sizes and forms

Because of their influence on the initial result, some other factors have to be taken into account such as:-

1. Size and location of package face, back, or any other surface to be printed.
2. Size and extent of folds, diecut slots, perforations, seams and other important features where they exist.
3. Exact print areas
4. Product- fill height and/or contours whenever they are essential to form
5. The names, color matching numbers and/or color swatches related to color preference or standards where they exist

Initial graphics can be roughed- in on the substrate or some similar material. The graphics plan is finally cheeked against machine drawings and other technical aspects. Substrate & Materials

Since flexo printing is done on paper board, film, textiles, foil and many other materials. The comprehensive design can often be prepared on these surfaces.

As per as substrate are concerned, the designer must be alert to possible printing problem . Some common concerns are as follows:-

1. Images printed on thermoplastic should usually be kept away from heat seal areas.
2. Packages for certain food items frequently require special inks.
3. Most plastics have a non absorbent surface and may not readily accept and retain printing ink

4. Design fine details to overprint solids (usually white) rather than have them print on the bare plastic surface.

After the design is ready for detailing in its final form, the following flexo requirements should be maintained.

1. Typography

If type matter is used, six-point type for positive and eight-point type for reverse copy are generally for wide-web . Again a four-point type for positive copy and six point for reverse copy is commonly seen in the narrow web field.

2. Line illustrations

Line drawings, made up of solid lines are very often used. If shedding or gray values from light to dark are called for, tints or halftone screens are used to achieve any desired gray or tonal effect. This can be done in the camera department or on computerized color separation equipment.

3. If illustrations or line copy on embossed, drawing surfaces or paper or board are required, should be mentioned.

4. Line count

If single-color reproduction of continuous tone copy such as photographs or wash drawings are needed. Halftone reproduction must be fully understood and right halftone screen count specified. For printing either tints or halftones on corrugated board, 45 to 65 line screens are fine. Wide - web package printing on film go anywhere from 65 to 133 lines screens, while narrow-web printers typically go from 120 to 150 line screen counts. But 200 line screen printing is also possible in narrow-web.

5. Color reproduction

when continuous tone, full color reproduction from original copy is required, as with color photographs and transparencies, oil paintings, reflective art, water colors and illustrations-a full understanding of three and four color process printing is mandatory.

6. Color selection

Where color overlap to maintain register, related colors are preferable to complementary color, because the latter may produce an undesired third color in the overlapping area.

7. Register

Although to-day 's sophisticated presses are able to maintain fairly tight register, it is still a good policy to avoid hard less register situations. In wide web 1/32 " is the accepted tolerance, if a design is prepared for central impressions press (C1). For in line stack types press, 1/16" is preferred, corrugated printers look for 1/4" whenever possible. While narrow-web printers often work with less than 1/64".

8. Rendering

The layout can be rendered on the actual material to be printed say either paper, or film or foil.

9. Transfer Materials

Many design can be rendered and made quite presentable by organizing the graphic design in a black and white format with lines, mass, illustrations and typography, the typematter can either be obtained from out side or prepared in-house from electronic desk top publishing (DTP). Anyone of the prepress transfer methods can be used, to come up with an attractive colored proof, with all the colors in perfect register. Additional colors can be added if necessary as the very design concept is developed. The transfer components can also be used in conjunction with hand-rendered elements and/or airbrush work in developing the design.

Now once accepted, the design is ready for the production artist to convert into finished mechanicals with little possibility of error. The production artist needs to follow guide lines set up by the chint, the art director and the printer -converter as work begins on the camera ready art.

Electronic imaging and computer graphics

In recent past many conceptual and mechanical aspects of design for flexographic printer have changed dramatically. Computer generated graphics, have touched virtually every aspect of production. More and more design studio realize the profit potential and enormous power of computer graphic systems now available.

Computer graphics workstation

Suppose a designer has the title, copy matter and pertinent legal description of a new product about to debut. The client wants decoration sign on the label and market research agrees. The designer pores over the file full of photos, from many international magazines and selects some. Then the designer sends an assistant out for the decoration sign.

With decoration and picture in hand, the designer moves to a copy stand with light and a small video camera. The decorations are arranged nicely and same colored paper is fixed up as a contrasting back ground. The computer scans the three-dimensional arrangement using the camera, and captures, the scene on the computer screen, keeping all options open, the designer runs, all the decoration signs through the scanner, where the image is stored as a "hard disk".

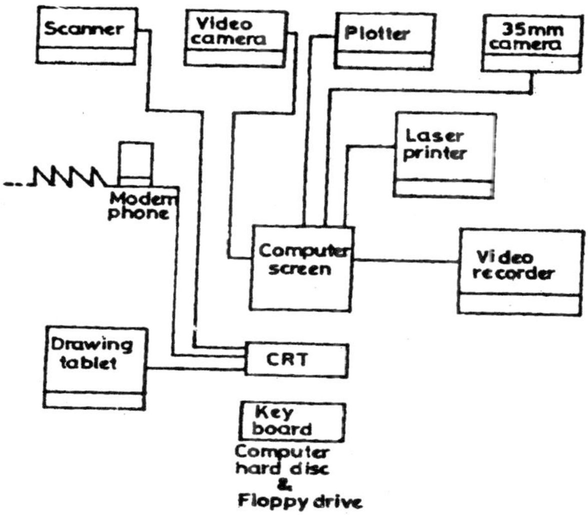

Computer Graphic work station

Computer graphic work station

The design's next step is to put the title and copy into the system.

Computer graphics provide almost immediate and unlimited variety. At every step a new view can be saved for comparison. Images & type matter can be twisted, stretched, turned and other wise modified in minutes, compared with hours and hours using traditional method.

Now the client's approval simply tells the designer to print out a final high resolution set of a negatives from an " image- setter" to be delivered to the printer. This branch of the computer system usually is at a printing facility. It does all the color separations which require only a final review by the art director.

Flexography Printing Press

There are so many kinds of presses, auxiliary equipment and in-line operations. The flexo press has four major components as follows-

1. The unwind with infeed
2. The printing section
3. The dryer
4. The rewinder with out feed.

The auxiliary equipment, such as web guides, web viewers, powder spray units, air shafts is important in the overall operation of the flexo press

Recently, more and more flexo presses (imported) have been supplied with in-line operations such as laminating and /or coating.

There are following three types of flexographic printing presses-

1. Stack
2. Central impression cylinder
3. In-line press.

In general these presses are used for printing flexible packaging materials, but they also do narrow-web corrugated and publication printing operations. Regardless of the end product, the printing principles remain fundamentally the same.

The stack press

In the stack press, individual colour stations are stacked one over the other, one on both sides of a main press frame. Each color station is driven through gear trains supported by the main press frame.

Stack presses are made with one to eight color stations, although the most common is a six color press.

This stack presses have following three main advantages:-

1. Can reverse the web, to allow both sides to be printed during one pass through the presses. Various web threading arrangement allow complete ink drying before the reverse side is printed, if enough drying capacity is designed into the area, where the web presses between the two series of stations.
2. Color stations accessibility, which facilitates change over, washup etc.
3. Ability to print large repeats.

The stackpress has found wide applications and has been used to print on almost every types of substrate. It lends itself well to applications such as printing in-line with other types of machinery such as bag machines, sheeters & laminators etc.

The central impression press (C1 press)

The impression cylinder supports the web, which is thereby locked to the cylinder as it passes all color stations. The C1 press supports all of its color stations around a single steel impression cylinder, mounted in the main press frame.

Since the chief advantage of the C1 cylinder press is its ability to hold excellent register, it has become the mainstay of the converting industry.

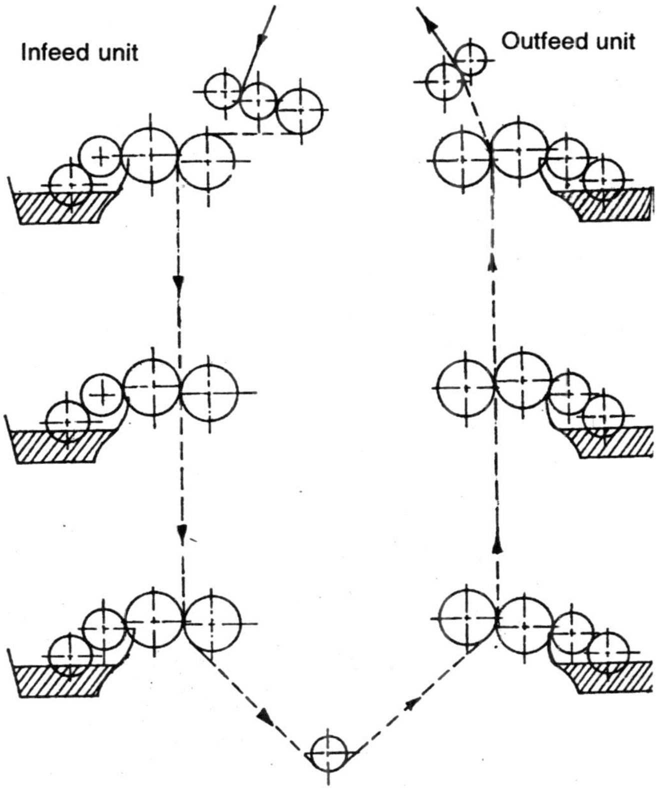

Figure : Stack Press with infeed & out feed units

The most common press is the six color central impression press. There have been many four color presses and designs for presses that include upto eight colors.

Impression cylinders of various diameters have been used. At first, four-color presses were the most common and they generally used 30" or 36" diameter impression cylinders. To get better speed impression. cylinders up to 60" are used for four-color presses. The first six-color central impression cylinder presses used 83" diameter cylinders.

As drying techniques improved and less required for drying between colors, smaller impression cylinders came back. The most common six color single impression cylinder press has a 60" diameter cylinder. There are still 83" diameter C1 six color presses as well as intermediate 72" diameter six color units available.

The C1 press- has found limited use when it comes to printing both sides of a web during one pass through the press.

The in-line press

Its color stations are separate, complete units horizontally mounted one to the other, and are driven by a common line shaft. In-Line presses can be manufactured with any number of colors. This type of press can be easily designed to handle extremely wider web widths, single frame need not support all colors. They are commonly found in folding box, corrugated and multi-wall beg presses, as well as other special applications. In line presses are also common in narrow-web widths for printing pressure sensitive and standard label stocks, where they offer the advantage of quick setup & accessibility. This design approach also holds true in those specialized areas, where a specific product line may need a short-run type press design. As in the case with stack presses, they are limited to printing designs that are not critical with respect to register.

However, on very narrow-web presses, where the web distance from one color station to the next is not great, the closeness of register which can be held is improved and maintained.

The in-line press has the versatility to print on both sides of a given web, by either turning the web over with the turning bar systems or using alternate threading.

Web Tension Control is a very important function of any web process machine, because it determines, in large part, the product's quality and efficiency.

Inadequate tension control can severely limit the performance of new machines Web breaks and wrap-ups around driven rolls (caused by slack web) are only the most obvious consequences of inadequate tension control.

Tension Zones A typical flexographic machine has more than one tension zone. This is because the process in any individual zone may require a different tension level or pattern than the processes in other zones.

A tension zone is that length of web, which extends from one tension-affecting device (TAD) to the next. Typical TADs (tension affecting device) are as follows:-

(a) unwind or rewind core shaft with attached more

(b) clutch or brake

(c) driven rolls

(d) braked rolls

(e) nip rolls where at least one roll is driven or braked

(f) drag bars or any other device that may add or substract a significant amount of tension to or from the web.

Unwind constant tension from full roll to core is desirable here. A significant deviation from constant tension, may be reflected in the next tension zone causing problem there. Tension level should be equal to or less than the tension used, while winding the roll, Greater tension can cause the roll tighten on itself. This is more serious with smooth or slippery webs than with rough or sticky webs.

Extensible webs such as polythylene and unsupported vinyl are run with much lower tension than nonextensible webs such as paper or foil to prevent wrinkling, stretching & reduction of width.

Rewind Either constant or tapered tension is used. The choice is determined by the web material, the build-up ratio (full role diameter divided by core diameter) and the tension capability of the rewind drive.

Tension is often used to correct web-handling problem. e.g. the web may have a loose edge, so the machine operator increases tension to stretch the web and eliminate the looseness. Or the web may not track properly through the machine, so again tension is increased to correct the problem. Unfortunately, this creates other problem such as web breakage, stretching, wrinkling and print length variation. It would be more beneficial to correct the causes of the web-handling problems rather than to create more problems by increasing tension.

Brakes Clutch & Motors Brakes are generally used to create tension in the unwind zone. There are several different kinds like manually actuated friction devices, pneumatic brakes with either single or dual disks, electric friction brakes and electric magnetic-particle brakes.

Web-Tension Control Systems

(a) Manual tension- control systems require the machine operator to judge the tension in the web and make appropriate adjustments by hand to the brake, clutch or motor torgue or speed.

(b) Roll followers are an improvements on manual control system. There are three basic types: follower arms, sonic rang finders and diameter computers.

(c) Draw systems Draw controls are used only in intermediate zones. The nip rolls at the upstream end of the zone are slaved to the main drive either mechanically or electrically. The nip is oversped a small amount, stretching the web and creating tension (draw) in the zone.

The most common mechanical draw system has a continuously variable speed transmission, driving the nip with a shaft from the main drive gear train. The transmissions output speed is adjusted manually a hand wheel.

The speed range is usually +- 0.5 % of the input. Speed electrical draw systems, eliminate the drive shaft, instead the nip is driven by a speed- regulated DC motor that follows a speed- reference signal from the main drive. The draw is accomplished by an operator adjustment that overspeeds the nip motor.

The correct draw is arrived at by trial & error and for best results must be reset each time a change is made in web thickness, width, speed & material. Moisture content of paper web must be considered and the draw adjusted accordingly. Excessive draw will cause web breaks, streching and wrinkling.

Automatic controls : This systems relieves the operator of the need for constant adjusting, hedging tension level, skill and experience and freez him for other things. There are several kinds of automatic tension control systems, each with its own advantages & disadvantages. However, no automatic system will eliminate tension problem, caused by mechanical deficiencies or poor quality web material. To get the most from an automatic system, the machine must be properly designed and in good condition.

Dancer Roll system : A dancer is an idler roll that is free to move in a straight line or under the influence of web tension

A counter force created by a weight or air cylinder opposes the tension force. A sensor is connected to the dancer to detect its position, and the position signal is fed to a regulator where it's compared to a desired-position signal, usually representing the mid-point of the dancer travel, set by the machine operator. If tension increases, the dancer will rise, moving the saucer and signaling the controller to reduce torque, allowing the dancer to return to its original position. If tension decreases, the opposite sequence occurs.

Tension is determined by adding or removing weights to the roll or by varying air pressure to a loading cylinder.

Dancers also are mechanically complicated and require at least two other property positioned idler rolls (one before and one after the dancer) to operate. This requires extra space in the machine. A properly designed dancer is lightweight so it can react quickly but strong, so it won't deflect and steer the web to either side. Its mechanism must also be designed with very low friction in its moving parts so it can react to small changes in tension.

Dancers usually begin to have difficulty maintaining control at web speeds over 500 rpm because of the in inherent friction and interia of the dancer itself and relatively low gain (sensitivity) of its controller. Low tensions and wide tension ranges are a problem for dancers, again because of friction, inertia &. low gain, the dancer roll will oscillate or haunt along its travel path, causing tension variations, web length variation and shifting to the side. If the haunting is severe, the dancer may reach the ends of its travel and cause web breaks or slack in the web, resulting in wraparounds, all of which disrupts the printing, coating or other process taking place in the machine and causes waste & loss of quality.

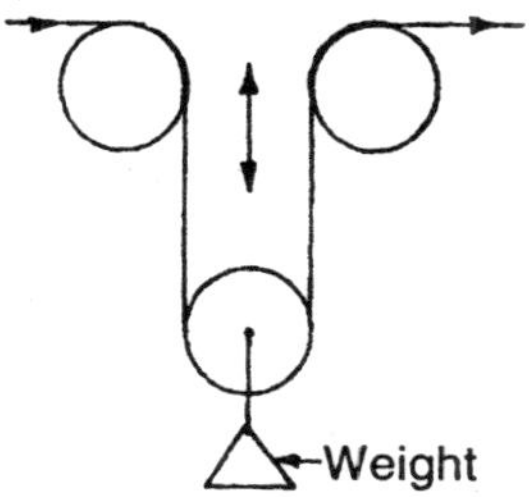

Linear Dancer

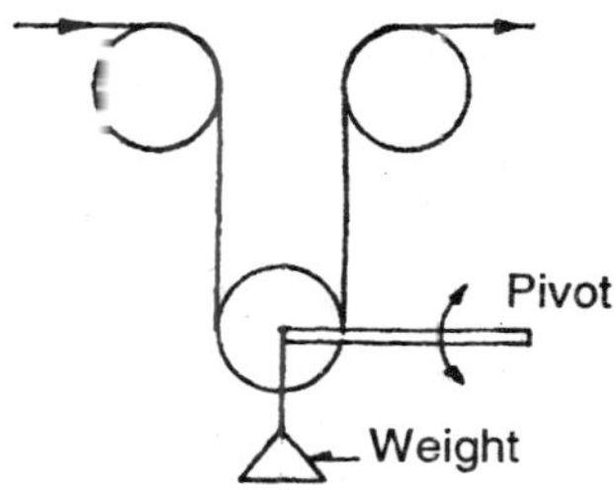

Pivoting dancer

Tension transducer systems : This systems use specially designed force transducers to measure actual web tension. They are normally used in pairs, with one installed on each end of an idle roll. Sometimes a single transducer is used, but accuracy is very poor because transducer output becomes dependent on web width, the placement of the web relative to the transducer and the location of the tighest part of the web, which can change continuously throughout a roll of material.

Dual transducer system are not subject to these factors because the transducers are electrically connected so that their outputs are averaged.

The appropriate load rating depends on the weight of the idler roll mounted on the transducers, web tension and wrap angle.

The indicator is the simplest kind of transducer system and it consists of a pair of transducers, an enclosure, with display meter on front, containing a circuit card to execute the transducers and amplify their output, and a pair of interconnecting cables. The circuit card usually has voltage and current outputs proportional to tension that can be fed to variable-speed drives, recorders or computers. It does not directly control tension by itself. The regulator outputs a voltage or current to a servo valve, motor brake or. clutch to automatically control tension in a "closed loop" control scheme.

Closed loop systems are very accurate because actual web tension is measured continually and compared to the desired tension, set by the operator. The regulator circuit automatically adjusts its output to eliminate any difference between actual & desired tensions.

The term closed loop comes from the fact that the output of the system (tension) is fed back to the input.

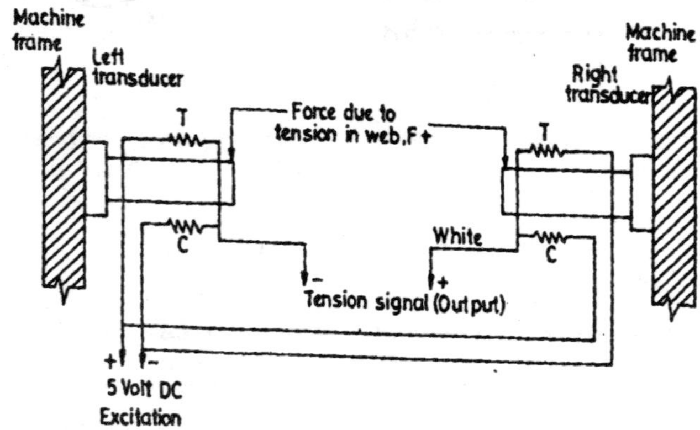

Closed loop tension control

Advantages—

1. It measures actual web tension.
2. The meter gives a display of actual web tension
3. It eliminates all the guess work
4. It gives accuracy of 2% to 3 %
5. Setting tension is very simple.
6. It gives consistency
7. Ability of transducer systems to operate successfully over a wide range of tension.

The term closed loop comes from the fact that the output of the system (tension) is fed back to the input.

Unwind Equipment : The unwind stand, which can be simple or complex plays an important part in the proper operation of the press.

1. Single position unwind
2. Flying-splice stands.

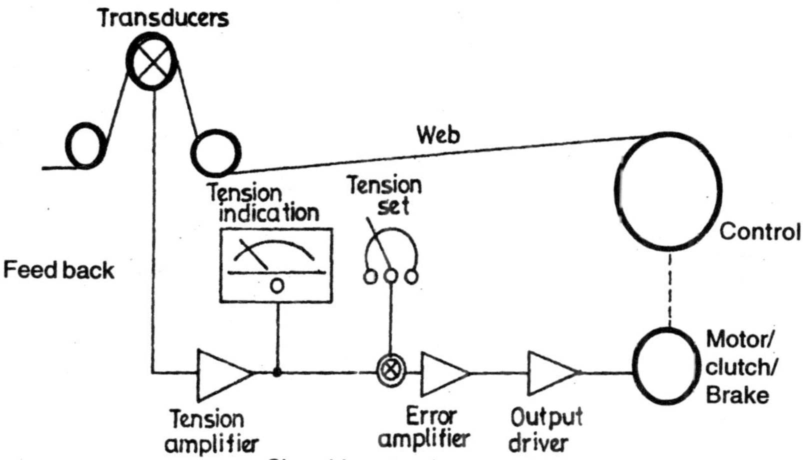

Closed loop tension control

Most single roll and older drop slice unwind stands are non-driven, while more modern flying splice unwinds are of the driven type.

Single- Position unwind is the simplest and most common in the single-position stand. It can be manufactured to handle a great variety of roll widths and diameters to make the stand more usable, designers normally incorporate a side shift adjustment so the core shaft may be moved by the operator either way from its counter position. This allows the operator to locate the web with respect to the printing, without having to shift the roll on the core shaft. For wide & large diameter rolls, loading apparatus may be used. Hydraulic arms may be added to the stand to facilitate the loading of rolls. Most important function of an unwind stand, is uniform tension controls during, the unwinding of the material to the printing sections.

The unwind stand can be automatically guided to provide lateral web position control at the press infeed

Another common single-position unwind stand is the shaftless type. This stand uses the roll core as a support for the material. The advantages of this system is that the operators does not have to shaft and unshaft rolls of material. Even with pneumatic shafts, this core becomes very difficult when handling wide rolls

Generally single position rollstands are used where large diameter mill rolls contain sufficient footage to keep the press running for a reasonable time between changes.

When changes become frequent consideration must be given to flying splice systems that although more expensive, increase the printing output.

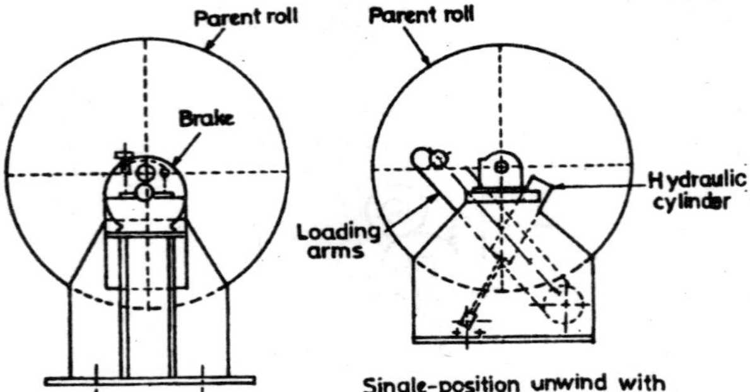

Single-position unwind

Single-position unwind with loading arms

Flying splice unwind

There are many types of flying-splice stands available. They are classified with respect to their roll diameters say from 24" to 72".

But 24" diameter, simple drop-splice unwind stand for common packaging materials such as polyethylene and other extensible films. This simple unwind is semi-automatic in operation, since it requires the operator to drop the splice into the splicing rolls. Drop slices are made by placing the pre-taped leading edge of the new roll into the puenmatic or hydraulically loaded marriage rolls that press the tape on to the expiring web. Operator skill is required to determine the precise movement to make the splice, in order to keep the tail of the exporting roll to a minimum, the operator also has to index the turret into the proper spacing position. Most modern stands are made to accept both " underwound" and "overwound" rolls.

24" diameter flying splice unwind

Larger roll diameters and the need for more uniform color through out the roll, leads to a need to drive the full roll during splicing. Rolls weighing 500 to 800 pounds are very difficult to move and making successful "drop" splice at running speeds where matching web speeds is required is virtually impossible.

The splicing operation whether 24" diameter or 72" diameter is normally the same. The leading edge of the new rolls is pre-taped but it is held in place by two or three pieces of break away tape" to keep it from unwinding as the parent roll is brought up to speed. When the roll reaches the proper speed, an air-loaded rubber roll forces the expiring web against the new parent roll. As the pre-taped leading edge of the new roll comes into contact with the expiring web, the break away tape releases and the transfer is completed. As with the" drop" method the transfer is at the operator's discretion and the amount of tail left on the expiring web depends on the operator's skill.

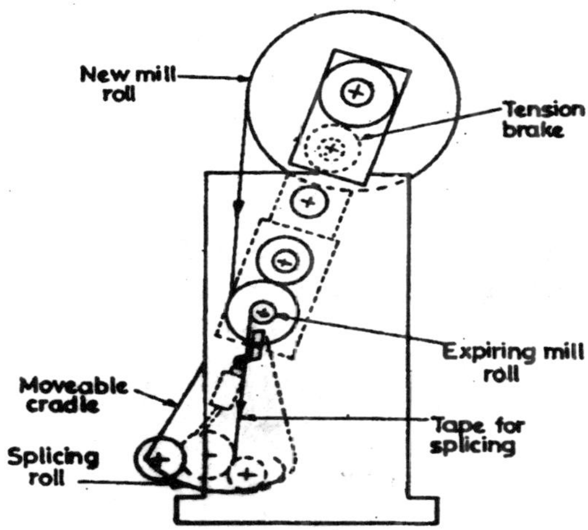

24" diameter Flying splice unwind

Unwind tension systems

generally there are two basic types of unwind tension control system, each type can use the same tension-sensing devices. The most common tension control is the unwind braking device, employing air electric or manually adjustable brakes. The other system uses either air, electric or hydraulic motors to drive the unwind shaft to pay off a predetermined amount of material into a tension-sensing device.

Unwind tension control is necessary for good register. This is especially true on stack presses and just as true in respect to repeat variations on CI presses.

Sufficient unwind web tension must be applied to maintain an even flow of material into the printing section. The tension value must not be so high that it can a cause slippage in the infeed draw roll section, or so low that there is not sufficient tension to properly track the web. In general when unwinding materials, the tension values are roughly half the values listed in the tension chart, as given below in the rewind section

For laminated web, sum the tensions for the individual webs and add 0.10 lb. per inch of width.

Approx. Tension for Web materials

Meterial	**Tension**
Acetate	0.50 lb.
Foil	0.50 lb.

Cellophane	0.75 lb.
Nylone	0.25 lb.
Polythylene	0.12lb.
Polyster	0.75 lb.
Polypropulene	0.25 lb.
Styrene	1.00 lb.
Saran	0.15 lb.
Vinyl	0.05 lb.

Paper : (based on 3000 sq. foot ream)

15 lb.	0.40 Per inch of width
20 lb.	0.50
30 lb.	0.75
40 lb.	1.25
60 lb.	2.00
80 lb.	3.00
100 lb.	4.00

Paper board :

8 pt.	3.0
12 pt.	4.0
15 pt.	4.5
20 pt.	5.5
25 pt.	6.5
30 pt.	8.0

For limitation webs, sum the tensions for the individual webs and add 0.10 lb. per inch of width.

In the unwind braking system, the braking power is decreased as the material unwinds from full roll to core, to maintain the desired web tension. A controlling device such as a dancer or a load cell may be used to automatically regulate any type of brake.

Small tension values must be held on narrow webs, overcoming cover-shaft inertia and gearing friction loads may take all of the brake's sensitivity and hinder its proper control of web tension. If the press speed is high and the core shaft and brake gear have a high inertia value, then as the roll decreases, the brake may be turned to a zero setting and the tension value in the web can still be too high , thereby stretching the web to overcome the high inertia value.

Thus the basic purpose of the unwind braking system, is to apply just enough hold-back force, to the expiring roll, to maintain a constant tension into the printing section.

Infeed unit : The function of the infeed, is to present the web to the print section at a constant rate. It is normally located prior to the first print deck as a stack press and is usually a three-roll system, comprising of two driven steel rollers and a rubber pressure roller, which is air or hydraulically loaded. It pulls the web from the unwind roll and helps establish the first tension zone on the press. Tension is established by braking the unwind roll. It also serves to isolate the tension in the print section, from the unwind and becomes the nip point for a new tension zone, through the print and drying sections.

On central impression presses the CI drum becomes the pulling element and the rubber covered nip roll, which secures the web to the drum, comprise the infeed. The infeed plays a vital roll in maintaining tension and the controlling register.

Outfeed unit (cooling drum) the function of the outfeed unit, parallels the function of the infeed. It provides an even constant pull through the print section and also serves to isolate the print section tension zone, from the rewind tension zone. It consists of one or more chill rolls, with a rubber covered nip roll, to secure the web to the chill roll and to prevent slippage. The drive is variable through a variable speed unit or, on the latest presses by means of an independent DC drive motor.

Rewind equipment - The rewinder is usually driven with a variable speed drive to set the basic tension pattern. Normally one roll is also driven slightly faster than the other to provide a differential surface speed, creating a hard roll. The winding drums can also be fluted to reduce wrinkling and give additional roll hardness. The roll can be wound with the printing in or out, depending on the method of threading. The centre of the roll, the core shaft, moves vertically upward as the roll diameter builds, the increasing weight of the roll as the winding drums assist in obtaining a lightly wound roll.

The single drum rewinder uses the same surface-winding principles, as the double-drum unit, but has a single drum to impart rotation. It is also driven with a variable-speed drive to set the basic tension pattern.

The roll being wound, moves horizontally rather than vertically during buildup. Hydraulic or pneumatic pressure, is applied to maintain a regulated roll pressure between the roll and winding drum.

A rider roll can he supplied with the double-drum reminder. The rider roll can be rested on top of the roll being wound & it can be puenmatically or hydraulically loaded to provide a greater nip pressure between the roll and the winding drums.

Hydraulic or pneumatic loading of the roll core shaft is another accepted method. The hydraulic cylinders act to maintain a uniform or steadily decreasing pressure relationship between the roll being wound and the winding drums.

Roll unloading equipment is available on both single and double drum reminders.

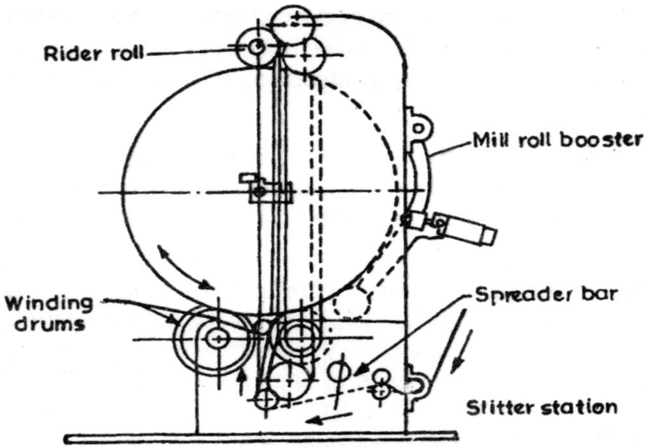

Fig. : Double drum rewinder

Centre winders : Centre winding gets its rotary movement through the core shaft. The centre winding units are available both as core-shaft and a shaft less types. On flexo presses the most common is the core-shaft type.

A common centre shaft winder is the single-position type.

This design employs one core shaft mounted inframes/ The shaft is driven by an electric, mechanical, hydraulic or combination drive, with same means of adjusting the drive speed to vary the roll hardness. The single position type must be stopped for roll changes.

The rider rolls is also driven on more modern design. The flying-splice rewinder uses the same basic centre winding principle but has two core shafts mounted as a turret.

Roll unloading equipment is available in the form of hydraulic or pneumatic roll-lowering arms for the single-position winder.

Sliting on center winders has grown in importance. Score & razor blade slitting attachments are available as auxiliary equipments. Shear slitting is also used in some instance where materials call for it. A spreader bar, driven expander roll or slatted expander rolls used after slittng attachments to prevent web interleaving.

Rewind jusion systems - The basic requirement of a good rewind tension system is to wind rolls with straight edges and uniform density, where preserving the accuracy of register & repeat length.

To do this, the unit must be able to rapidly and accurately follow the accleration & declaration of the printing press, compensate for changes in the diameter of the rewinding roll and, when used with flying splice rewinders, be able to make high-speed transfer. The drive must be able to handle wide or narrow webs, flexible, stiff or stretchable materials and in some cases perforated webs.

Web Guides

Edge guiding is the simplest approach to controlling a web and is generally applicable in cases, where it is desired to maintain the edge, in constant reference to the press and where the sensor can be repositioned in accordance with changing web widths.

Centre guiding is used in situations where web widths vary during operation and it is desired to keep the material, centered relative to the press rather than referenced to one edge.

The typical web guding system consists of several components integrated into a closed control loop. In the typical closed-loop position control system, the set point, or command is the prepositioning of the sensor on the press to determine the guide point.

Various auxiliary control systems are available for use, with the automatic guiding systems. With intermediate guides, the most common auxiliary control system is a feature which allows the operator to disable the automatic guiding system and to center the intermediater guiding device in its travel.

With unwind and rewind guides, the most common auxiliary control system, is a feature which enables the operator to mannually control the system, as well as to operate in the automatic guiding mode.

There are three locations on flexographic presses where guiding is normally applied. These are 1 at the unwind 2, prior to printing 3, at the rewind.

The unwinding roll of material is shifted laterally and the sensor is fixed so that the edge of the material is aligned to the desired position relative to the press. An idler roller should be provided with shifts lateral with the unwind and provides for a constant plane at the point of sensing.

Since the web plane between the tangent to the unwinding roll and the first idler in the press is constantly changing, it is not practical to sense in this area.

The use of the shifting idler serves to establish a fixed web plane, suitable for sensing. This eliminates undesirable time lags in the sensor seeing the output of the control cylinder.

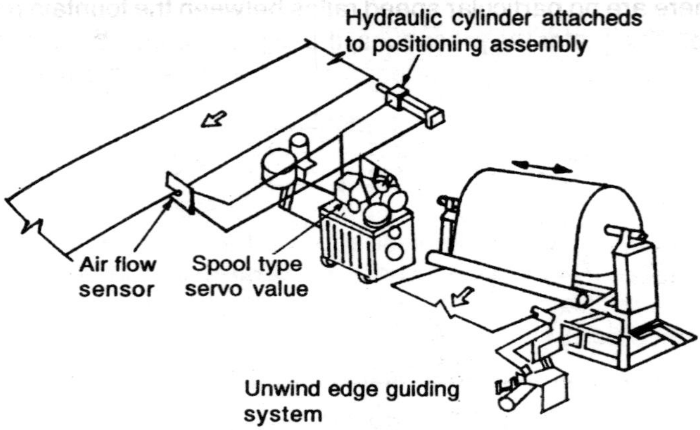

Printing stations

Two roll - An ink metering system, is a means of controlling the amount of ink being presented to the plates and subsequent substrates. Although great advances have taken place with the development of blade-system, many presses still operate today, using a roll system. Basically this system consists of a fountain roll and an anilox roll.

Anilox roll - The purpose of the anilox roll, is to transfer a measured amount of ink to the plate cylinder. In the two roll system, ink is transferred to the anilox from the rubber covered fountain roll.

Fountain roll - This is the roll that actually the excess amount of ink from the anilox roll. It is subjected to fairly high nip pressure, either from the roll loading system or from the hydraulic pumping action, caused by the excess of ink at the nip between the fountain roll and the anilox roll.

Fountain- Roll speed- In flexographic printing press, the web is being printed, travelling downward, that the ink puddle between the fountain roll and the anilox roll, is on the top or front, while when the web is being printed in the upward direction, the ink puddle is on the bottom, or back. Therefore, the direction the printing station is rotating will depend a great deal on how fast the fountain roll should travel.

On most modern presses, the fountain roll is independently driven either with a hydraulic drive, or some other type of independent drive. Therefore, the speed is adjustable to suit the condition of the printing station.

On the up station, since the puddle is on the bottom, the excess ink will not accumulate as much as it does in the downward station. Therefore, the speed of the fountain roll should be slightly faster.

There are no particular speed ratios between the fountain roll and the anilox roll. The faster the press runs, the higher the ratio becomes. It's only a question of maintaining an adequate supply of ink at the nip.

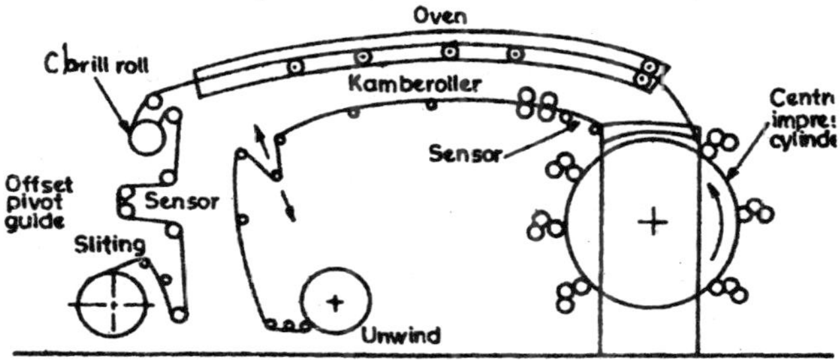

Figure-: Edge guiding prior to printing

Nip Loading- Once the speed has been established, set the loading of the fountain roll against the anilox roll. There are three different ways to load the fountain roll, against the anilox roll mechanically, pneumatically & hydraulically.

If the fountain rolls is mechanically loaded against the anilox roll, the fountain roll hardness plays an important part in the doctoring system. It means softer the rubber, the more ink will be carried. The pneumatically & hydraulically loaded fountain rolls, are far more popular. Here the load is applied directly to the fountain roll and latter's hardness is less important in doctoring the ink, since the loading of the roll will simply deflect softer rubber more than harder.

Deflection- When the fountain rolls are loaded against the anilox roll, the nip loading pressures and hydraulic pumping press wires, cause the fountain rolls to deflect. When deflection takes place, the roll will tend to transfer more ink in the centre of the press than at the end. This is particularly true of wide presses.

Because of roll deflection and the hydraulic pumping action of ink in flexo's two-roll system, the system is called speed sensitive. This means that as the press speed increases, the hydraulic pumping action of the ink, between the fountain roll and the anilox roll increases and the amount of the ink transferred to the plate will increase also. In view of this, the production speed of halftone work is limited. As speed increases, the amount of ink being transferred to fine screen printing plates may eventually cause dots to join. This is known as "bridging" and gives the appearance of dirty print. This is why reverse- angle doctor blade systems were devised.

Reverse-Angle Doctor Blade System-

In order to eliminate a number of the so called deficiencies with a fountain roll system, press manufacturers have developed a number of doctor blade devices, to assist the operator to control the distribution of ink.

The purpose of the doctor blade, is to remove excess ink (or fluid) from the surface of the engraved anilox or transfer roll, allowing better control of ink transfer to the plate cylinder. The device is particularly helpful in printing process colours.

Ideally, the doctor blades should make contact at a 30 degree angle with the tangent point of the anilox roll. At this angle, the blade shaves or doctors off the excess ink, leaving the precise amount of ink contained in the engraved cells of the anilox rolls e.g. for any given use, the substrate on which the printing is done, the type of work (solid, type, halftones etc.) and type of ink will be factors in the selection of the engraved transfer roll. To obtain good doctoring in any application a number of requirements must be satisfied.

1. The doctor blade, should be set at the minimum blade pressure to accomplish its task.
2. The doctor blade, must be accurately aligned and adjusted to the anilox roll, in the designed location
3. The anilox roll must be in reasonably good condition to start with. It should be manufactured for use with a doctor blade.
4. The doctor blade anilox roll, must be given sufficient maintenance to prevent deterioration & misalignment.
5. The doctor blade must be designed and manufactured for the specific application, taking into account machine width and speed, the function and location of the roll being doctored, and the location of the doctor blade on the roll.

With the advent of the reverse-angle blade for the down side and up side, flexography was given a tremendous boost in print quality.

The introduction of the double bladed holder, presented to the operator a tool, which not only made the job easier, but reduced clean-up time as well.

The double bladed holder is commonly used by the major press manufacturers but will differ in design in reference to the area of ink supply and return.

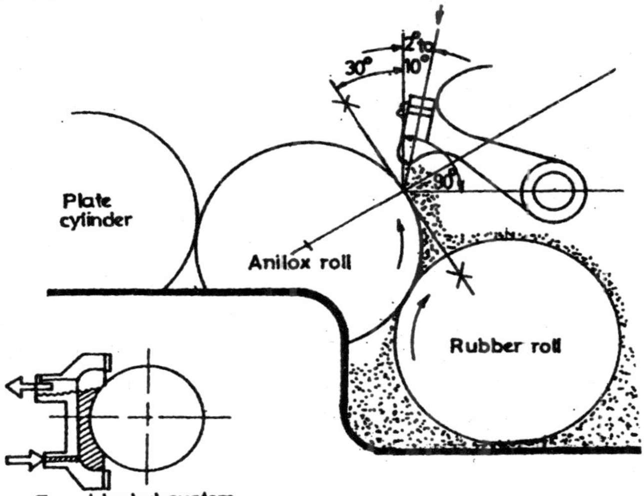

Deck control Press manufacturers are constantly required to design printing decks, that allow the operator to rapidly change from one job to another. As a result, today's presses, feature a number of devices connected to press retraction gearing to greatly decrease the time required to manually move the deck. Hydraulic or electric motors connected directly or through a gear box, are very common for moving decks in & out.

To make things easier, press manufacturers have developed systems that automatically position the printing & deck. These system cut the time required to change jobs by automatically setting the impression, within a given number of the calculated printing positions. The objectives of automatic deck positioning, are to set the impressions, engage the gears & register the printing decks to each other.

Generally this is accomplished by setting the full gear, plate cylinder gears and anilox roll gear in known positions, then using a programmable controller or dedicated computer to activate either hydraulic or electrical motors, that actually move the plate cylinder and anilox roll carriages into the printing position. Also each anilox roll is preset to a specified angle, so the gears mesh with those of the plate cylinder. This is accomplished with a special hub and degree scale on the gear side of the anilox roll, or in some cases the position may be pre determined with magnetic and non-magnetic sections located on the hub.

Continuous Inking

Since the flexographic inks are fast drying, the anilox roll in the ink distribution system must continue rotating, when the press is in a non-printing mode. If not, the inks will dry in the cells, and controlled transfer will no longer be smooth. Therefore, when the press is idling, if the fountain roll and anilox roll are to rotate, the anilox roll must be separated from the plate cylinder. Otherwise, the anilox roll will wear the plate along the line of contrast.

If the plate cylinder is not separated from the web, the ink from the plate will dry on the web, and when the press is restarted the web will stick to the plate and may break. So it is important to separate the plate cylinder from the web when in the stop mode.

If ink has been applied to the plates before the press is stopped, it may be necessary to clean the dried ink from the plates before starting the press again. After the plates have been cleansed, the press can be restarted and the impression control energized. After this, the anilox roll comes in contact with the plate cylinder and the plate cylinder in contact with the web and the printing resumes.

Side and circumferential Register control

A number of devices are presented to the operator by press manufacturers, to adjust side and circumferential registration. Most common is a mechanical apparatus, such as a hand wheel, which, when connected in number of ways to the plate cylinder, will cause the plate cylinder to be laterally moved. Circumferential register can also be furnished by a simple hand wheel by allowing the plate cylinder and associated gearing to be connected to a helical gear.

The male spline clamped to the plate cylinder and associated gearing to be connected to a helical gear. The male spline clamped to the plate cylinder journal in conjunction with a female spline, will allow the plate cylinder gear to slide forward and back on the male spline to affect the circumferential register, without affecting the side register of the plate cylinder.

A number of hydraulic or electrical devices can be designed and installed to allow the operator not only convenient access to the register controls but allow a degree of automation.

Dryers Flexographic presses have integratd drying system. There are individual dryers after each print station except the last one, which has much longer, dryer. These dryers are known as either between color or interstation dryers. The dryer after the last print stations is called the main tunnel or overhead dryer.

Interstation dryers remove a sufficient amount of volatiles from the ink, so that the next print station may apply another colour without altering the previous one. On a four-colour stack press, there are typically three interstation dryers. A colour CI press usually has five. In either a stack press or CI press, there is one main tunnel dryer.

Again, flexographic press dryers use hot air to remove the volatile portion of the ink. This air is ducted and directed to the substrate, ink solids ink volatiles are raised to a temperature that will do the job. The important considerations for any dryer are:

1. Air temperature
2. Air velocity
3. Air volume
4. Time

Generally, the supply fan takes air from either the roof or inside the plant, past a gas-fired burner and directs this air to the substrate. The heated air is pushed on to the substrate through a series of air nozzles, narrow slots running perpendicular to the substrate travel, across the widths of the substrate. After vaporizing the volatiles, the exhaust fan pulls this air from the dryer and sends it to either the atmospheric or pollution-control devices.

Now that pollution-control equipment is mandated, flexographic press dryer air-flow schemes have been altered to reduce the volume of air to be processed by pollution control devices.

Most press manufacturers offer a recirculation air-flow scheme that recycles air with volatized solvents

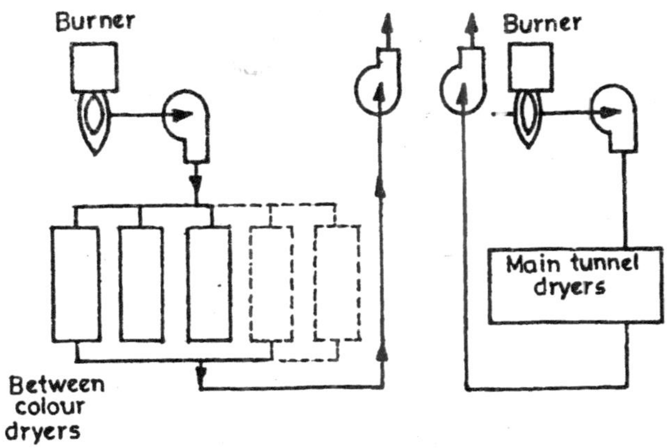

Mechanical components

CI drum — Drum, construction can be of double-wall steel or cast iron. In either case, the drum will be temperature controlled by a heating /cooling device. As market advanced in flexography and the product expected from a CI press improved, to a high degree of excellence, press manufacturers were compelled to a high degree of excellence.

Very commonly employed today is the use of digitally controlled heating/cooling elements, which guarantee to hold drum temperature within +1° F. The close tolerance reflects directly to achieving printing excellence, being demanded of press manufacturers & converters today. If the press experiences variation in drum temperature, which reflects directly into CI drum size variation, the operator will be continually attempting to compenstate for this condition, causing missed impression or over impression.

Plate cylinders The plate cylinder shaft and cylinder wall, may be of integral construction similar to the roll body of the fountain roll and anilox roll.

Also the roll diameter should be of such size, that when the stickyback and rubber or photoplymer plate is applied, the printing plate will run at the proper gear pitch line.

There are several different types of demountable plate cylinders. They are commonly used where a converter may have several different flexographic presses of about the same width, but which require different types and lengths of journal construction.

Anilox rolls This roll is often referred to as form roll, meter roll, knurled roll, engraved roll, ink-applicating roll, ink transfer roll and of course anilox roll. It is a very important link in the complete roll chain, used in successful flexographic printing.'

The ink-metering roll in the flexo printing units has changed a great deal in the past decade and in the next decade it will change more than any other component.

The main purpose of the anilox rolls is to supply a predetermined and uniform volume of ink to the surface of a rubber or photopolymer printing plate. This is done either by transfer from a rubber roll or directly to the plate with a doctor blade system, to wipe the roll's engraved surface, so that only a controlled and measurable volume of ink to the printing plate.

Construction

The engraved ink-transfer roll must be constructed properly to give good printing results. Very often economy gets ahead of quality when a replacement anilox roll is purchased.

The engraving process used in making rolls require soft, maleable and nearly homogeneous material. When the roll is engraved, it must still be chromium plated to present a wear surface, hard enough to give satisfactory service life.

Selecting the proper anilox roll to achieve the best printing, the longest wear and lowest cost has become more difficult because of the large variety of screens and engraving methods now available.

Fountain Rolls

Most rubber rolls is simply a core or shaft, metal or otherwise with a rubber or elastomeric coating.

The coating may vary in thickness from a few thousandth of an inch to an inch or more of covering material. The rolls themselves may vary in size, from fractional dimensions to an excess of 6' in diameter and 30'in length.

To day modern rubber roll suppliers use a scientific, computerized approach to solving formulation problems.

Cooling Rolls

Cooling rolls can be devised by using other rolls as substitutes such as uncovered fountain rolls or plate cylinders, impression cylinders, in positions following the dryers between the last dryer and rewind or where a temperature reduction is necessary.

The heat transfer through the wall of the cooling roll is also by conduction. However the heat transfer from the cooling roll wall to the coolant is by convection.

With the ever-increasing cost of energy, cooling roll heat transfer efficiency will become an ever increasing cost problem.

Balancing Flexo Rolls

The increase of printing speeds and quality standard in the flexo industry to day makes it necessary to consider the importance of accurately balancing the various rolls. Balancing is a process. whereby the distribution of mass in a rolls is altered to eliminate vibration at the supporting bearings.

In general the need for balancing a flexographic roll is evident when we consider that the mass centre of a roll will not necessarily coincide with the rotational axis as determined by its supporting bearings. A roll which is not restrained by bearing, will rotate about its mass center. Balancing is only an operation, which eliminates vibration by redistributing the mass of the cylinder, so as to cause its center of gravity to lie in the same axis as that determined by the supporting bearings.

Gears- Power is delivered from the main motor to the impression rollers or drurm. All impression printing & inking rollers must have their diameters match the pitch diameter of the gear that propels them. Pitch diameter is the diameter, at which the gear operates, not their internal or external teeth diameter.

It is very important that the gear always operate on their pitch diameter, because otherwise they can create a speed mismatch which will cause print quality problem and excessive plate wear.

Flexo gears are precise parts of the flexo printing system, and if they become damaged or worn, they will affect the print quality & register. Most flexographers mount the gear temporarily on the plate cylinder shaft and remove it when the job is finished. They do this simply because they think they are saving money. If they have several sets of similar size (diameter) plate cylinders, they can purchase one set of gears for all the sets of plate cylinders, that will match it.

However temporary mounting of plate cylinder gears creates faster tooth wear & increases the possibility of damaging the gear, because they are used frequently. Some flexographers mount the gear permanently on the plate cylinder shaft and that gear is used only when that plate cylinder is printing a job. This reduces the wear & potential damage to the gear. While it costs more money to purchase a gear for each plate cylinder, it also saves money in the long run, by improving the print quality and colour register. It is common to see temporarily mounted gears with broken teeth, nicks and other damage marks. Therefore, it is suggested that the gear which drives the plate cylinder be permanently mounted. A second gear on the opposite side of the plate cylinder delivers power to the anilox roller.

Web viewers

Right from the beginning of flexographic printing, press manufacturers have constantly made improvements in their design & process techniques, to reach higher optimum printing peeds for improved profitability. As web speeds increased, many process functions became more critical, requiring greater operator attention & skill, to obtain maximum efficiency, quality assurance and waste control.

As web speeds exceeds the 250 to 800 feet per minute (fpm) range, above which operator could no longer accurately inspect the printed web, web scanners became a necessity for quality assurance.

Today with the greater demands for more sophisticated print requirements, the selection of a suitable web viewer is of prime importance.

The degree of efficiency of obtainable is greatly dependent upon the ability of the web viewer to provide a stable image of such quality that precise definition of detail, registration etc, are readily visible to the operator at all printing speeds. When this is accomplished the operator can constantly monitor the print quality and make the necessary press adjustments, as well as observe the adjustment results, while the press is running at optimum web speeds. This minimizes material waste and press downtime.

In-Line Laminating

To day in-line laminating incorporated a separate and a self-contained laminating machine in line with flexo press. It is independent of the press except that it and the press use the same rewind and the electric drives of the two machines are connected and employ necessary web tension controls.

Following precaution must be taken while using in -line laminations

1. To prevent loss of web control, the web travel between the two machines and to the reminder should be as short as possible.
2. The electrical & mechanical drive systems, that synchronize and laminator to the press, must be of the highest quality with adequate web tension controls to maintain, precise control of the web throughout the operation
3. The adhesive applicator station must be tied into the press function, so that when the press stops the nip opens and the applicator roll continue to rotate. Automatic or manual controls can be incorporated for engaging the adhesive applications station.
4. The above function must also be incorporated into the laminating machine drying tunnel controls. When the press stops, the drying temperatures must be reduced to prevent damage to the web in the tunnel.
5. The laminating station must also automatically open at press stop to prevent damage to the rubber impression roll, as well as the web. Independent control for closing nip at press start-up is acceptable. For the converter, who has a large volume of laminating with printing, the press and the laminator are completely separated, except for electrical and mechanical controls, which can be operated individually or in tanderm.

This arrangement allows the converter to laminate in-line with printing, printing without laminating, laminate without printing or use both machines at the same time but independent of each other.

Folding Carton Presses

Folding carton have been printed by flexo for many years, but it has only been about a decade that the industry has tried to upgrade flexo and start using state-of- the -art central impression presses.

Folding carton equipment differs as a results of the substrate. Heavy board stock is processed, which requires higher tension levels, larger idler rolls and more sophisticated splicing & control units.

The unwinds must handle 72 inch diameter rolls and automatically splice them without an overlap. Idlers must be of large diameter to reduce the risk of breaking or crating the board.

22

Gravure & Packaging

Printing

Gravure, is a photographic version of the original 'intaglio' copperplate etching or hand engraving process. In this technique the image is cut or etched into the metal plate, mostly copper, then rolled and filled with a viscous ink. Surplus ink is wiped from the surface, a sheet of paper laid over the plate and both plate and paper subjected to pressure in the intaglio press. Ink from the recesses is transferred to the paper according to the width and depth of the evgraved lines. The result is a unique image with its own characteristics, which make it distinctive from the other printing processes widely used for long run publication and packaging work.

Using the photographic principle to the intaglio process, meant the introduction of a halftone screen and the replacement of continuous etched or engraved lines by a pattern of minute recesses cells. Flat plates are now replaced by cylinders, a liquid ink replaces the stiff copper plate in and hand wiping is replaced by mechanical 'doctoring' with a metal blade. Instead of solid copper, the cylinders are electodeposited with a 'skin' of copper which can be stripped at the end of the run and start for the next job.

Basic methods of producing the gravure image are as follows:—

* 1. Etching
* 2. Engraving

Etched cylinders are the original form of photogravure and fall into following two types:—

1. Conventional gravure
2. Halftone gravure

Conventional Gravure - This is mainly manually operated but requiring a great extent of skill and experience to achieve expected very good results as follows:—

1. A light sensitive pigment or carbon tissue, comprising a coating of dyed gleatine on a stable paper base is exposed in vacuum frame, to a gravure screen, comprising a patte of opaque squares separated by transparent cross lines. The exposure hardness the pigment paper in the transparent cross line areas and this will form the cell walls of the gravure image.
2. The carbon tissue is then given a second exposure. This time to a continuous tone positive. This exposure hardens the gelatine selectively to form an acid test.
3. The tissue is transferred to the prepared printing cylinder by wetting and rolling under pressure to ensure thorough adhesion.
4. Etching with a ferric chloride solution, determines the depth of each cell, depending mainly on the previous exposure. Cell wall areas will remain enetched due to the primary exposure to the gravure screen.

Conventional gravure gives the nearest to a continuous tone effect as the printed ink film is more or less continuous and the tonal effect depends on the depth of the ink film. However, it does have some problems in that, the very fine highlights may be difficult to hold and the slightest wear will effect them. This is mainly important when superimposing multicolour ink film.

EXPERT Four Colour Rotogravure Printing Machine.

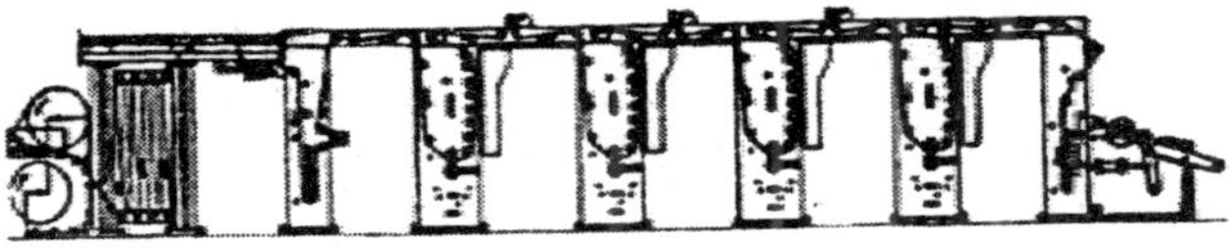

Data subject change without notice to incorporate further developments

Halftone Gravure— Halftone gravure plus latest technique has largely eplaced the conventional method mainly four colour production. This echnique have the following features:-

1. Pigment paper is replaced by presensitized film or by direct coating of the cylinder, with a light sensitive emulsion.
2. Double exposure for screen and tone is replaced by a single exposure to a litho—type screened positive to produce dots of varying diameters rather than cells of constant areas.
3. Automatic etching machines and sequential production lines, have replaced the traditional hand bathing method. Flow rate, acid value, temperature, viscosity and other variables are accurately controlled and monitored during the precessing cycle.
4. Depending on the particular system, dots may also be of varying depth as well as varying area, to give a complete and subtle range of tone complete with a full depth of colour.

Electronic engraving— Etching of gravure cylinders either by automatic means is a time consuming series of operations. For high volume work, requiring speedy production of large numbers of cylinders may be preferable to use a one stage electronic engraving systems, which will incorporate the following features:—

1. A scanning unit will examine the copy via pulsed electronic light beams and transmit digitized information to a central computer console.
2. A visual display station will enable the operator to examine the copy and modify the images to be engraved.

The modified information will be used drive an engraving unit in which diamond tupped styluses cut into the copper surface of the printing cylinders at more than 4000 cuts per second.

This diamond stylus is designed to cut an "inverted pyramid" cell into the copper surface of the cylinder. This is an ideal feature in many ways in that, the highlight tones will have a small area and shallow depth, while solids will have a maximum area and full depth.

Computerized control also enables the electronic engraving system to save information for repeat work, cut multiple image from one master copy and make seamless images for continuous pattern work.

Any method is used, the continuous copper cylinders are capable of long runs in the order of several hundred thousand impression. For very long runs, say in the order of millions, the cylinders may be chromium plated to give indefinite life.

Proofing presses for gravure printing therefore highly specialized widths and diameters and to print on a wide variety of substrates. For staps and lables, the cylinders may just a few centimeters in circumference and width, whereas larger publication presses can use cylinders over a metre in circumference and up to three metre wide.

Cylinders are normally proofed one at a time with modified inks, to allow for the difference in press speed and the time interval between printing. The proofs can give a good indication of the main image properties such as image location, register relative ink thickness etc. Sufficiently to determine whether any modifications are necessary before sending the cylinders to the pressroom. Correction can be made either by further etching to deepen cell volume.

Gravure printing units comprise the printing cylinder running in a trough of ink, a rubber impression roller and a 'doctor' blade to remove surplus ink from the surface of the cylinder.

Printing cylinders are generally one piece continuous cylinder containing the recessed, screened image which has been produced either by etching or engraving. The cylinders are designed to easily removable for quick change overs and to enable the cylinders to be proofed, prior to printing and whenever necessary, stored for future use. Many gravure press can accomodate cylinders of different diameters to allow for the production of work with varying print repeats, such as packing work.

Ink container may be trough in the elementary types of press, but it is more likely to be part of a complete circulating system where ink is pumped to the tray from a bulk supply and surplus ink returned, filtered and adjusted for solvent balance etc.then repumped to the cylinder. The entire system should be enclosed as far as possible to ensure cleanliness of the ink and the avoidance of turns escaping into the atmosphere.

Doctoring of the ink is by means of a flexible steel blade, which is prepared to ride on the surface of the printing cylinder and remove surplus ink, without damaging the surface of the printing image.

The 'doctor' is usually located as near as is practicable to the printing nip, in order to minimize the possibility of the volatile ink evaporating and drying in the cells between doctoring and impression. Adjustments can generally be made to the pressure and the angle of the doctor inrelation to the cylinder, to ensure the cleanest and most efficient wipe of ink.

Impression is usually by means of a hard rubber covered roller, in firm contact with the web and printing cylinder. Firm pressure is essential to ensure maximum transfer of ink from the minute cells of the gravure image. The impression roller is often backed up by a steel bow roller to reinforce its pressure and prevent deflection under the high stresses involved.

Electrostatic may be used to ensure efficient ink release. The impression roller is electrostatically charged, so that its electrostatic field attracts the oppositely charged ink in the cells of the printing cylinder. Electrostatic of 'cell skip', where some cells fail to print and at the same time permits printing with less pressure and consequently less wear on the image.

Arrangement of printing units for gravure is almost always in-line with the web being transferred from one unit to the next, via a series of path rollers. Drying units are generally placed abve each unit to ensure that it is sufficiently dry for overprinting or backing up, mainly on non porous materials. Following three are main ingredients of gravure inks:—

1. Colouring
2. Resins
3. Solvents (Liquid Phase)

Colouring is generally via dyes or pigments of very small particle size to enable formulations, which will easily enter and leave the cells of the gravure image.

Resins, are synthetic polymers which mix with the colouring matter to give the ink its 'body'. The ratio is like this. The higher the production of resin, the stiffer the ink, the formation of the final ink film, depends largely on the physical properties of the resin content. Therefore resins are chosen largely for their ability to dissolve in the solvent, carry the pigment and key to the substrate.

Solvents are normally based on aromatic hydrocarbons such as toluene and xylene, which can dissolve most resins used in gravure inks and will evaporate quickly once the ink is transferred to the substrate, other solvents may be included to either accelerate or slow down the drying rate and specific alcohol may be incorporated, to lower the surface tension of the ink, which will assist flow into and out of the cells.

Hydrocarbon inks are ideal for printing on paper stock, which initially absorbs a high proportion of the liquid in. However, where gravure is used for printing of packing materials such as wrappers, bags and other containers, the residual odour of hydrocarbon solvents may be unacceptable and their evaporation rates may be unacceptable and their evaporation rates may be too slow for films and foils where is no absorbency to assist the drying. In these instances, inks based on a specific range of commercial alcohol will be preferred, due to their faster drying rates and lower residual odour.

WEB CONTROL DURING PRINTING

Web must be kept under positive control before, during and after the printing process. The reel brake, web sensor and infeed system all do their job prior to printing by ensuring that there is constant tension in the unwound reel and the feed, through the printing units should ensure a similar consistency of control.

At each revolution a specific length of material will be fed through the press, at a rate equivalent to the circumferential speed of the printing cylinders. If plate or blanket calipers are inaccurate, this will affect the passage of the web, causing problems such as overfeeding or underfeeding, wrinkling, creasing, stretching, wandering or web breaks.

Web register - Register rollers are a simple mechanism, in which the web is passed over an intermediate rollers which can move from its original position. Movement of the roller will alter the length of web, between one unit and thus control the register.

Manual adjustment - This is simple and economic method for comparatively slow running presses or smaller, compact machines, where all units and controls are within hand reach. The operator will observe a copy at the delivery of the press, note the feed for adjustments, move to the appropriate stations, make the corrections, and repeat the sequence until an acceptable state is attained. This adjustment is done by hand wheels connected to the register rollers or differential gearing arrangements as suitable.

Remote control is more desirable for larger and faster presses, where the delay in travelling from one station to another station will result in an unacceptable waste level. Register controls will be extended mechanically or electromechanically to the central area of the press, where the print is first inspected and corrections initiated.

Automatic correction is by means of photoelectric scanning heads situated between the units. This requires particular register marks to be included in the printed images and positioned on the running web, in an appropriate manner for inspection by the scanning heads. Accurately with this technique will be finer than the earlier methods.

Web turning- The main reasons for turning a web are either to meet a space problem or to make the press more versatile. The most common arrangements are as follows:—

1. Right angle turns via bangle bars.
2. Lateral displacement via turner bars.
3. Inversional of the web via turner sections.
4. Exit and re-insertion via 'bay-window' devices.

Right- angle bars simply comprise a bar placed in the web path at 45 degree so that the web can be wrapped around it and turned at the right angle. Reel stands can be position, at right angles to the printing units, as a multiweb press to save space, or a web printed on the one side may be turned over for perfecting on a second unit at right angle to the first. This enables to build presses, in which the web will 'turn corners'.

However, where the inked side of a web is required to contact the angle bar, there will be a risk of ink smudging as its rubs across the bar.

This usually avoided by providing an 'air cushion' between the web and the bar by pumping low pressure air through apertures, in the wrapped section of the bar.

Lateral displacement via turner bars comprise a pair of bars at 45 degree to the web path in such a manner, that they will take a half web on a publication press and displace it laterally to run in line with the other half. This enables to build 'double-width' presses i.e. with webs twice the width of a standard newspaper, thus doubling the page capacity but allowing the slit and turned webs to come to a single width folder.

Inversion of the web via turnover sections comprise a pair of bars, one at 45 degree and another at 135 degree to web travel plus an intermediate, roller at 90 degree. The web is thus turned over three times in all to finish up inverted, but travelling in the original web line.

Air cushioning may be incorporated to avoid marking up, and mobile web turnovers can be used on multiunit presses to increase versatality. e.g. a four-unit press may print two colours on each side or three colour on one side plus one on reverse etc.

Exit and re-insertion via 'bay-window' devices comparison arrangement of angle bars and rollers designed to take a web or one strip of a web out of the press and re-insert it at another position. A basic method would be to pass the web over a 45 degree angle bar which takes the web to the side of the press. A 90 degree roller will then divert the web towards its new location and another 90 degree roller, will redirect the web back into the press, where a second angle bar will re-align it in its original direction, but with a different relationship with the other web strips.

Web drying - Once the ink is deposited on the web, it needs to change quickly from its free- flowing paste or liquid state to a plastic or solid form, firmly anchored to the substrate.

Liquid inks used in flexographic and gravure printing will also be reduced to thin film. Failure to dry thoroughly may give rise to a few problems. Tacky, undried inks may cause layers of material to stick together, which is undesirable where work is to be rewound. Residual odour from unevaporated solvents may be offensive. Trapped solvent may cause dyes and pigments to bleed in to the substrate, causing unpleasant halo effects around the image areas. The drying process can be accelerated in a number of ways including forced air, heating and radiation curing.

Liquid Inks - used in flexographic and gravure pain will also be reduced to thin film. Failure to dry thoroughly may give rise to a few problems. Tacky undried inks may cause layers of material to stick together, which is, undesirable where work is to be rewound. Residual odour from unevaporated solvents may be offensive. Trapped solvents may cause dyes and pigments to bleed into the substrates,causing unpleasant halo effects around accelerated in a number of ways including forced air, heating and radiation curing.

Heating of the environment around the web will also accelerate, the evaporation of solvents and at the same time speed up the polymerization of residual resin-based ink film. Radiation curing is used to convert resinous ink films, from a semi fluid or viscous condition to a stable plastic or solid state, by means of selective radiation. Infra red, ultra violet and electron beam radiation, may be employed on the particular need.

Inter colour drying units are used on some simple type gravure and flexo presses, particularly when printing on non porous film, and foils. These are generally hot air blowers, employed at a sections of web between two printing units and the aim is to drive off sufficient solvent to enable the first ink film to set to a plastic stage, sufficient to receive the next layer of ink. Controls of temperature and all will be included to allow for variations in substrates, press speed etc. so that each ink film successfully 'traps' the following layer. A final dryer may be included, when all colours have been printed, to ensure that the overlaid films are sufficiently dry, before finishing operations such as folding, cutting and rewinding. Drying hoods are often preferred for the larger gravure and flexo presses. Here the drying equipment is located above the printing units. Immediately after printing, the web is passed around either a large drum or a bank of roller and subjected to high velocity pre heated air. The air is then extracted, with its evaporated solvent content, and the web continuous to the next station with the ink in virtually dry condition.

Web cooling - Web cooling, is essential after heat assisted drying, as the web emerging from the drying system will contain hot, tacky ink, which could cause marking up problems on press rollers and other moving parts and cause the web to stick to itself or other items with which it comes in contact. Cold air blowers are sometimes used but 'chill rollers' are considered more efficient, where the web is passed around a series of hollow rollers of drums, containing continuously circulating refrigerated water.

This is the final operation to ensure that the ink on the substrated to the web finishing operations.

Length folding in standard news paper production, is most commonly performed by a vertical, triangular plate, around which the web is wrapped and pulled by drawing rollers to the cross fold and cutting sections.

Cross folding ereally requires the web to be pushed into the nip of a pair of rollers by a metal blade. The web will by this stage, have been cut into its final print length. More than are cross fold may be made in order to achieve magazine, pocket book, digest or diary page sizes.

Zigzag folding - is a technique for cross folding uncent webs for such items as business forms (continuous stationery) pack label and computer listing papers. The web is usually preperforated laterally as an aid to precise and positive folding.

Combination folding - Combinations of these technique are necessary for newspapers or magazine. Newspaper folder may start with a set of feed rollers, which propel the combined webs to the former plate. A second set of feed rollers, at right angles to the first, will draw the web around the former, to make the length fold on a standard newspaper.

The webs may then be cut to length (therefore the term'cutoff') and passed to one more sets of folding cylinders or rollers for cross folding.

Finally the completed papers are delivered via an endless belt delivery system for counting, batching and despatching to their destinations.

Tabloid newspapers are reduced in the same way except that the webs are slit before folding, in order to produce the final page size.

An additional 'Quarter page' fold may be provided to produce a magazine size product, cutting the web to its final size is generally by means of rotating blade cutting against either a counter cylinder or another blade.

Rewinding - When the end product is required in roll form like adhesive tapes, wallpaper, snap-off bags, wrapping papers etc, it is essential to ensure that the ink is dry before rewinding to avoid the problem of blocking where successive layers or rewound web stick to each other.

Batch counting devices may be included to identify quantities of newspapers or magazines in bundles. These are quire count arrangements, whereby a kick copy or copy kicker is slewed or displaced every so many copies to indicate the completion of a quire. More sophisticated arrangements are available if required, where papers can be batched according to the quantities required by the eventual destination such as distribution or retailer.

Bundling devices - Automatic bundling is very common for modern high volume production to wrap either standard size bundles or variable quantities, by means of programmed control.

Labelling devices - Ink jet or laser printing are commonly used, as they can be computer programmed with each label printed and production on site at the point of application. Preprinted labels may be used where there is a standard pattern and where the publication is to be individually wrapped and mailed.

Inserting - Inserting is the technique, of placing loose sheets cards, leaflets or other items of promotional material into each publication before bundling, labelling and despatch. Automatic equipment is used for high volume, fast turnaround work.

Ink container may be a trough in the elementary types of press but is more likely to be part of a complete circulating system, where ink is pumped to the tray from a bulk supply and surplus ink returned, filtered and adjusted for solvent balance etc. then repumped to the cylinder. The entire system should be enclosed as far as possible to ensure cleanliness of the ink and the avoidance of fumes escaping into the atmosphere.

Doctoring of the ink is by means of a flexible steel blade, which is prepared to ride on the surface of the printing cylinder and remove surplus ink, without damaging the surface of the cylinder or its image. The slightest nick, burr or other imperfection in the face of the blade, will either prevent efficient shearing of the ink from the cylinder surface, or damage the fine cell structure of the printing image.

The 'doctor' is usually located as near as is practicable to the printing ink, evaporating and drying in the cells between doctoring and impression. Adjustments can generally be made to the pressure and the angle of the doctor, in relation to the cylinder to ensure the cleanest and most efficient wipe of ink.

Surface tension of the ink, which will assist flow into and out of the cells. Hydrocarbon inks are ideal for printing on paper stock which initially absorbs a high proportion of the liquid ink. However, where gravure is used for printing of packaging materials such as wrappers, bags and other containers , the residual odour of hydrocarbon solvents may be unacceptable and their evaporation rates may be too slow for films and foils where there is no absorbency to assist the drying. In these instances inks based on a specific range of commercial alcohol will be preferred, due to their faster drying rates and lower residual odour.

Heating of the environment around the web, will also accelerate the evaporation of solvents, and at the same time, speed up the polymerization of residual resin-based ink film.

Radiation curing is used to convert resinous ink films from a semi-fluid or viscos condition to a stable plastic or solid state, by means of selective radiation. Infra-red, ultra-violet and electron beam radiation, may be employed depending on the particular need.

Intercolour drying units are used on some simple type gravure and flexo presses, particularly when printing on non-porouns films and foils. These are generally hot-air blowers employed at a section of web between two printing units and the aim is to drive off sufficient solvent to enable the first ink film to set to a plastic stage, sufficient to receive the next layer of ink. Controls of temperature and air will be included to allow for variations in substrates, press speed etc., so that each ink film successfully 'traps' the following layer. A final dryer may be iuncluded when all colour have been printed, to ensure that the overlaid film are sufficiently dry before finishing operations such as folding, cutting and rewinding.

Drying hoods are often preferred for the larger gravure and flexo presses. Here the drying equipment is located above the printing unit. Immediately after printing the web is passed around either a large drum or a bank of rollers and subjected to high velocity pre-heated air. The air is them extracted, with its evaporated solvent content, and the web continues to the next station with the ink in virtually dry condition.

Web cooling - Web cooling is essential after heat assisted drying, as the web emerging from the drying system will contain hot, tacky ink which could cause marking— up problems on press rollers and other moving parts and cause the web to stick to itself or other items with which it comes in contact. Cold air blowers are sometimes used but 'chill rollers' are considered more efficient, where the web is passed around a series of hollow rollers of drums containing continuously circulating refrigerated water.

This is the final operation to ensure that the ink on the substrate is sufficiently solidified to enable it to be subjected to the web finishing operations.

Web Finishing - These operations may include folding, cutting, rewinding and a variety of ancillary operations.

Length folding in standard newspaper production is most commonly performed by a vertical, triangular plate, around which the web is wrapped and pulled by drawing rollers to the cross fold and cutting sections.

Cross folding generally requires the web to be pushed into the nip of a pair of rollers by a metal blade. The web will by this stage have been cut in to its final print length more than one cross fold may be made in order to achieve magazine, pocket book, digest or diary page sizes.

Zigzag folding is a technique for cross folding uncut webs for such items as business forms (continuous stationery) pack label and computer listing papers. The web is usually pre-perforated laterally as an aid to precise and positive folding.

Combination folding - Combinations of these technique are necessary for newspaper or magazine. Newspaper folder may start with a set of feed rollers, which proper the combined webs to the former plate. A second set of feed rollers, at right angles to the first, will draw the web around the former to make the length fold on a standard newspaper.

The webs may then be cut to length (Therefore the terms 'cut-off') and passed to one more sets of folding cylinders or rollers for cross folding.

Finally the completed papers are delivered via an endless belt delivery system, for counting, batching and despatching to their destinations.

Tabloid newspapers are produced in the same way except that the webs are slit before folding, in order to produce the final page size.

An additional 'quarter pager' fold may be provided to produce a magazine-size product.

Cutting the web to its final size is generally by means of rotating blade, cutting against either a counter-cylinder or another blade.

Rewinding - When the new product is required in roll form like adhesive tapes, wallpapers, snap—off bags, wrapping papers etc. or it may be that rewinding is an intermediate stage for further operations as in the case of preprinted colour work for newspapers, overprinting, varnishing, laminating bag and sack making etc.

It is essential to ensure that the ink is dry before rewinding to avoid the problems of blocking, where successive layers or rewound web stick to each other.

Batch counting devices may be included to identify quantities of newspapers or magazines in bundles. These are quire count arrangements whereby a kick copy or copy Kicker is slewed or displaced every so many copies to indicate the completion of a quire. More sophisticated arrangements are available if required where papers can be batched according to the quantities required by the eventual destination such as distribution or retailer.

Bundling devices - Automatic bundling is very common for modern high-volume production to wrap either standard size bundles or variable quantities by means of programmed control.

Labelling devices - Ink jet or laser printers are commonly used , as they can be computer programmed with each label printed and produced on site at the point of application. Pre-printed labels may be used where there is a standard pattern and where the publications is to be individually wrapped and mailed.

Inserting - Inserting is the technique of placing loose sheets, cards , leaflets or others items of promotional material in to each publication before bundling, labelling and despatch. Automatic equipment is used for high-volume fast turnaround work.

Rotogravure is a form of "intaglio" (cut-in or sunken) printing and prints directly from unconnected cells engraved into the plate cylinder. The true intaglio or steel-die process prints from sunken lines or grooves that are connected and cross each other.

In gravure, cells are not interconnected. The print cylinders in gravure are machined and electroplated with copper, ground & polished. The cylinders are coated with light -sensitive emulsion. After drying, negatives are contacted completely around the cylinder and exposed to light. The sunken cells are etched into the cylinder with an iron chloride solution. To increase the run length of the copper cylinder, chrome plating to protect and harden the surface, is applied over the copper for short runs, the copper cylinder can be used without chrome plating. Today the use of laser etching is growing too.

Screen rulings can be very high. Rulings of 600 lines per inch and above have been achieved using waterless plates, although 150 - 200 lpi is more usual in commercial printing on coated papers and 100 - 133 lpi as un coated stock and newsprint.

Pressed are built in configurations from single- colour sheetfed machines upto very large web units that are combined together to print publications with multiple pages. The paste ink used in offset contain catalyfic drying agents, that cause the ink to dry slowly in normal room conditions. UV curing systems are widely used to give instant drying and high glass levels.

Infrared systems are the main alternative method of accelerating the drying process.

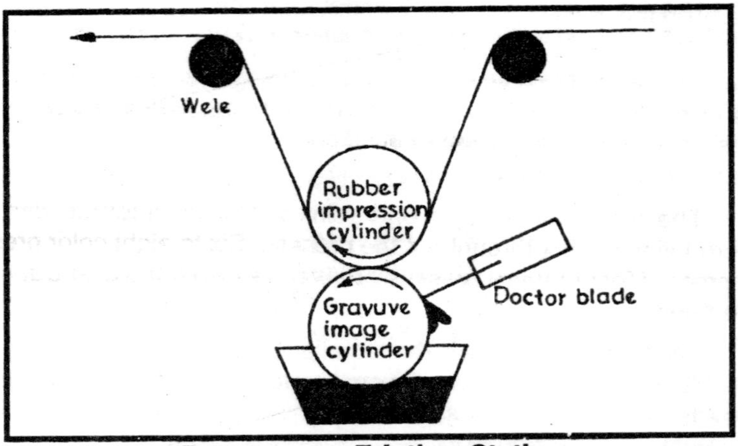

Rotogravure Printing Station

On web offset pressed coated papers are printed with inks that set hard after being heated and then chilled. Web presses heated with this equipment are referred to as heat set and other web presses are called coldset. Heatset presses are necessary for web printing. Heatset presses are necessary for web printing on coated papers.

Modern multicolor CNC (CRC) presses have computer controls, that accelerates function like ink setting and registration control, side ways or clock and anticlock wise. Some press manufacturers have developed systems that are fully automated control of quality by on-line measurement, as it is printed and automatic follow-up adjustment. Due to alcohol dampening printing & dampening is possible, with out extra impression & without scumming problem.

The cylinder's print surfaces are etched as microscopic, cup—like cells, while nonprint areas remain untouched. The larger & bolder the copy the larger & deeper the cells. Fine tonal areas have a smaller cell size and depth.

Gravure inks are fluid and have very low viscosity they are formulated of resins dissolved with solvents, pigments and additives. On press, the image-bearing cylinder is either flooded with an applicator or rotates in an ink pan or fountain in order to fill the cells with ink, excess ink on the surface of the cylinder is wiped off with a steel doctor blade. As the cylinder makes contact with the substrate, ink leaves the cells by capillary action to transfer the image to the substrate. A rubber roller provides an electrostatic assist to attract the ink from the cells to the substrate. Often the ink will not release from the cells to the web substrate, causing print "skipping." Electrostatic assistance helps eliminate skipping by allowing the electrostatically charged ink in the cells to be attracted by an opposite charge in the roller.

Gravure inks must be free of foreign particles. These can cause streaking on the cylinder surface and can cause doctor blade streaking on the web. If streaking does occur, the cylinder has to be removed from the press and refinished. If the doctor blade has nicks or other defects, the blade must either be replaced or reground to a smooth finish.

The web travels from one print station to another with wet ink overprinting dry ink throughout the process. Six to eight color presses are common. Most gravure presses are web-fed (rotogravure) But some are sheet-fed and have a flat plate, that clamps to the plate cylinder. Other gravure systems use removable sleeve-type cylinders.

Gravure is used to print line work and fine half tones, at relatively high speeds and their print runs go into millions of impressions. Run length depends on the condition or wear of the print cylinder.

Gravure is used for packaging, magazines, newspapers and other specialty printing applications.

Ideal substrates for gravure are smooth-finish, clay—coated papers, super calendered papers, rigid film & foils. Since effective ink transfer depends on thorough cell contact with the substrate, irregular or "toothy" rough surfaces are generally not printed gravure.

Gravure has been on outstanding choice for printing process color for mass circulation magazines and newspaper. Gravure, printed postage stamps are another example of the fine print results of rotogravure.

OFFSET GRAVURE

Offset gravure is a combination of offset & rotogravure. Because the flat offset plate image has a limited life & can wear off during long runs, a longer wearing gravure cylinder can be used instead. A gravure cylinder transfer its image to an offset blanket with excellent fidelity. The image on the gravure cylinder must be positive, so it can transfer a reversed image to the offset blanket. In turn the blanket's image prints positives on the substrate.

The disadvantage of gravure's need to print on machined or coated papers, with or without electrostatic help, is offset by the way the ink easily transfer's its image to the smooth rubber blanket. In doing so the image from the blanket faithfully delivers its minute dots to the substrate, or even woven fabrics are printed with surprising fidelity.

Gravure cylinder preparation is time-consuming and costly, making gravure competitive only on longer runs. The image on a gravure cylinder lasts for many millions of impressions without wear. Colour variation is virtually eliminated by the design of the inking system, making the process highly suitable for printing packaging where consistent color is very important.

The gravure cylinder is normally solid steel plated, with a skin of copper and the image is formed by etching or engraving a pattern of cells with a variable depth. After proofing the copper may be plated with chromium, to harden the surface and extend its life.

Cylinders can be imaged either photographically or electronically. Photographic imaging involves exposing a film positive to tissue with acid resistant properties. The tissue is attached to the surface of the cylinder and chemically etched.

Electronic engraving is similar to analog scanning, A positive is prepared and mounted on an analyze scanner, which sends image data to an engraving device as which the cylinder is mounted.

A stylus scans the cylinder, "pecking" cells to the required depth. There can be upto twelve scanning heads on the analyze drum and a corresponding number of engraving heads on the cylinder.

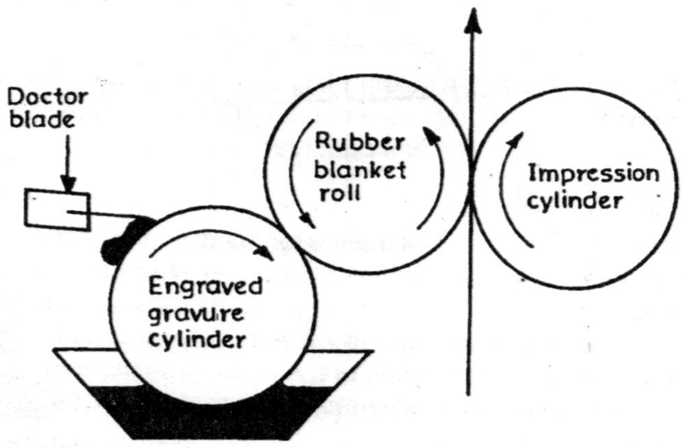

SHEETFED GRAVURE

Today printers are opting for sheetfed gravure is because of a growing demand among advertisers and customers for even better quality. Gravure printing also allows. The ink to be modified to suit the substrate or type of application (e.g. food packaging). Sheetfed gravure is basically simple process from the construction of the printing units. Another reason which must not be ignored is the growing environmental awareness concerning solvents in packaging for foodstuffs and luxury goods.

Applications : General applications for sheetfed gravure presses are as finishing units in a single-or two-colour gravure/offset duplex, or for multicolour finished products. An additional application would be e.g. for the cigarette industry. Where sheetfed gravure presses could be used for test runs and reproduction before transferring to webfed votogravure presses which can run long runs more economically. There is currently a trend towards smaller production runs, Product-specific application are is to be found where a high print quality is of paramount importance i.e. for foodstuffs, luxury goods and confectionery (on paper, board, foil or laminates), or cosmetics (on board). A further application is the printing and /or finishing of glossy catalogues.

Print -Specific applications : The print specific applications are obvious; highest reproductive quality of halftones, solid preprints, solids printing with negative cut-outs, varnishing (solids), varnishing with negative cut-outs, substitute for bronzing, substitute for hot-foil stamping. Printing with special inks: metallic inks (gold, silver etc.), luminous inks, nacresous/ iridescent pigments,

Those who are well converscent with international Market or world market, will agree that while writing about sheetfed gravure, automatically the name comes in our mind is KBA REMBRANDT (142/104).

This new generations of KBA Rembrandt (142/104) presses are developed to meet the following demands; case of operation, as in offset, while conforming to regulations an accident prevention and explosives, preparation of the forme and plate cylinder outside the press, precise infeed register, high register precision between the printing units, adjustable circumferential cylinder, adjustable side by register, shrink wheels, cocking at the printing unit, immaculate, cylinder inking via a molleton roller, doctor blade as close as possible to the printing zone , easy access to the doctor blade, hand-wheel adjustment of the doctor blade, lateral reciprocation of the doctor blade, easy access to all sheets guides, long drying paths i.e. drying without heat , drying with heat for special applications, use of IR or UV dryers optional, smear and scratch-free sheet run through the press, high output.

Technical features and function— The sheet feeder and delivery are the current state-of-the- art and can incorporate non-stop facilities for packaging printing. A double-pile delivery is also available and will be filled on the new machine. The sheet is passed from The feeder to the infeed via a suction -tape feed table. The press incorporates state-of-the art systems for double-sheet, missing sheet and overshoot protection. As is customary in sheetfed gravure printing the sheet is passed to the feed table "face down", i.e. on the side which is to be printed. The feed table and sheet infeed fulfil the criteria for a scratch-free sheet run, to protect vulnerable substrates. After reaching the front lays as the pre-infeed the sheet aligned by pneumatic side by . It is then accelerated to machine speed by suction drums and passed to the transfer drum. As an option the circumferential and diagonal registers on the transfer drum can be adjusted on the run for thé transfer of preprinted offset sheets. A photoelectric cell in the transfer drum activates impression thrown-on the printing couple.

The printing couple basically comprises on impression cylinder, forme cylinder, doctoring system. It should be noted that on the impression cylinder filled with cylinder grippers these are located below the cylinder surface Printing pressure is adapted to substrate thickness by using a blanket and underby sheets. For KBA the engraved forme cylinder is still the basic printing forme for sheetfed gravure, particularly for multicolour gravure printing. For certain specialized applications they are currently developing a plate cylinder which can also take copper or nylograu plates. The machine and can be inserted into the printing unit via support rails. The sliding bearings pivot into the relevant position. The forme cylinder is inked via an ink-feed incorporating a molleton-covered forme roller an ink trough. Ink is supplied to the trough via an ink circulation system. Viscosity is measured manually but can be regulated automatically as an option. The ink-feed system and ink trough are designed so that a minimum of inks is in circulation at any one time. The ink-feed system can be easily swung out or removed for washing and make ready. The doctoring system-a key component of the press wipes the surface of the forme cylinder absolutely clean. There are quick action blade locks for a fast and easy change of blade holder.

The doctoring system is positioned to allow easy access and viewing. The blade angle is adjusted via a hand -wheel according to a scale. Throw-on pressure is controlled pneumatically (manometer display). Lateral reciprocation of the blades is adjustable.

Register setting for the forme cylinder is also by hand-wheel the circumferential, sidelay and diagonal register can be adjusted on the run. Particular attention was paid to ensuring a register-true transfer between the individual printing units. The sheets are guided from one printing unit to the next, and from the last printing unit to the delivery, by register gripper systems. An effective system of cetcher prongs ensures the accurate positioning of the gripper system at the moment of transfer. The sheet is always transferred in precise register, as if by a transfer gripper system. There are shrink wheels to counteract sheet distortion.

The dryers overlap like roof files to make maximum use of the space available. The drying paths, however, have suffered no reduction is length.

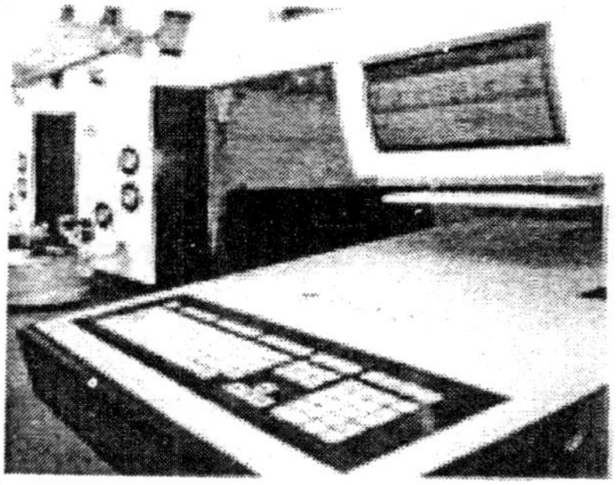

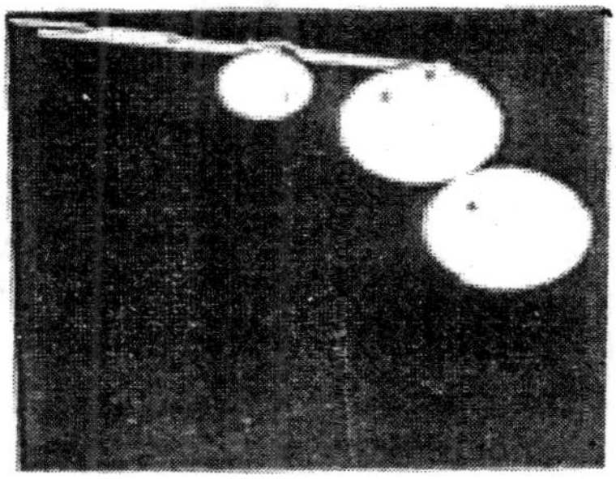

Five-colour sheetfed gravure press KBA Rembrandt 142 with modern control console-the data and control centre for the printer.

Experienced has shown that with this length of drying path-plus the correct ink setting-standard packaging materials (carton board) can be printed without the need for additional heaters. For specialized applications the circular dryer air can be heated via heat exchangers to improve drying. In this case the freshly printed sheet passes through a cold-air zone to a warm-air zone and then back into the cold-air zone again.

The single sided printing sheet is guided through the dryer using suction air i.e. air nozzles in the outer wall of the dryer hold the sheet by suction as it is toward along the gripper system. This ensures a smear free passage through the dryer to the next printing unit. However, sheetfed gravure can also be used to print foils and aluminium coated materials, surfaces on both sides are extremely vulnerable to scratching. In order to eliminate the danger of damage to such vulnerable surface the KBA engineers developed a new type drying system called air-cushion operation. The sheet passes over the guide plates on a cushion of air so that there is virtually no contact on its path through the dryer and from the last printing unit to the delivery.

The fact that the cost of and time involved in cylinder production have both been dramatically reduced by the use of electronic gravure has certainly contributed towards this revival of interest.

KBA REMBRANDT 142(40" × 56") upto 8 colours at 8000 sheets per hour speed-standard equipment is a high-performance stream feeder, with a suction-tape feed table available as an option the sheet transfer system is one which has already proven in its effectiveness. The new sheet is conveyed into the feeder and aligned while the preceding sheet is still covering the infeed. The patented sheet feeder with suction drum before the first printing couple allows a high machine output with register-true sheet infeed.

Another standard feature is photo-electric monitoring of the sidelay at the infeed. If sheet alignment deviates by more than 0.1 mm an acoustic warning signal is activated. Circumferential and sidelay register can be adjusted on the run.

Make ready times have been reduced to a minimum through the use of pneumatically swivelling doctor blades with quick -action tensionning devices, and of pivoting journals for rapid forme-cylinder changes. The lateral movement of the doctor blade is infinitely variable, the throw-on angle adjustable between 50 degree and 90 degree. For cylinder changes the doctor -blade holder can be swung down pneumatically. Before inking. The forme cylinder is immersed is a height-adjustable ink trough made of refined steel. A cloth covered inking roller improves ink saturation.

Chain gripper systems ensure a register -true sheet transfer between the printing couples. The sheets are adjusted in catcher prongs at sheet transfer and delivery. Each gripper change is thus effected in absolute register at all speeds, as if with a transfer drum. A long drying section (approx 12 metres between printing units and 14 metres between the last printing unit and the delivery) with extractors & blowers for warm air drying followed by cold-air cooling ensures optimum intermediate drying and a smear-and scratch-free sheet run. For continuous production there is the option of non-stop infeed and delivery and /or double pile delivery.

The printing couples are built to last the solid side frames have gearboxes cast on for optimum stability. All essential drive elements run in oil baths, with chain guides lubricated via a central pump. The impression cylinders are nickel -plated as standard. The machine is enclosed in sound proofing to minimise noise emission.

KBA REMBRANDT 104 (28" x 41") upto 8 colours at 10,000 sheets per hour-a new model in the popular medium format in continuous operation. Sheet turning is available as an option.

Environmental factors— Recently there has been a definite revival of interest in modern sheetfed gravure. The reason for this are two fold: a heightened ecological awareness, which has led to new regulations concerning non-toxic inks for food packaging; and a growing trend towards metallic inks for luxury packaging.

In gravure, conventional toluene-based inks can be replaced by alcohol-soluble water based inks and varnishes, which are not only virtually odourless but rub-proof and non-fade. In fact it is not only eco— friendly but also user friendly.

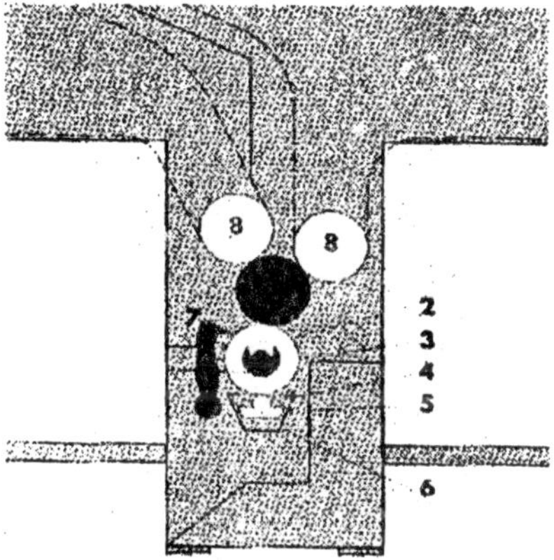

5 Printing unit with built-in forme cylinder, doctor blade in working position

1 = Impression cylinder, 2 = forme cylinder, 3 = pivoting bearing, 4 = ink-forme roller, 5 = overflow trough, 6 = collecting trough, 7 = doctoring system, 8 = sprocket wheels with chain gripper system.

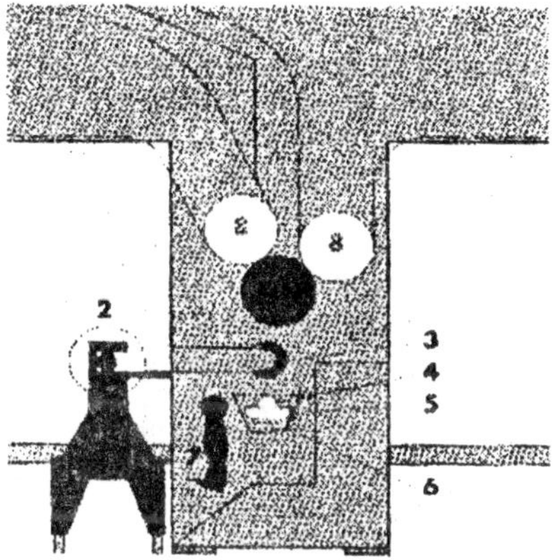

6 Printing unit without forme cylinder, doctor blade thrown off.

1 = Impression cylinder, 2 = forme cylinder, 3 = pivoting bearing, 4 = ink-forme roller, 5 = overflow trough, 6 = collecting trough, 7 = doctoring system, 8 = sprocking wheels with chain gripper system.

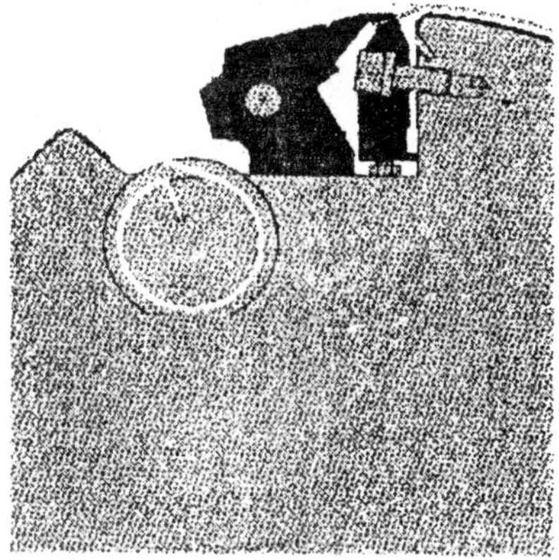

9 Impression cylinder with gripper system.

Advantages of sheet-fed gravure

1. Dry printing, not wet-on-wet, and therefore no risk of doubling, no ink/water balancing, no emulsification-and no fluctuation in tonal values.
2. Quick drying inks and varnishes allow immediate finishing e.g. cutting, embossing, stamping punching etc.
3. No powder spraying, no set-off, virtually no inking/colour fluctuations.
4. Easy printing of all gold shades, silver, luminous inks, opaque white, nacreons pigments, mau colours and varnishes as well as all the customary solvent-or water- based varnishes.
5. Problem-free printing of all substrates from a weight of 70 gsm (45 lbs) to a thickness of 0.75 mm (30 thou)
6. Processing of almost all critical substrates such as aluminium, steamed paper and metallic or synthetic foil.
7. With a judicious choice of screen sizes and etching depths a perfect ink coverage can almost always be obtained which in offset would usually require two plates.
8. A virtually unlimited choice of printing sequences, e.g., gold coloured inks; varnish or coloured inks-a goild, varnish or coloured inks-varnish-gold.
9. Accurate colour matching, since in gravure (unlike offset) there is no ink absorption to cause alterations in colour shades (provided coloured inks are varnished immediately, which is almost always the case).
10. No ink fade, no ghosting on large solids
11. Lower printing-forme costs because existing cylinders can be re-etched/rechromed
12. Print runs of millions per cylinder
13. Straightforward printing couples-no inking rollers, no ink keys.

Advantages in Finishing

Sheetfed gravure also makes finishing much easier. The sheet is absolutely dry and smearfree so that it can be finished much sooner. There is no interference from powder dust.

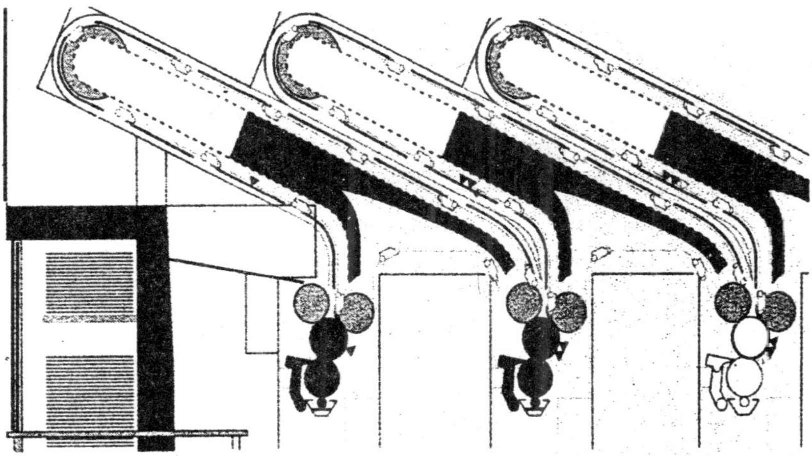

11 KBA Rembrandt 142: drying path with warm and cool air zones.

23

Screen Printing

Screen-process printing can be accomplished with only a minimum amount of equipment. Basically all that is required is a press, a squeegee, and a place to dry your prints. While you can buy a hand-operated screen-printing press for relatively little money, even house wives can also easily make one.

The Screen-Printing Press

A hand-operated screen printing press. The basic parts of (as shown in the Figure A) are the frame, base, screen fabric, and squeegee.

The *Frame* serves as a support for the screen fabric. It can be made

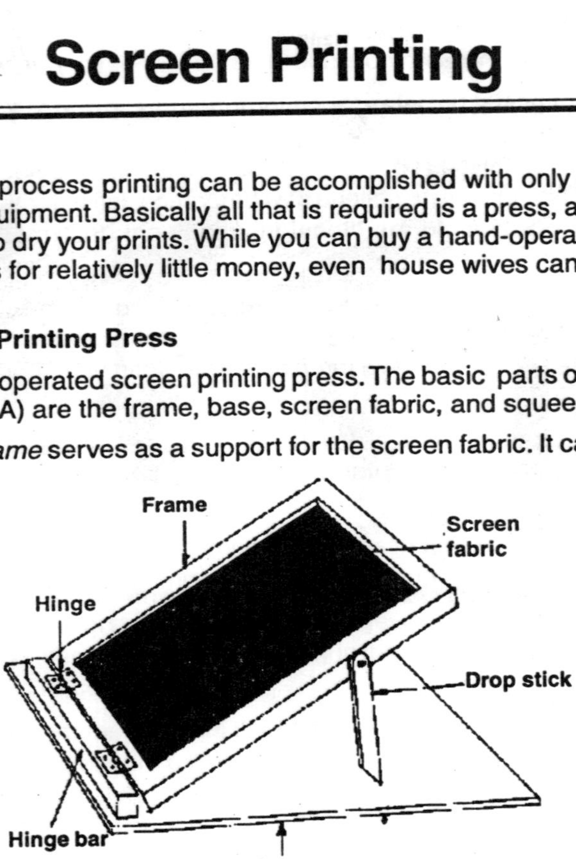

Fig. : A

from wood, metal, or any other rigid material. However, a soft, straight grained wood such as pine or popular is generally utilized for this purpose. Miter, end lap, or spline joints may be used to fasten the corners of the frame together. Each of these joints are illustrated in Fig. B. Angle and corner irons are sometimes used to reinforce the corners of a large screen-printing frame, Fig. C

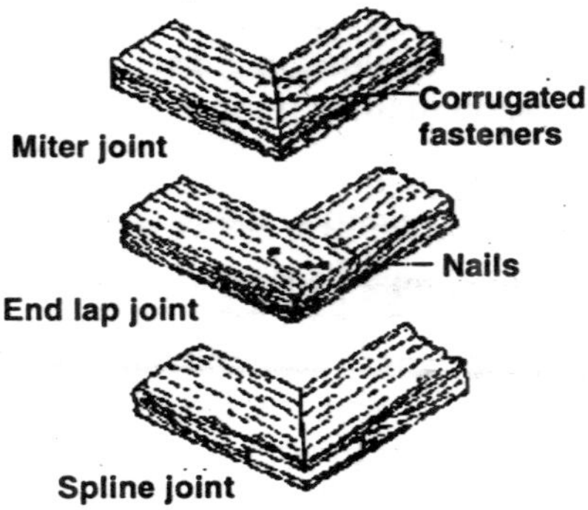

Fig. : B

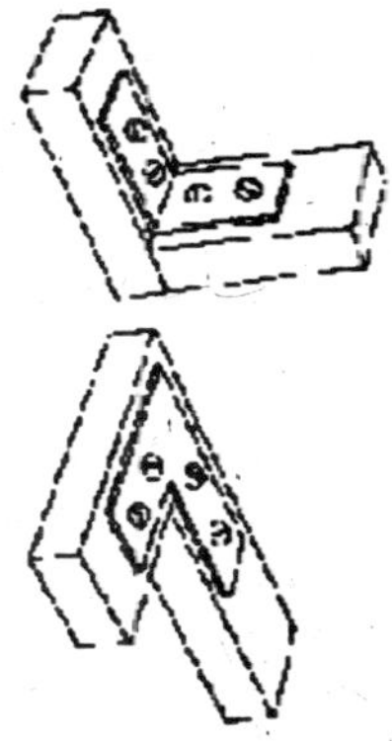

Fig. : C

The base of the press is the surface upon which the substrate (material to be printed upon) is positioned and held during printing. It is usually made from a thin sheet of plywood or hardboard that has been cut slightly larger than the outside dimensions of the frame. Loose-pin butt hinges serve to hold the frame and base together. Common methods of attaching the frame to its

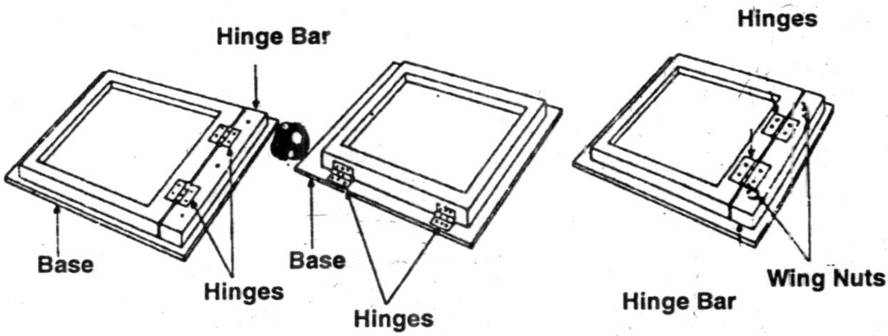

Fig. : D

The *screen fabric* is a piece of woven material, that is stretched tightly across the frame. It serves to carry the stencil throughout the printing run. Three important factors must be considered when selecting a fabric for your screen-printing press. These include the type of material from which the fabric is made, how the fabric is constructed, and the size and number of openings in the screen fabric through which ink can pass.

Types of Fabrics

Silk, polyester, nylon, and metal wire are the materials, that are most often used to make screen fabrics. Silk is a natural fiber produced by the silkworm as it spins its cocoon. Polyesters such as Dacron. Terial, and Polylast are synthetics. They are man-made from materials containing cellulose, resins,and hydrocarbons. Polyester fabrics are characterized by their uniform weave and good dimensional stabilty. They are also extremely strong and can be used for long runs even on abrasive substrates. A major disadvantage is that indirect photographic stencils will not adhere as well to polyesters as they will to silks.

Nylon is also a synthetic. Like polyesters, nylon fabrics are uniformly woven from man-made fabrics that are both strong and durable. Because of their durability, nylon fabrics can be used for long runs on rough and, or, uneven surfaces. Because of their uniformity of weave, nylon fabrics are capable of laying down an even layer of ink on these surfaces. Unlike polyesters, however, nylon fabrics lack dimensional stability. They tend to stretchwith use. Nylon fabrics will also absorb moisture and react to temperature and humidity changes. To maintain print registration and screen life, nylon fabrics must be very taut on the screen-printing frame. It is a good idea to first wet a nylon fabric before stretching it on the printing frame.

Fabrics woven from copper or stainless steel wires are available for printing long-run commercial work while maintaining accurate registration throughout the run. Metal fabrics are expensive. They are also difficult to stretch. Use caution when working with metal fabrics. These fabrics are easily deformed or kinuked as a result of too much pressure or mishanding.

Construction of Fabrics

Fabrics may be woven from either multifilament or monofilament threads. Multifilament threads are made from very fine fibers that have been twisted together. Both silk and poleyster type fabrics are offered in multifilament construction. Hand-cut and indirect photographic stencils adhere well to fabrics made from coarse, multifilament threads.

Monofilament threads are made from single strands of manmade fibers or from metal wires. Nylon, stainless steel, copper, and some of polyester fabrics are of monofilament construction. To obtain good stencil adhesion on this type of fabric, it is generally necessary to roughen the surface of the threads. Various types of abrasives are available for this purpose.

Mesh Count, Mesh Strength, and Mesh Opening.

The coarsness or fineness of a screen fabric depends upon its *mesh count* or the number of threads per inch of fabric. Multifilament fabrics such as silk and some polyesters may have as few as twenty to as many as 200 threads per inch. Monofilament fabrics may have upward of 400 threads per inch. The mesh count for multifilament fabrics is generally indicated by a number ranging from 1 to 25. The higher the number, the greater the mesh count and the finer the fabric weave.

Number 12 silk or its equivalent can be used for most general purpose screen- process work. A number 14 or finer screen should be employed when printing detailed line illustrations. For coarse work, a number 8 or10 mesh screen will suffice. Monofilament fabrics should be used when reproducing halftones by the screen process. The mesh count needed will depend upon the halftone screen ruling that was used to prepare the required film positive. Fabric with a mesh count that is at least 3-1/2 to four times that of the halftone screen ruling is recommended.

STRETCHING THE SCREEN FABRIC

Stretching involves tensioning and attaching the screen fabric to the printing frame. The optimum tension that is required will vary with the type of fabric used. An airoperated fabric tensioning device is designed to stretch all types of screen fabrics, including stainless steel mesh. While such a device is helpful, it is not required for most work. Drum-tight screens can also be stretched by hand.

The screen fabric may be attatched to the frame, with staples or by inserting heavy cord into grooves that have been cut into each side of the frame. If staples are utilized, attach the fabric as shown in Fig. E. Use a heavy-duty staple gun to first staple the center of each side of the fabric to the center of each side of the frame. Then, while pulling a corner of the fabric taut with one hand, begin to staple along one long side of the frame. Work from the center, outward. Locate a staple approximately every half inch until you reach the corner of the frame. Now go back to the center of the same side of the frame and repeat the procedure. Then staple the other long side from center to ends. Finally, repeat the procedure along the short sides of the frame.

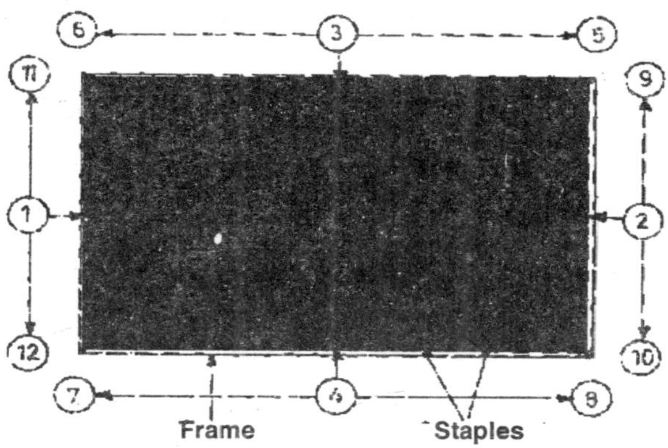

Fig. : E

To attach the fabric with cord one must cut a groove down the center of each side of the frame. Cutting may be done with a table saw, or frame stock may be purchased with the groove already cut. The relationship between the diameter (thickness) of the cord and the depth and width of the groove that is required, is illustrated in Fig. F. Stretching is accomplished by first covering the groove side of the frame, with a piece of screen fabric that has been cut to the outside dimensions of the frame. Then a dull-edged tool is used to press a continuous length of cord down into the grooves. It is advisable not to press the cord down completely the first time around. Instead circle the frame with tool two even three times, each time press somewhat harder on the cord. This will help to ensure a more uniform tension on the screen fabric.

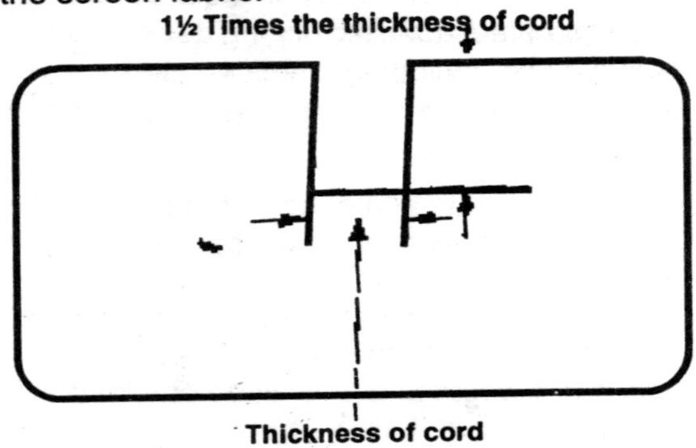

Fig. : F

The *squeegee* is used to spread ink across the screen and to force it through the stencil onto the substrate below. It consists of a hard rubber blade held in a wooden handle, Fig. G. The blade is generally between 3/16 and 1/2 inch thick. Blades shape will depend upon the requirements of the job to be printed.

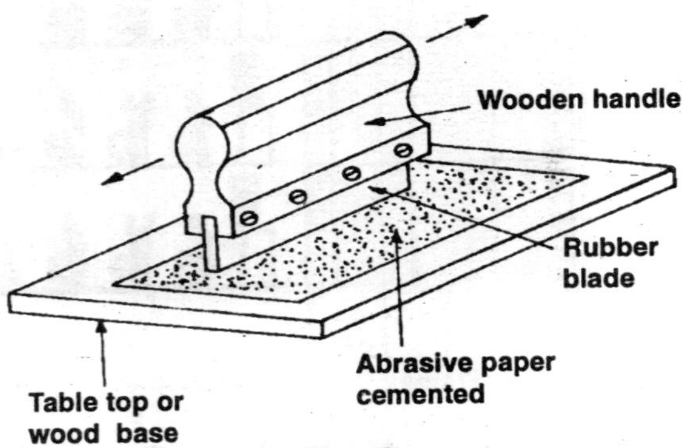

Fig. : G

How to Build A Screen-Process Press ?

Instruction for building the screen-process press. These dimensions, of course, will depend upon the sizes of the jobs to be printed. (Suggested inside frame dimensions for a general purpose screen process press are 14" x 21").

Step 1 : Assemble Needed Materials

Gather the materials as specified in Fig H

BILL OF MATERIALS

Frame, clear pine of similar. 1" x 2" x as desired
Hinge bar clear pine or similar 1" x2" x frame width
Drop stick, clear pine of similar 1/2" x 3/4" x as needed
Base, hardboard or plywood 1/8" - 1/4" thick
Screen fabric and stretching cord for staples)
Loose-pin butt hinges (2 required) with screws
Corner irons with screws (if needed)
Gummed paper tape
Corrugated fasteners (or nails)
Glue (Elmer's or similar)
Shellac
Abrasive paper (medium grit)

Fig. : H

Bill of Materials

Frame, clear pine of similar 1" x 2" x as desired
Hinger bar clear pine or similar 1" x 2" x frame width
Drop stick, clear pine of similar 1/2 x 3/4" x as needed
Base, hardboard or plywood 1/8 " - 1/4" thick
Screen fabric and stretching cord for staples)
Loose-pin butt hinges (2 required) with screws
Corner irons with screws (if needed)
Gummed paper tape
Corrugated fasteners (or nails)
Glue (Elmer's or similar)
Shellac
Abrasive paper (medium grit)

Step 2 : Construct the Frame

Constructing a grooved frame with mitered corners can be used to fasten frame corners together.

Cut a groove down the center of the frame stock with a table saw.

Now cut the frame stock into the strips that are required to form the frame .Two long strips and two short strips are needed. Use a miter box to cut the ends of each strip at a 45-degree angle.

Glue one long strip to one short strip, as pictured in Fig. 1, then reinforce the joint with nails or with a corrugated fastener. Repeat this procedure until all four sides of the frame are joined. Now that angle irons can be used to rein force the corners of large frames complete the frame by smoothing and removing all sharp edges from the wood with a piece of abrasive paper.

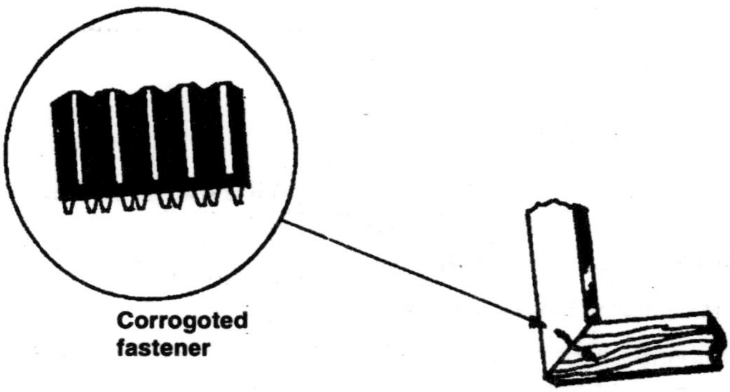

Fig. : I

Step 3 : Attach the Screen Fabric

Cut the screen fabric to the outside dimensions of the frame. Then stretch and attach it. Be sure the fabric is tight on the frame before proceeding.

Step 4 : Tape and Seal the Screen

Taping and sealing serves to extend the life of the screen. It also simplifies cleanup after a press run. Taping is accomplished by applying gummed paper tape to both sides of the screen , as illustrated in Fig. J. Wet the gummed side of the tape with a glue- water mixture before applying. The tape should extend onto the screen approximately 1/2 inch.

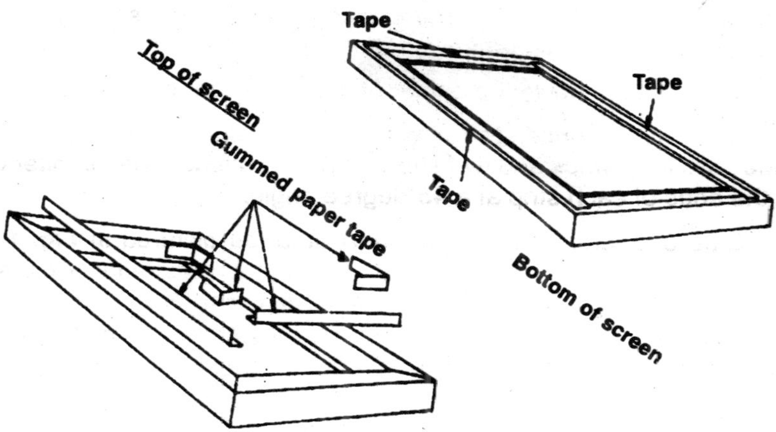

Fig. : J

After the gummed tape has dried, coat the tape and frame with shellac. Allow the shellac sealer to extend onto the screen fabric about 1/8 to 1/4 inch. Avoid getting shellac on any other part of the fabric.

Step 5 : Prepare the Base

Cut the base board slightly larger than the outside dimensions of the frame. Also cut the back bar to length. Then attach the back bar to the baseboard with glue and nails, or with screws.

Step 6 : Hinge the Frame to the Base

Attach the frame to the base with loose-pin butt hinges. Hinge clamps are sometimes substituted for loose-pin butt hinges. Such clamps may be attached to the baseboard or fastened directly to a table. When fastened to a table, the table top serves as the base of the screen-process press.

Step 7 : Add a Frame Support

A Frame support is used to keep the screen away from the base, while loading and unloading the printing stock. Three methods of providing this support are illustrated in Fig. K. The use of a drop stick represents the simplest technique. Counterweighted and spring-loaded supports, however will lift the frame, automatically, after each print is made.

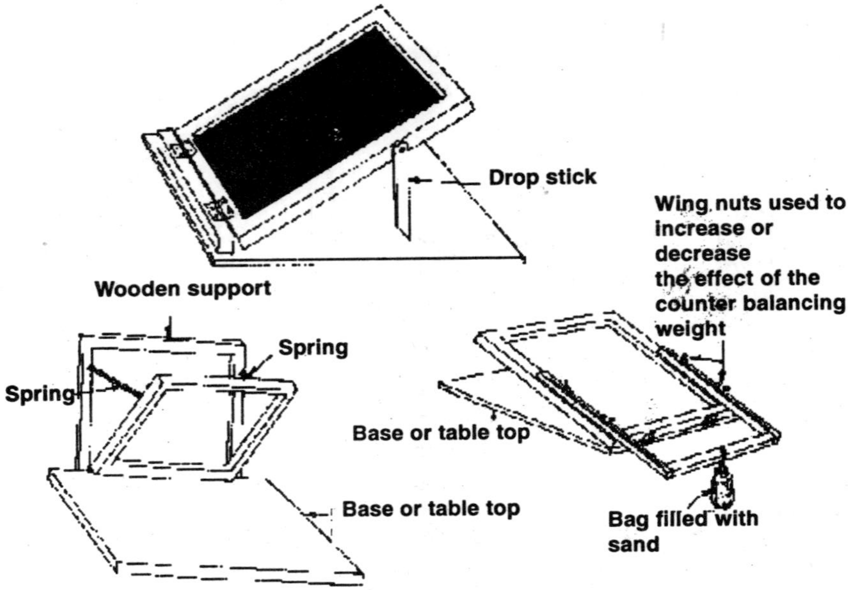

Fig. : K

Print Dry Equipment

Wet prints may be conveniently dried on either a homemade rack or on a line. An easy to build *drying rack* is shown in fig. L. It consists of individual shelves made from corrugated cardboard and narrow strips of wood. Two strips are nailed together along each end of the cardboard. Fig. M. To use, simple pile the shelves, one upon another, as each is filled with wet prints.

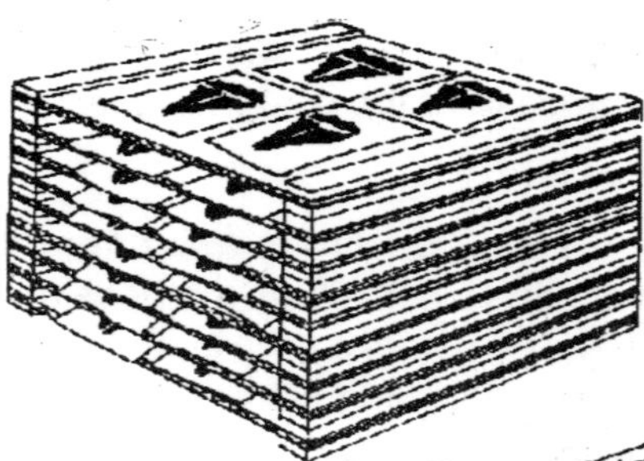

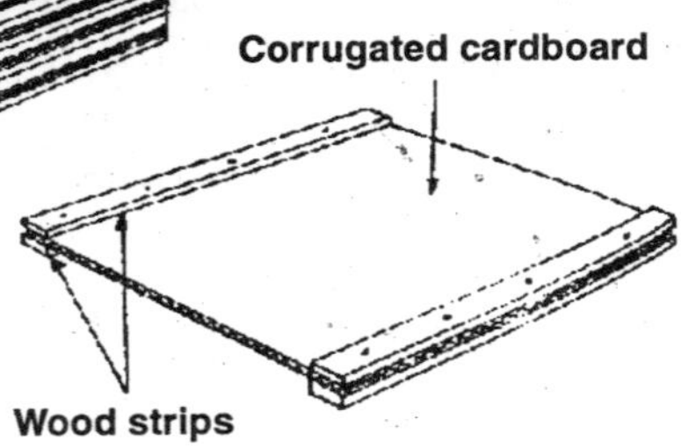

A *drying line* is pictured in Fig. N. It is made by drilling holes in the flat ends of clothespins and stringing them on a length of strong rope or wire, Fig. O. Each clothespin can hold two wet prints, if the prints are placed back to back.

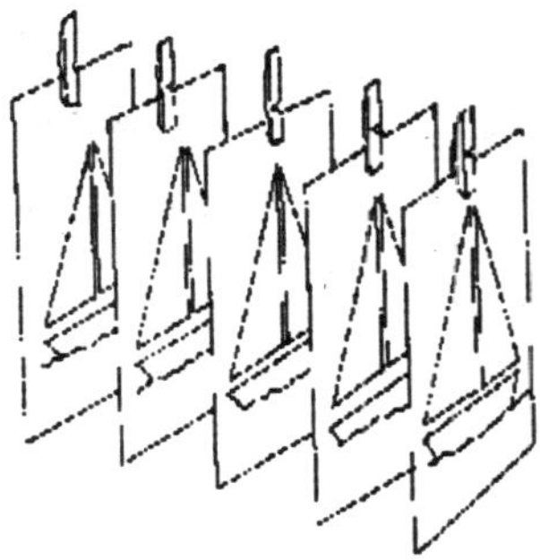

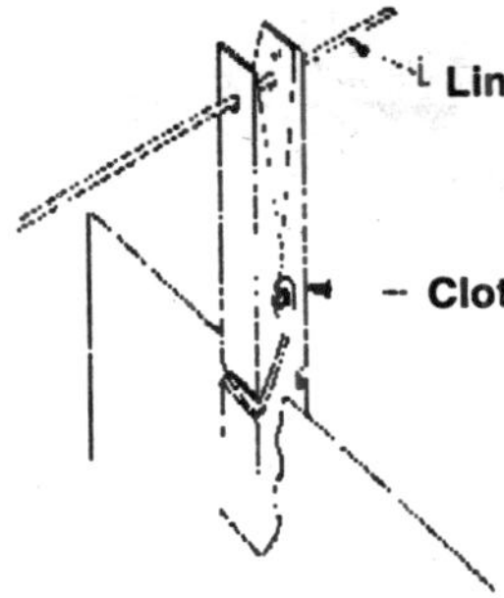

Constructing Specialty Equipment

Specialty equipment for printing on cylindrical and other irregular shaped surfaces are available commercially. Such equipment may also be constructed by the screen process printer. A homemade apparatus for printing on boxes and other thick objects is picture in Fig. P.

The operating principle of a press designed for printing on cylindrical objects is presented in Fig. Q. Note how the squeegee is held stationary, while the screen is able to move back and forth. Note also how the round object will rotate because of the movement of the screen during the printing operations.

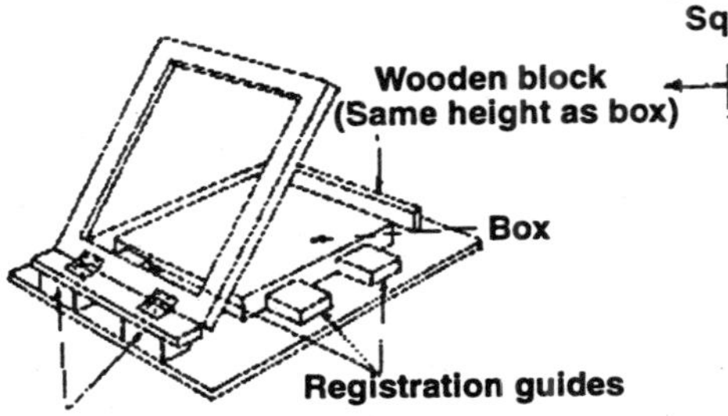

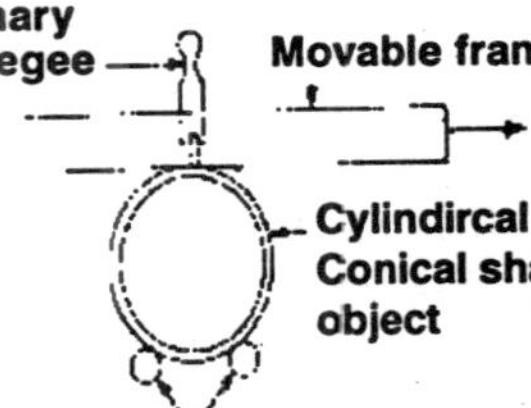

General plans for constructing a screen-process press than can be used to print on cylindrical objects are provided in Fig. R. It consists of five major parts. These include the housing, carrier, carriage, screen and squeegee.

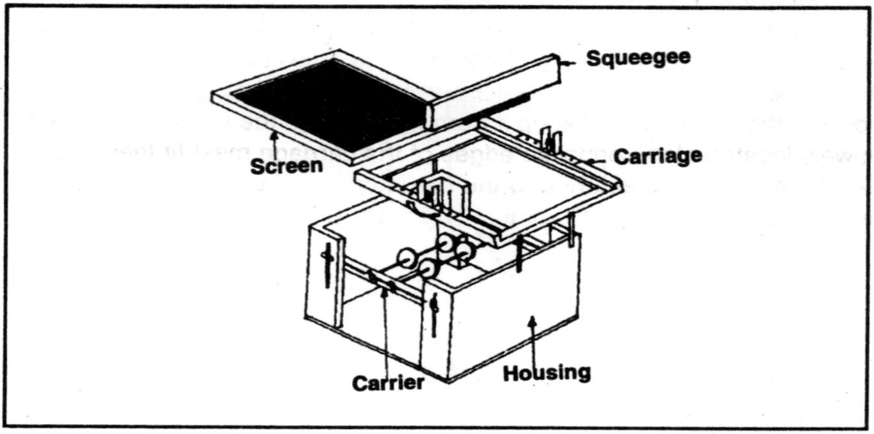

The *housing* may be made from solid lumber, plywood, or particle board. Three - quarter-inch thick material is suggested. Fasten the parts of the housing together with glue and a nails or screws. Once completed, the housing will be used to position and hold the carrier.

The *carrier* may be from threaded steel or brass rods, that measure at least 1-4 inch in diameter. Two rods are required. At least two wheels are placed on each rod. The wheels may be made from wood (for example, from a Tinkertov set). Metal (for example, from an Erector set), rubber (for example, wheels from a toy car), or some other material. The wheels must be able to turn freely on the rods. Roller skate wheel are ideal in this regard. The end of the rods are festened with wing nuts to slotted strips of metal that are positioned across the two open sides of the housing. Note that each of the two strips of slotted metal can be moved vertically within the housing. This makes it possible to print on different diameter cylinders. Note also that the slotted strips move independently of one another. Because of this it is possible to print on tapered as well as cylindrical surfaces.

The *carriage* may be hinged to the housing with strips of band iron as shown. Hinging the carriage in this manner allows for the easy insertion and removal of cylinders. The carriage may also be fastened to the housing with regular hinges. Construct the carriage from solid lumber. It is a good idea to reinforce the corners of the carriage with corner irons. Position the corner irons so that will not interfere with the movement of the screen.

The *screen* is free to slide within the track formed by the sides of the carriage. Procedures for constructing the screen are the same as those given previously.

Finally, the *squeegee* is constructed and positioned over the screen, Fig S. Note that blade is narrower than the inside dimension of the frame and that it has a beveled edge. Note, also the holes that have been drilled through either end of the squeegee handle. These holes are needed to position the sqeegee and keep it from moving with the screen. Screws or dowels located along opposite edges of the carriage must fit these holes exactly. Angle irons fastened to the carriage on either side of the handle help to keep the squeegee vertical during the printing operation.

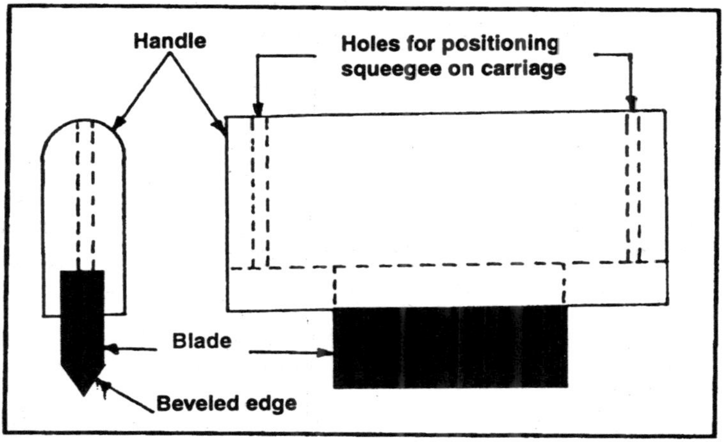

Fig. : S

24

Screen Printing on Paper, Textiles and Other Printing Substrates

A *substrate* is the base material upon which printing is done. The substrate receives and holds the screen- process printed image.

A wide variety of materials can be used as substrates in the screen process. By far the most important of these materials are papers and textiles, other common substrate categories include plastics, metals, woods, and ceramics.

Type of Paper

The screen process can be used to print on all papers and paperboards (cardboards), including those that offer a problem to letterpress, gravure, and lithography. Papers that are commonly printed by the screen process are listed in Fig. 2.1.

Textiles

No one knows exactly when textiles were invented or by whom. However, archeological evidence suggests that out ancestors were producing and using textiles many thousands of years ago. Indeed, strands of linen fish nets have been found in Swiss lake dwellings thought to be at least 10,000 years old.

Originally the term textiles only applied to woven fabrics. Today, however, it generally includes any type of cloth made from fibers, whether woven, knitted, felted, or interlaced in some other way.

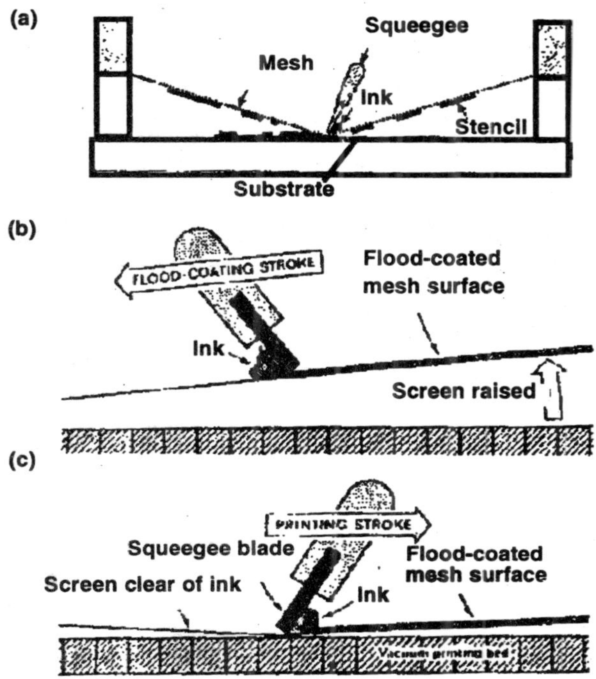

Fig. : 2.1

PRINCIPLE OF SCREEN PROCESS PRINTING
COMMON TYPES OF PAPER

Type	General Uses	Basic Size (Inches)	Basic Weights (Pounds)
Bond	letterheads, direct-mail advertisements	17 x22	9, 13, 16, 20, 24
Bristol (Card)	Tickets, postcards, covers, menues, novelty items	22 ½ x 28 ½	67, 80, 100, 120, 140, 160
Bristol (Index)	Index cards, folders, records	25 ½ x 30 ½	90, 110, 140, 170
cover	covers, binders, menus, posters, tags	20 x 26	Coated : 60, 65, 80, 100, 125 Uncoated : 50, 65, 80, 130
Label	Labels	25 x 38	50, 60, 70
Mimeograph	Mimeograph duplicating	17 x 24	16, 20, 24
Paperboard	Advertising, displays	22 x 28	

Color of Stock

Most papers are available in a choice of colors. Indicate the precise color that is required.

Fibers constitute the basic raw materials from which textiles are made. They are first spun into yarn, then woven or interlaced in some other way to form a fabric.

Fibers may be either natural or synthetic. Natural fibers come from animals and plants. Common animal fibers include wool (from sheep and other animals) and silk (produced by the silkworm as it spins its cocoon). Cotton and flax (used to make linen) are the most important of the vegetable fibers, synthetic fibers are man-made. They are made from materials containing cellulose, resins, hydrocarbons. Nylon, polyester, rayon, acetate, and acrylic are the names of some popular synthetic fibers.

Type of Fabric

Fabrics that are commonly printed by the screen process. Uses for each type of fabric are also provided here. Fabrics made from natural fibers are the easiest to print on. They are the most ink receptive. Synthetic fabrics are more difficult to print on. Synthetic fibers are less absorbent than natural fibers. Printing on synthetics such as nylon and polyester require the use of specially formulated inks.

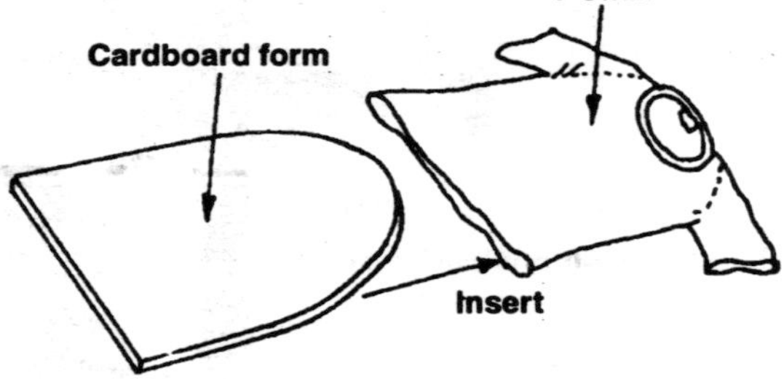

COMMON TYPE OF FABRICS

TYPE	General Uses
NATURAL	
Cotton	Napkins, T-shirts, textiles, towels
Felt	Pennants
Silk	Ribbons, textiles
SYNTHETIC	
Acetate	Drapery and upholstery
Acrylic	Added to other materials to impart crush resistance, warmth, etc.
Fiberglass	Draperies and insulating materials
Nylon	Jackets, dress fabrics, curtains
Polyester	Wash and wear apparel, curtains

Printing on T-shirts

The screen process is often used to print designs directly on T-shirts, shirts made from cotton or synthetic fibers may be imaged in this manner.

For best results the T-shirts must be free of starch or synthetic sizing. Therefore, you should prewash all sizes fabrics prior to printing. Also use a cardboard form to stretch and hold each T- shirt in position on the screen-process press, Fig. 2.2. The form, which is inserted between the front and back layers of the fabric, will prevent the inked design from bleeding through to the other side of the shirt.

Plastics

Chemists produced the first plastic material little more than 100 years ago. Since than, and particularly since World War II, many new plastics have been developed. Today, products manufactured from plastics are all around us.

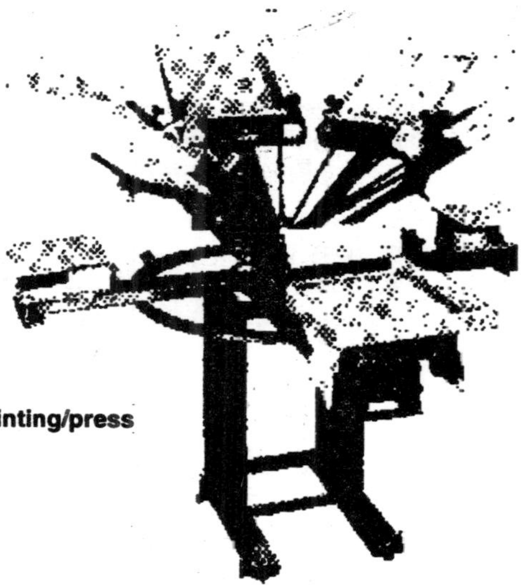

Six-colour T-shirt screen-printing/press

Fig. : 2.2

Plastics are a family of synthetics made from common natural materials such as wood, water, air, petroleum, salt, and natural gas. Heat, pressure, and chemical action serve to separate the atoms and molecules that comprise these natural materials and then recombine them to form the long molecular chains that characterize all plastics. Different types of plastics are created by slightly modifying the processes and ingredients used during manufacture.

Types of Plastics

All plastics may be classified according to their thermal properties into two groups : thermoplastics and thermosets. **Thermoplastic** materials become soft when exposed to heat and harden when cooled. Thermoplastics can be repeatedly reshaped or reformed by alternately heating then cooling the material. *Thermosetting* materials, once formed by heat and pressure, cannot be returned to their original state.

Metals

Printing on metals presents no special difficulty to the screen process. The surfaces to be printed, however, need to be completely free oxidation, corrosion, grease, and oil.

Metal traffic and street signs are often screen printed. Circuit boards that consist of metal foil laminated to a plastic base are also prepared by screening a resist on the metal surface of the board.

Woods

Woods, inculding plywoods and hardboards, are relatively easy to print upon using the screen process. Best result are obtained by first filling and, or, sealing the surface of these materials.

Fillers serve to plug the grain on open-grained woods such as oak, mahogany, and walnut. Commercial wood fillers are available for this purpose. A brush or cloth is used to first coat the wood with filler. The fillers is applied in the direction of the grain. Then, before the filler hardens, it is removed from the woods' surface. This is done with a coarse rag by wiping across the grain of the wood. The filler should be allowed to dry completely before proceeding.

Sealers are applied to woods in order to seal their surface pores. The sealer keeps the surface of the wood from being penetrated or affected by subsequent applications of printing inks. Both open-and closed-grained woods are sealed prior to printing. A mixture of one part shellac and two parts alcohol is commonly used for this purpose. The sealer should be applied with a brush. Let dry twenty-four hours before proceeding.

Solid woods are sold by the *board foot*, which is a piece of material 1 inch thick ,12 inches wide, and 12 inches long, or its equivalent. A board foot is actually a measure of volume. One board foot equals 144 cubic inches, Fig. 4. The following formulas are used to calculate board feet :

Thickness (in.) × Width (in) × Length (in.)/144= Board Feet

T" × W" × L"/144= Bd. Ft. **Or**

Thickness (in.) × Width (in) × Length (in.)/12= Board Feet

T" ×W" × L"/12= Bd. Ft.

Plywoods and hardboards (for example. Masonite) are sold by the square foot or as a standard sheet. Standard sheet size is 4 by 8 feet, or 32 square feet.

Ceramics

Ceramic materials are manufactured from earthen particles such as clay and sand (silica). Since the dawn of civilization, people have formed clay particles into useful shapes, then fused them together by baking and firing. Pottery, tiles, plumbing fixtures, dishes, and even the nose cones of some spacecraft are produced in this manner. Glass, which is one of the oldest materials, is made from melted sand. Drinking glasses, bottles, mirrors, light bulbs,and fibers for insulation and reinforcement are examples of products made from sand.

Ceramic materials are often decorated using the screen process. For example, glaze like inks can be applied to clay tiles before they are fired in the kiln, then ,when fired, these inks bond themselves to the tiles. Special inks are also available for printing on tiles after they have been fired, as well as on glass panels and mirrors. It is even possible to print on the curved surfaces of pottery and glassware using special apparatus.

25

Screen-Process Stencils

The stencil serves as the image carrier (plate) used in screen-process printing. Openings or holes in the stencil allow ink to pass through to the substrate below. The shape of the stencil openings determines the shape of the printed image. Stencils for screen-process printing may be hand cut from paper or film, painted directly on the screen, or produced photographically. Procedures for preparing each type of stencil are provided in this chapter.

Hand-Cut Paper Stencils

Paper stencils may be hand cut from just about any kind of thin paper. Kraft wrapping paper is recommended, however, because it is relatively strong and easily cut. While paper stencils are economical, they are also less durable than other types of stencils. Use them for short printing runs only. Runs greater than 100 should not be attempted with a paper stencil.

To Prepare a Paper Stencil

Step 1: Image the Paper

Cut a piece of kraft paper slightly smaller than the outside dimensions of the screen frame. Then draw the desired image directly on this piece of paper. The image may also be transferred to the kraft paper with carbon tissue.

Step 2: Cut the Stencil

Place the imaged kraft paper on a piece of smooth cardboard or a glass plate and proceed to cut the stencil. Use a sharp knife or a single edge razor blade to cut completely through the paper.

Designs may also be created by folding and cutting or tearing the kraft paper. The unfolded cut-out or the paper remaining, is then used as the stencil.

Step 3: Adhere the Stencil

Carefully transfer the cut stencil to the base of the screen-printing press. Then lower the screen almost to the base and position the stencil. After positioning , lower the screen completely and run a bead of ink across the screen with a squeegee, in Fig. :3.1. The ink acts as adhesive and will hold the stencil to the screen fabric. Finally open up the stencil image by peeling away the appropriate cut-outs from the bottom of the screen. Fig. :3.2. The stencil can now be printed.

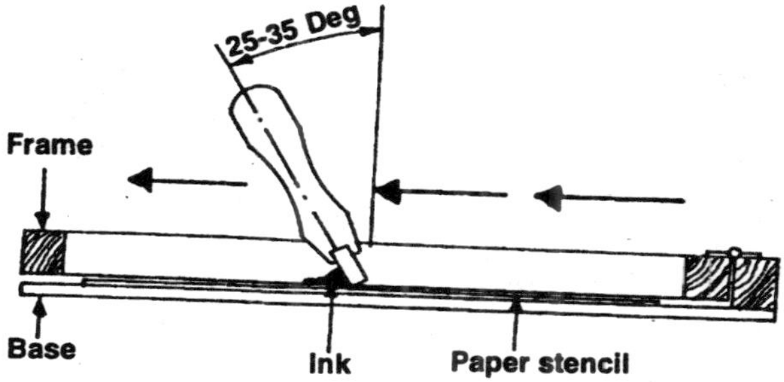

Fig. : 3.1

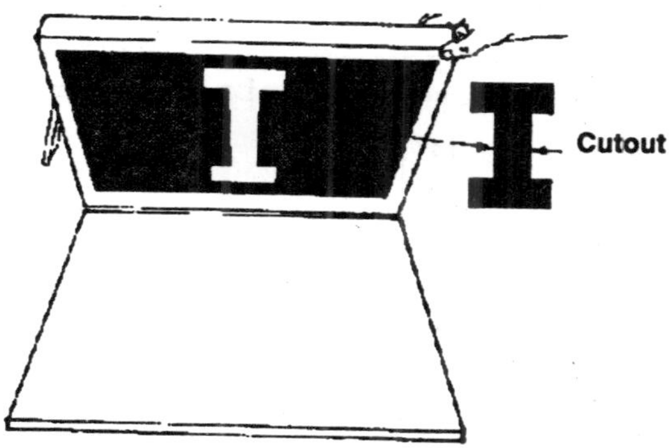

Fig. : 3.2

Hand-Cut film Stencils

Film stencils are ideal for reproducing poster size lettering and illustration, that contain many loose floating parts.They are capable of producing printed images with very sharp corners and edges.These stencils are quite durable and, unlike paper stencils, can be left on the screen after a press run. Many thousands of impressions are possible from a single hand-cut film stencil.

A cross-section of a piece of stencil film is shown in Fig.3.3. It conists of a thin layer of lacquer or water-soluble gelatin material, on a paper or plastic backing sheet. Lacquer and gelatin coated films are prepared in much the same manner.The solvents used to adhere each type of film to the screen are different, however. Lacquer films are adhered with lacquer thinner. Gelatin-coated films are adhered with water. Because lacquer films are soluble in lacquer thinner, they cannot be used with lacquer-base inks. Similarly, gelatine-coated films cannot be used with water-base inks.

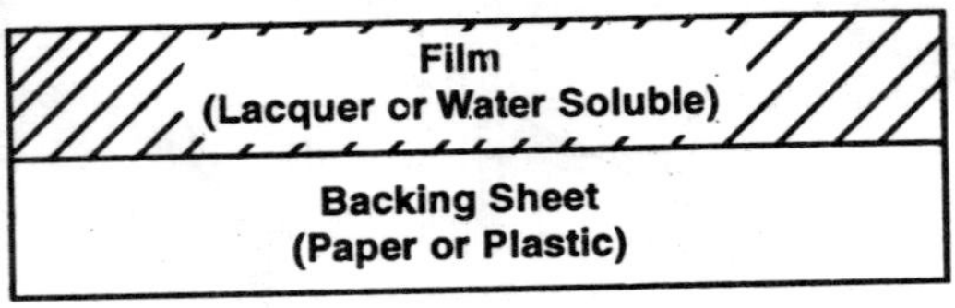

Fig. : 3.3

To Prepare a Film Stencil

Step 1 : Prepare for Cutting

Tape the design that you wish to produce to smooth, hard surface.Then cover the design with a piece of stencil film, film side up. Center the stencil film over the design and tape it in place. The design will be visible through the stencil film, the stencil can now be cut.

Step 2 : Cut the Stencil

The stencil knife is the most important. It may be used in conjunction with a metal straightedge to cut straight lines, or alone to cut curving lines thought the stencil's film layer. The stencil knife is also utilized when lifting and peeling unwanted lacquer or gelatin from the stencil. The blade in the swivel cutter is freed to revolve within its handle. This permits 360-degree manoeuverability with only the slightest touch.

Cutting must be accomplished with as little pressure as possible. The weight of the knife alone should provide sufficient cutting pressure. Do not cut into the backing sheet. The best results are obtained with an extremely sharp blade and a light, gentle touch. This excess solvent can dissolve or burn away the sharp edges of the stencil image.

Step 3 : Adhere the Stencil

Place the stencil, film side up, on a build-up board that has been cut slightly smaller than the inside dimensions of the screen-process printing frame. The build-up board should be between 1/4 and 1/2 inch thick. A sheet of glass or several layers of cardboard that have been glued together will serve nicely in this regard. Next position a clean screen over the stencil. The stencil is now ready for adhering, Fig. 3.5.

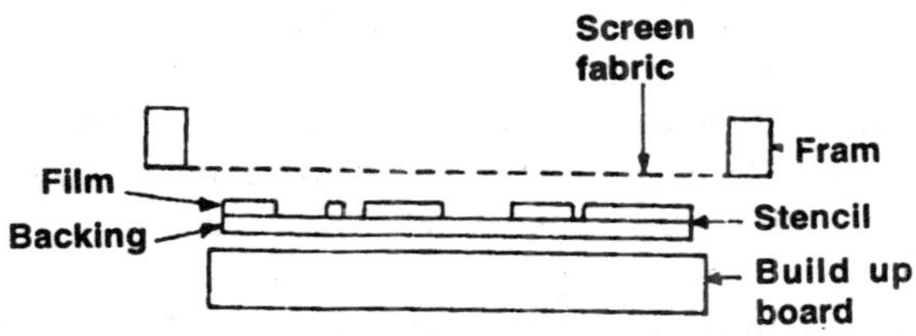

Fig. : 3.5

The overall dimensions of the stencil must be smaller than the inside dimensions of the printing screen. The stencil must not be allowed to extend onto the shellacked and taped areas of the screen.

Lacquer films are adhered to the screen with lacquer thinner. The strength of the required solution will vary with brand of film used. Gelatin films are water soluble. Some brands are adhered with plain tap water only. Other require a mixture of water and alcohol or vinegar and alcohol, as the adhering liquid.

To adhere a water-soluble stencil, saturate a sponge with the recommended solvent and move it over the fabric. Overlap your strokes until the entire surface of the stencil is wet. Aplly just enough solvent to uniformly soften the stencil film.Next blot the water-softened stencil surface Fig 3.6. Cover the screen fabric with a pad of newspaper and then roll over this pad with an ink roller. Use moderate pressure only. Repeat this procedure several times, until all of the excess water has been removed. Blotting in this manner will cause the stencil to adhere to the screen.

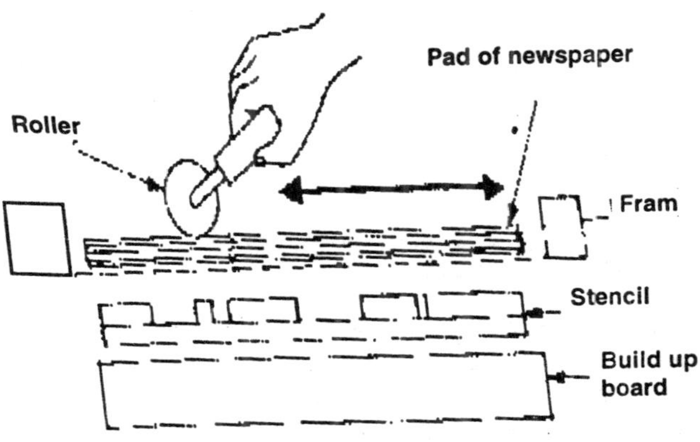

Fig. : 3.6

To adhere a lacquer stencil, dampen an absorbent cloth pad with the recommended adherent. Then, starting at one corner of the stencil, press the cloth pad against the film. The adherent will soften the film and cause it to stick to the screen fabric. Be careful not too use too much adherent or the film will begin to dissolve. Continue dampening until 4 to 6 square inches of the film's surface have been moistened. Now immediately bolt the moistened area until it is dry. Use a clean, dry rag for this purpose. Repeat the process until the entire stencil is adhered.

Too much adherent will cause the stencil film to dissolve, leaving a hole that is also a light spot.

Step 4 : Remove the Backing Sheet

Allow sufficient time for the film to dry before attempting to separate the backing sheet from the stencil. A fan or hair dryer may be used to shorten the drying time that is required. When the stencil is completely dry, use your fingernail to lift a corner of the backing sheet from the stencil. Then peel away the sheet, as shown in Fig. 3.7.

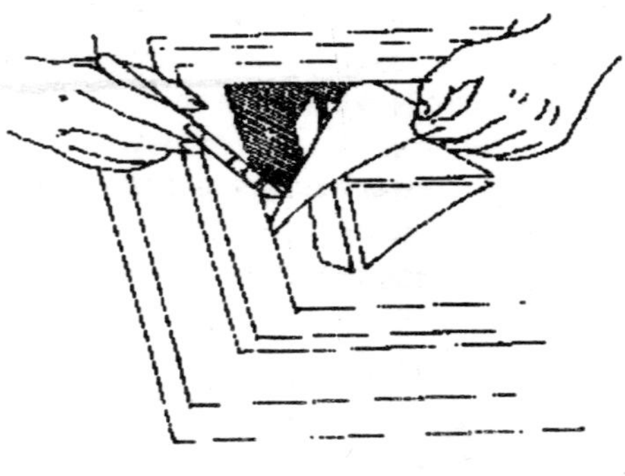

Fig. : 3.7

Photographic Stencils

Photographic stencils are used when it is necessary to reproduce fine detail by the screen process.

Positive transparencies are required to produce photographic stencils. Image areas on such transparencies must be completely opaque to actinic light. Actinic refers to the light energy that is used to harden a photographic stencil's emulsion, so that it will be insoluble in water. Non-image areas of positive transparencies may be either transparent or translucent. Non-image areas allow actinic light to pass through to the light-sensitive emulsion that forms the photographic stencil.

The positive transparency that is required can be prepared in several different ways. It may be made by drawing an India ink image directly on a piece of tracing paper or on a clean sheet of acetate. The procedures are similar to those used to prepare pictorial and technical drawing. The transparency may also be made by applying transfer type to thin plastic. Photographic processes may also be used to prepare the transparency. This is done by first preparing a line or halftone negative, then converting the negative to a positive by contact printing on process film.

There are three basic types of photographic stencils : indirect, direct, and direct/indirect.

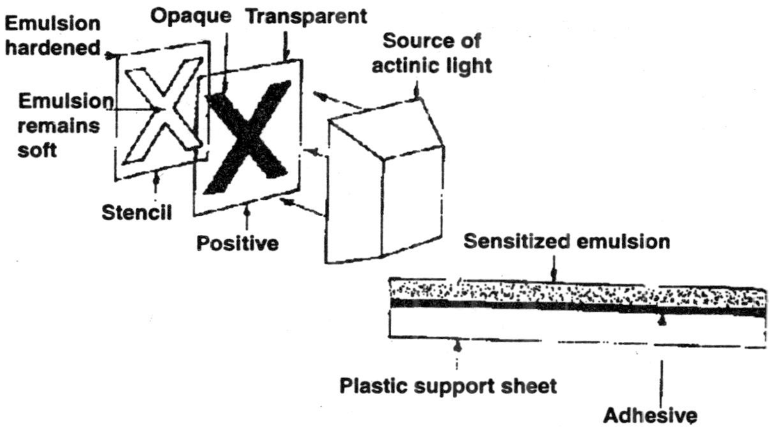

Fig. : 3.8

To Prepare an Indirect Photographic Stencil

A cross-sectional view of a piece of photographic stencil film is presented in Fig. 3.8. It consists of an emulsion layer on a plastic support sheet. The emulsion layer is sensitive to blue, violet, ultraviolet light. It is not sensitive to yellow and red light, therefore the stencil film can be handled and processed under yellow or red safelights or even under subdued incandescent illumination.

All indirect photostencil films are prepared in much the same way. They are first exposed through a positive transparency to a source of actinic light. Areas of the film that are struck by light are hardened, while unexposed areas remain soft. Processing follows exposure and serves to dissolve away the unexposed (unhardened) areas of the film's emulsion. It is these unhardened areas, that form the image during printing. After processing the film is adhered to the screen and allowed to dry. When the film is dry, the stencil's plastic support sheet is removed and the screen is ready for use.

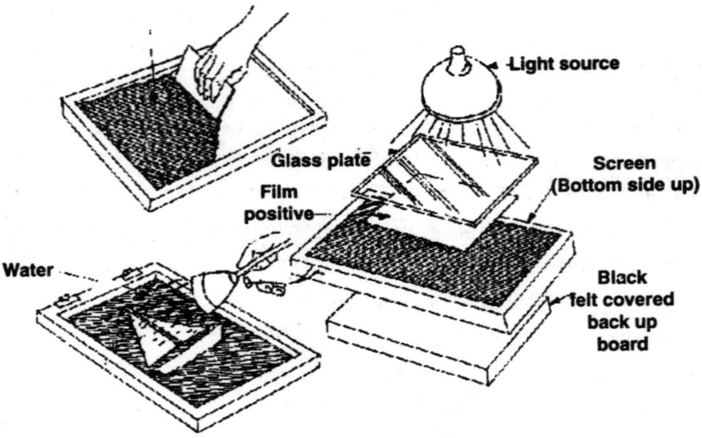

Fig. : 3.9

Step 1 : Prepare for Exposure

A vacuum-or contact-printing frame is needed to hold the positive transparency and stencil film in tight contact throughout the exposure process.

A source of actinic hight is also required. Carbon arcs emit a great amount of blue. violet, and ultraviolet radiation.

Actinic light sources emit ultraviolet light. Ultraviolet light can damage eyes. Do not look directly at the light that is used to burn (expose) a photographic stencil.

Step 2 : Load the Frame

Cut a piece of photographic stencil film so that it is at least 2 inches longer than the dimensions of the image that is to be reproduced. Place this is piece of stencil film, emulsion side down, on the rubber blanket of the vacuum frame. The emulsion side of Hi-Fi Green is its shiny side. The emulsion on most other Ulano stencil films is the duller of the two sides. Now cover the stencil film with transparent positive. The emulsion or readable side of the positive must contact the back of the stencil film.the unreadable side of the positive faces up. Next close the vacuum frame and turn on the vacuum pump. Once the air is evacuated, the positive and stencil film will be held tightly together. If a flip-top platemaker is used, turn the vacuum frame toward the light source. The stencil is now ready for exposure.

Step 3 : Expose the Stencil

Proper exposure is required if the stencil is to adhere to the screen and print correctly. Underexposure will produce an emulsion that is too thin to be readily adhered to the screen. Such an emulsion may begin to show signs of wear, for example, develop holes in non-image areas early in the press run.Overexposure produces an emulsion, that is thicker and harder than desired, such an emulsion is difficult to adhere, will not reproduce fine details, and may even crack under the pressure of the squeegee.

Correct exposure will depend upon the type of stencil film, the light source that is used, and the distance between the light source and the stencil film. Suggested starting exposure times for several films manufactured by Ulano are listed below. These are suggested starting exposure only.

Step 4 : Develop the Stencil

Ulano stencil films are developed in special chemical powders labled "A" and "B". Note that "A" and "B" powders can be bought in bulk or in premeasured foil packets.Mix 1 ounce (28 grams) of "A" and 1 1/4 ounces (35 grams) of "B" in 1 pint (0.5 liter) of water. The temperature of the water should be between 64° and 75°F (18°-24°C) Stir or shake the mixture until the powders are completely dissolved.Then pour the developer into a glass, stainless, steel, enameled, or plastic tray. The stencil can now be developed.

Quickly slide the exposed stencil film emulsion side up, into the developing tray and develop for 90 seconds. Rock the tray occasioally during the developing period. After 90 seconds the films is ready for washout.

INDIRECT PHOTOGRAPHIC STENCILS : EXPOSURE TIMES IN SECONDS*

	CARBON ARC LAMPS			***PULSED XENON***	***BLUE ACTINIC FLUORESCENT TUBES***		***No.2 Photo-Flood Bulb, 18"***
Type of film	15 amp 24" (60cm)	30 amp 24" (60cm)	40 amp 30" (75cm)	3000w 40" (110cm)	400w 40" (100cm)	built-in	
HI-FI GREEN*	336	168	198	330	250	3-5 minutes	7 ½ -8 min.
SUPER PREP*	306	153	180	330	250		" "
BLUE POLY*	272	136	160	270-300	200-225		" "
ULANO XPM*	244	122	144	270-300	200-220		" "
ULANOCRON RX*	244	122	144	300-320	225-235		" "

***Hi-Fi Green, Super Prep,Blue Poly XPM, and Ulanocron RX are registered trademarks of the Ulano Compainies.**

Step 5 : Wash Out the Stencil

When removed the stencil from the developer, the stencil image is not yet visible. Washout under running water makes the image appear. Washout temperature for several Ulano Photographic stencil films are listed here below. Set the water temperature as required before proceeding.

Place developed stencil film, emulsion side up, on a sheet of plastic or glass. The plastic or glass will serve as a support for the stencil during washout. Now place the support* on an incline in the sink basin, so that the washout water can flow over the stencil. Direct the spray across the entire surface of the film, not just the image areas. Continue washing until all of the unexposed (image) areas are dissolved from the stencil. Several minutes are usually required for complete washout.

After the image areas are completely clear, gradually lower the temperature of the washout water until it reaches the approximate temperature of the room. Chilling the stencil in this manner will harden the emulsion slightly and help to maintain image sharpness during the adhering step.

WASHOUT TEMPERATURE*

HI-Fi*	100°-115°F (44°-46°C
Super Prep*	110°-115°F (44°-46°C.
Blue Poly2-3*	97°-104°F (36°-40°C.
XPM 2*	97°-104°F(36°-40°C.
RX* 200, 300	97°-104°F(36°-40°C

Step 6 : Adhere the Stencil

The stencil must be adhered to the screen immediately after washout. Place the stencil, emulsion side up , on a hard, flat build-up board.

Adhering is accomplished by blotting the top surface of the screen with clean newsprint until the stencil' stacky emulsion meshes with the screen fabric. Do not use pressure. Simply cover the screen with newsprint and then gently wipe over the newsprint with a soft, clean rag. After blotting in this manner, change the newsprint and repeat the blotting procedure. Continue this process until color from the stencil no longer transfers to the newsprint. Normally four or five chages of newsprint are required

Allow sufficient time for the emulsion to dry before attempting to remove the stencil's plastic support sheet. A fan may be used, do not use heat, however. Heat can cause the stencil's edges to seperate from the screen. When the emulsion is completely dry, carefully peel away the plastic support sheet. The screen is now ready for printing.

To Prepare a Direct Photographic Stencil

Direct photographic stencils are prepared directly on the screen fabric. The fabric is coated with a light-sensitive emulsion; then it is exposed to a source of light through a positive transparency. Light passing through the clear areas of the positive hardens the emulsion below. Because no light passes through the opaque areas of the positive, the emulsion behind these areas remains soft, Fig 3.9. After the stencil is exposed, the unhardened ares of the emulsion are washed away with water and the stencil is ready for printing.

Direct stencil coatings are usually supplied in two parts-an emulsion and a sensitizer. The emulsion is not light sensitive until mixed with the sensitizer, which may be of the diazo or bichromate type.

Mixing, coating, exposing, and processing procedures will vary somewhat from one brand of emulsion to another. For best results, follow the instructions provided by the manufacturer of the direct emulsion coating that will be using.

Step 1 : Mix the Emulsion

The emulsion must be prepared under safelight conditions (no actinic light). Begin by dissolving the dry diazo sensitizer in lukewarm water. When completely dissolved, add the sensitizer to the emulsion and stir until a uniform color is achieved. The sensitized emulsion can now be coated on the screen or stored for future use.

Remember once sensitized, a diazo emulsion will begin to deteriorate. Refrigerated storage in a tightly caped, light tight bottle will increase the useful life of a sensitized emulsion.

Step 2 : Coat the Screen

It the emulsion has just been mixed , it should be checked for air bubbles. If none are found, coating may commence immediately. Otherwise allow the emulsion to stand until all of the air bubbles have disappeared. Air bubbles are to be avoided since they can cause an uneven coating to be applied to the screen's fabric. Also refrigerated emulsions should be allowed to return to room temperature prior to application. Emulsions thicken when they are refrigerated. Thick emulsions are difficult to apply correctly. Coating is accomplished under safelight conditions. A stiff piece of cardboard or a brush may be used, to spread the emulsion across the screen.

Step 3 : Expose the screen

Exposure requirements are similar to those described in Steps 1 through 3 of "How to Prepare an Indirect Photographic Stencil".

Step 4 : Process the Stencil

Processing is accomplished by spraying both sides of the screen with water at 600-100°F (15°-40°C). Gently spray the squeegee side of the screen, until the image areas open up,. Then intensify the spray and wash the printing side of the screen. Finally, reduce the water pressure and spray both sides of the stencil until it is completely clear and free of all foam and bubbles. The screen is now ready for printing.

To Prepare a Direct/Indirect Photographic Stencil

Direct/Indirect stencils are made from an unsensitized film that is adhered to and sensitized on the screen itself.The resulting stencil possesses both uniform thickness (characteristic of indirect stencils).

Step 1 : Sensitize the Coating Solution

The coating solution must be prepared under safelight conditions. Procedures are similar to those described in Step 1 of "How to Prepare a Direct Photographic Stencil". After the diazo powder is completely dissolved, however, it is added to a coating solution rather than to an emulsion.

Step 2 : Adhere the Film to the Fabric

Place the unsensitized stencil film, emulsion side up, on a hard, flat build-up board. The dimensions of the board should be slightly smaller than the inside dimensions of the frame. Now lower are properly roughened and decreased screen on to the stencil film.

After drying is completed (approximately 15 minutes is required when a fan is used) the plastic support sheet can be peeled from the screen.The stencil is now ready for exposure.

Step 3 : Expose the Stencil

Procedures are similar to those described in Step 3 of "How to prepere a Direct Photographic Stencil". Suggested starting exposure times for Ulano materials are listed below

Step 4 : Wash out the Stencil

Procedures are similar to those used to process a direct photographic stencils. After drying, the washed out stencil is ready for printing.

Ulano Fotocoat* Emulsion

Type	Stencil Color	Sensitizer	Description	Type of work
569	purple	diazo	unexcelled resolution : complete solvent resistance	fine details and halftones
569CL	undyed*	diazo	unexcelled resolution : complete solvent resistance	fine detailes and halftones: see -through for registration
569Fast	violet	diazo	excellent resolution : fast exposing; complete solvent resistance	fine detailes and halftones; ideal for high- volume production or weaker light sources
77112	violet	diazo	extra hardness; excellent resolution	large screens: ceramic printing: glass decorating
449	purple	diazo	complete solvent resistance wide eposure latitude	multi-purpose or special work including decals
339	undyed*	diazo	complete solvent resistance wide exposure latitude	general work: see-through for registration
TZ	light blue	diazo	water and plastisol resistant; decoratble; post-hardening unnecessary	sportswear, T-shirt and textile printing
TZ/CL	undyed*	diazo	water and plastisol resistant: decoatable ; hardening unnecessary	sportswear, T-shirt, textile printing; see-through for registration
TZ/HY-200B	undyed*	diazo	water and plastisol resistant; high viscosity	sportwear; ideal for coarse mesh; see-through for registration
FX88	blue	diazo	fast exposing; extra stencil hardness	large number of screen per day
FX88-CR	undyed*	diazo	fast exposing; complete solvent resistance	weaker light sources aggressive inks

*separate dye supplied with purchase

DIRECT PHOTOGRAPHIC EMULSIONS :
EXPOSURE TIMES IN MINUTES*

CARBON LAMPS	*ARC*	*PULSED*	*XENON*	*FLUORESCENT*	*MERCURY VAPOUR*
15 amp	40 amp	2000w	5000w	40w tubes	250w
20"	40"	20"	40"	6"	26"
(50cm)	(100cm)	(50cm)	(100cm)	(15cm)	(65cm)
1, 2/3-62/3	2,1/2-10	1-4	1 ½-6 ½	6-15	2-9

*Using screen made from white polyester fabric (305 threads per inch.)

DIRECT/INDIRECT PHOTOGRAPHIC STENCILS :
EXPOSURE TIMES IN MINUTES*

*TYPE OF DIRECT/ INDIRECT FILM ** *	*CARBON LAMP*	*ARC*	*METAL*	*HALIDE*	*PULSED*	*EXNON*
	30amp	40amp	2000W	5000W	2000W	5000W
200PR	8 ½	6 ½	2 ½	1	23	5 ¾
250PR	11 ½	8 ¾	6	2 ½	30	7 ½
300PR	14	10 ¾	9 ½	3 ¾	–	9 ½
200R	15	11	4 ¼	1 ¾	23	10
250R	22	16	6 ½	2 ½	–	13
300R	30	22	8 ½	3 ¾	–	16

*Exposure time using white fabric at 40 inch distance. For distances other than 40 inches multiply times in chart by the following factors :

20 inches-0.025

30 inches-0.56

50 inches-1.56

** Registered trademarks of the Ulano Companies.

26

Screen Printing Automatic Press

The Printing Forme makes it Possible

All printing methods is essentially differ by their printing formes, however, have in common the process technological fact-apart from the screen printing method-that the printing ink is transferred from the surface of the printing forme to the substrate, either directly or indirectly: This is quite different in screen printing, for in contrast to all other printing processes in screen the printing forme is porose and this fact is decisive for all the possibilities, capabilities and the nearly unlimited fields of application of this method. This is where the decisive strong points of screen printing originate from, i.e. the versatility with all its process technological chances, the flexibility and the adaptability, the innovative strength with its innovative capacity in many industries.

The decisive factor is the screen printing forme. It consists of a frame, formerly made of wood, now in most cases either made to metal, mainly aluminium, across which the screen, either made of polyester threads or metal wires, is stretched like over a drum. The printing subject is fixed in the mesh of the screen printing fabric by barrier layers which are produced in a photomechanical way. These barrier layers, also called stencils or screen fabric in form of direct stencils or also as directed stencils be fixed at the outer side of the fabric or they can be produced as combination. The screen fabrics themselves are available in very different degrees of fineness; the standard fabric polyester 120, 120 threads per cm, is now being supplimented by essentially finer, but also many coarser fabrics. The decisive influence is that by the combination of fabrics with different degrees of fineness with possibilities of the variations in screen stencil production, i. e. by different stencil methods, the thickness if the ink layer applied can be varied and controlled. The screen printing ink is applied across the printing screen with stencils by means of squeeges, and they will get into the substrate only at the permeable spots through the mesh of the screen wire.

The Screen Printing features and their singularity

From these features of the printing forme outlined above the essential characteristics of the singularity of screen printing will result; they can best be summerized under three focal points :

The screen printing forme in the mordern version is versatile considering both its size and the form of the surface of the substrate to be printed. The means the range of possibilities for the screen printing forme goes from few sqcm up to 30 or even more sqm, and it is clear that the large sizes will remain the domain of screen printing.

The thickness of the layer in screen printing is not only thicker than in all other printing methods, but can be varied by means of the characteristics, already mentioned. In accordance with the characteristics, the durability and the effects asked for, the thickness of the layer can be altered and if necessary be extremely increased. If in normal industrial screen printing the thickness of the dry ink layer is between 5 and 7 micron, it is possible to increase the thickness of the layers in industrial processes up to 100 micron. This way the coating technological criterium of screen printing has a decisive part in many fields of application typical for screen printing across many industries. Significant examples are the production of outdoor signs with durability, but also most of the technical tasks in the industrial field. Because of these coating properties of screen printing it is not only a printing method but at the same time a qualified coating technology! Especially in this function screen printing is needed for numerous technical applications in many industrial and trade branches and for even changing and new applications.

In screen printing the making of the printing inks is not only influenced by the printing method but also by the task to be done, that is the purpose it has to serve. This way a singular versatility is reached as shown in the wide range of programmes with the many ink types and products of the screen prinitng industry.

A screen printing ink essentially needs hardly more than the fine grinding of the pigments and the press-ready adjustment. This way a most diversified range of effects is realized according to the respective requirements, e.g. highly opaque effects with high gloss, transparent effect, fluorescent effect, technical effects, and all that with a great variety of inks, so to speak the language of screen printing. The unique diversity of inks with their light intensity and their lasing properties has been advanced by new ink mixing systems and colourimetrics system; especially since the creation of the computer assisted C-Mix-Comp a variety of millions of shades for all applications can be calculated fast, precisely and economically, and realized in practice. This increased use of the variety of ink shades in screen printing shows to an even greater degree the typical close relation between ink and printing and the expressiveness of the colour by means of screen printing !

This means that on the whole the almost unlimited variety of application is the special strong point of screen printing with reference to the different materials, surfaces, purposes, and tasks.

Apart from the variety of tasks it is the extraordinary flexibility in screen printing, this flexibility for shortest, short, medium length up to long and very long runs with various sizes, combined with the effects of economic efficiency and the special surface effects that can be realized. Basically the process technology in screen printing is relatively simple and this way opens up many possibilities. Today and in future the semi-automatic technology will be needed for certain tasks and full automation has spread for many years now in form of fast running flatbed screen printing presses and cylinder screen printing presses up to the monocolour and varicoloured rotary printing technology in screen printing. The printing output has been steadily increased and at the same time the specialization in screen printing has been advanced further and further. It turned out that in the graphic screen printing sector, i.e. mainly for the production of advertising media of all kinds, especially paper posters and adhesive films/foils, a new orientation in engineering had become necessary focussing on standardization, series production, systematic further development and research, concentration and tasks to come concerning the improvement and refining of the existing technology.

The development in technical screen printing in industrial applications was quite different; in this field screen printing is not an aim of the company but an important stage in the production process. Remarkable achievements have already been realized in certain fields of production, especially in case of high-capacity screen printing presses for printing packages and labels (rotary printing) and in other special applications.

In commercial screen printing field, screen printing is the purpose of the company and the main client is the advertising industry with its wide demand for advertising media of various kinds and with different effects. This sector can be assessed to cover a market share of about 30-40% at present. In contrast the technical screen printing sector covers about 60-70% and dominates the industry with unlimited variety of different applications which are outlined above.

Apart from the variety and flexibility the third strong point of screen printing is the innovative ability which actually is a conditon for the introduction of new processes and products. The capacity for innovation shown again and again this way has been proven by significant examples in the present-day production methods both in graphic-commercial and in technical-industrical screen printing. The examples range from the production of signs to printed circuit boards, from the printing of glass to textile screen printing, from package printing to membrane keyboards in order to outline the diversity of products that are common products now and owe their existence to the innovative capacity of screen printing.

Process-technological limits have been expanded in screen printing again and again and this process will go on. Limits are expanded as regards the fineness of line and screen, the machine output, the safeguarding of printing quality and its reproducibility, the environmental orientation, the ink technology with the creation and the further development of new systems as an alternative to the highly developed conventional ink series by UV-hardening screen printing inks and water-based screen printing inks, by the continuous further development of the diversity of inks by colourimentrical systems such as C-Mix-Comp.

It is not sufficient to look at the art of printing alone and to print the jobs taken over. The ingenuity in looking for new tasks is of increasing importance for screen printing. It will become more important to realize new possibilities needed, to develop new products, to participate in new developments in the fields of substrates, inks, but also materials such as metals and plastics for industrial use, to keep up relations with producers who not only need advertising media but also have to treat the surfaces of their products and need to add marks and decorations.

On the whole that means that a good development can be expected for the future in screen printing, if it is possible to unfold the strong points even more and to realize that in the market, that means to sell the potential included in the variety of tasks, the flexibility, the innovative capacity again and again and with increasing intensity.

The graphic-commercial field of screen printing profits from the alliance with the advertising industry, which is marked by a certain degree of interdependence. Advertising is mainly depended on screen printing as regards the not extremely short-term media and their efffects, the durability, the variety and the dimensions and materials, but also as regards the flexibility when having to meet the requirements and the length of runs. The advertising industry and screen printing need each other, look for each other, depend on each other. This means that is not only important partners and clients in the advertising industry what its strong points are and to sell its effeciency effectively. The near future of screen printing will mean increasing communication tasks in the advertising industry, especially concerning the European Single Market on the one hand and the development in the East on the other hand, and the overall increase in advertising activities that is to be expected at the places where the goods are sold and probably an increasing demand for display, increasing labelling with better quality, but also more flexible fields of application in outdoor advertising, especially as regards paper poster.

The development will be quite different in technical screen printing and the development reached there in the past years with steady growth rates and increasing variety of tasks leads to the conclusion that the future has only just begun.

Within the diversified tasks in the field of electronics and production, screen printing will be needed to an increasing degree even if the economic development is not that favourable as regards the demand; the many articles and apparatuses in the field of consumer goods that will even increase in number will have to be marked, decorated, and the metallic and plastics surfaces will have to be treated. Screen printing with its suitability for printing and at the same time coating is needed to an increasing degree in those fields where large -scale production takes place. The decisive factor for the future will be, however, that the advantages of screen printing will be realized to an increasing degree and made use of.

27

Choosing A Printing Process

1. LETTER PRESS

Advantage :

Faithful Reproduction versatile & late correction or type matter even on the printing ink

Operation such as numbering & perforated can be carried out while printing..

Limitations :

Good coated paper reqd. for half tone printing inpractile for large illustrations i. e. poster, wall charts etc.

omparatively low running speed storage of type & forms expensive & space consuming.

2. LITHOGRRAPHY

Advantage :

Wide range of paper board qualities. Intricate matter can be easily handled by using art work or by stripping negatives or pos. Fast ink speed No limitations to size of illustration (posters etc.) convenience of storage o plates or films. Letterpress books may be reprinted by photographing original. Making typesetting un-necessary.

Limitation :

Alterations to plates inpracticable.

Inferior lithography printing appear flat & dul with weak bad printing

3. PHOTO GRAVURE

Advantage :

Rich tonal effects, in full colour mono-chrome, ideal for very long runs in colour, in expensive paper coated or un-coated (women magazine, prepress for newspaper advert etc.)

Very fast running speeds.

Limitations :

Preparatory cost expensive. Alteration to cylinder inpracticables. Cylinder storage expensive. Type matter &fine line detail becomes somewhat broken by screen pattern.

4. SCREEN PRINTING

Advantage :

Low preparatory cost suitable for short runs in multicolours, light colours & fluoroscent ink widely used.

Light colours can be printed on dark materials. Versatile use of process for printing on paper &board, fabric, plastic, wood, china pottery, glass& may be printed on curved surfaces, also used for posters, show cards, bus sides, Aeroplane body.

Limitation :

Printing speed comparatively slow. Halftone screen is limited to fairly course work.

5. COLLO TYPE

Advantage :

No screen reqd. beantiful colours, very true to two original & fairly inexpensive. Printing reels of paper, foil, plastic, film, rubber stereo long lasting.

Limitation :

Suitable for short run only (upto 1000 copies only). Printing must be colour for colour (i.e. No 4 colour process).

Only one printer commercially. Only suitable for packaging type work such as wrapping paper etc. Not suitable for type matter excepting the small quantities in large sizes.

6. FLEXO GRAPHIC

Advantages :

Fast running speed, good colour areas, suitable for printing reels of paper, foil, plastic, filn etc.

Limitation :

Only suitable for packaging type work such as wrapping paper etc. but not suitable for type matter.

What Process to use :

Making to decision each printing process involved cost in various stages. Preparatory cost are the same whether the quantity of printed matter is 15000 copies. And this must be borne in mind. Litho is the most widely used process at present because the overall cost is lower than letterpress printing & this only overtaken by Gravure when very long runs are involved. The high preparatory cost of preparing gravure cylinder is soon outweighed when a very long run is envisaged.

At the other end of the scale screen printing has low preparatory cost but the printing is either manual or semi automatic & then drying is either achieved by using racks or by passing under an infra-ray drier. Which involves a further handling charge, cosequently long run becomes costly.

28
Security Printing

Security printing encompasses areas including bank notes, stamps, credit cards, cheques, postal orders and many types of top secret Government information.

There are following three main concerns for security printing:-

1. To secure the confidentiality of the documents, through origination and production, from the preparation of artwork to proofing, platemaking and printing.
2. To ensure the authenticity of the documents and make them readily identifiable by all concerned.
3. To make copying, duplicating and forgery of the documents as difficult as possible.

Currency printing : Bank notes are printed on a specially tough rag-based paper produced under conditions of high security. The paper contains it own individual watermark, which could possibly be different in design and position for each denomination. Some bank notes comprise more than one layer of paper which enables metal or plastic strips and other devices to be incorporated to aid identification and deter forgery. The design of a bank note is normally based on hand drawn or engraved image, which is difficult to *copy* or photograph. These are complicated line patterns, produced by mechanical or computer-driven plotters, often in several interwind colours and difficult to copy or reproduced by conventional printing techniques. Micro images and subtle hidden or secret patterns may also be included to aid weak detection of forgeries.

Cheque printing : Unlike bank notes, cheques themselves have no particular value-it is the personalized information written on the cheque which gives it value. The security factors for cheques printing are, therefore, centered more around the problem of fraudulent issue of cheques rather than copying and forging.

The information to be printed on a cheques are as follows:-

1. The name of the bank or other financial institution.
2. The address of the individual branch.
3. The code of the branch and bank.
4. The account number.
5. The name(s) of the account holder (s).
6. The individual cheque number.
7. The amount in numbers and words.
8. Pay ....

Provision must also be there for the individually typed and/or written instructions.

The printing process may vary e.g. the basic cheque form may be required by the million and could be printed by web-fed presses . But more localized information such as the branch, account and individual cheque number, may be superimposed in batches via either sheet-fed printing or individual imprinting machines.

The hand -filled or typed individual instructions are usually contained within a background tint area, often comprising the name of the issuing institution, printed in minute lettering and repeated in closely spaced lines.

The ink used for this is usually of a fugitive nature in that it is water or solvent based and easily disturbed if any attempts are made to erase, obliterate or modify the information typed or written on it.

The individual cheque number and other coded information may be printed with an ink containing minute ferrous particulars, which can be detected by a magnetic ink character recognition (MICR) system. To aid rapid recognition the characters are printed in a distinctive, stylized form so that the magnetic reading head can distinguish each character by recording the levels of ferrous material present, as the characters pass below the reader . Ink film thickness, character size and spacing plus the position of the coded information on the cheque must be precisely controlled to ensure that MICR cheque sorting and data recording function efficiently.

Credit cards

Plastic cards were originally used for obtaining or transferring money from one account to another, or to obtain credit within predetermined guidelines.These basic functions are still central to the plastic card economy, although they are now capable of recording far more information and performing a greater variety of functions.

Usually the cards are multiplier laminates on a polyvinyl chloric (pvc) base, printed on both sides, with a protective film layer and containing additional material such as magnetically-encoded strips, embossed information and holograms.

Caliper and dimensions

Cards must conform to the appropriate ISO standards in order to function correctly with the relevant machinery and processing equipment. Printing may be by screen process or litho depending on the quality and effect required.

Screen will be preferred for solid lay down and thicker film of ink, whereas litho may be used for more subtle images and halftone reproductions.

Protection Of the printed image may be by means of a fine coat of varnish or lacquer, or for more complete protection a plastic film may be used to complete protection, a plastic film may be used to complete the lamination on each side.

Embossing is by physical deformation of the laminate, requiring an individual master die to be made for each card. The embossed image is then formed by heat and pressure in a stamping press. The relief height of the imager needs to be precisely controlled in order to register correctly in the recording equipment. Very shallow embossing will mean that the digits may not be fully recorded, whereas very high an image could mean that the card may not be received, or premature wear and distortion of the image could occur.

Magnetic strip is generally bonded to the card prior to final limitations. This is subsequently encoded with invisible magnetic information, which can only be decoded in an appropriate reading device. Information from the magnetic strip can be used to activate automated machinery in cash dispensers, credit recorders, communication devices, security entry doors etc.

Signature panels are a feature of many plastic cards. These panels or strips are bonded to the card after final lamination. The card holder is required to make a signature across the panel, which generally has a printed background image to avoid the possibility of an altered or substituted signature.

Holograms are often included in plastic cards as identity markers and security devices. An original hologram is produced by photographing an object with laser light from different angles and superimposing the image optically to reproduce the original with a three-dimensional effect. There are different types of hologram, but for printing purposes the most commonly used is called embossed hologram

An embossed hologram is a lamination of several film layers, including a base of stable polyester material and two or more layers of metallized foil containing the printed and embossed images, plus a protective film layer or coating.

Metallization is the deposition of vaporized metal on to a plastic base to give a pure, even film of metal, just a few microns in thickness.

Embossing of the foils is normally from a metal master derived from the original hologram or three-dimensional design. It is to be noted here that embossing means more in the form of a subtle variation in the surface structure of the metallized films, which are only a few microns in thickness.

Superimposition of the printed and embossed films will give the unique three-dimensional effects obtainable only by embossed holograms.

Holograms are very difficult to photograph, which makes them ideal for security purposes. The specialized techniques required to produce holograms in the first instance makes forgery an equally unlikely possibility.

All types of postal stationery printing are being done by Gravure process in Nasik & Hyderabad and currency note are being printed by offset Intaglio process in Nasik and Dewar .

The concept of railway tickets printing have been changed, almost all tickets are being printed by computers today.

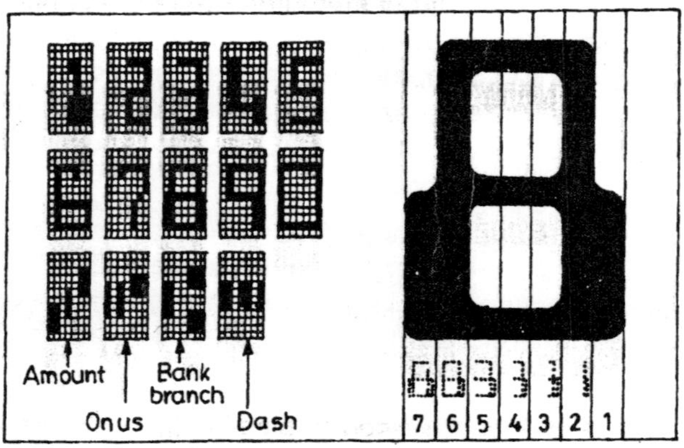

Figure : Magnetic ink character recognition (MICR). The varying thickness in the strokes of the character will result in differing wave forms as the character pass below a reading head. This coded information can then be used computer sorting of cheque and other items.

HOLOGRAM

Holography is a process that creates the illusion of three-dimensionality on a two dimensional surface. Though it was discovered in 1947 by Nobel prize winner physicist Dennis Galor but holograms were limited to the laboratory until the early 1960's development of the laser, a device instrumental in creating a split beam of light.

During this period, holograms largely remained a scientific curiosity, simply because images could only be produced one at a time and recorded only to the emulsion of photographic film in a time consuming process.

Then by the early 1980's technology was developed that enabled holographic film images to be transferred to a metallic plate and subsequently mass-produced onto the mirror -like surface of hot stamp foil. This progression allowed a holographic image to be adhered to any surface that could be foil stampea, creating a new market for foil stampers that has now come into its own.

Today, nearly any image or model can be transformed into holographic foil.

HOLOGRAM TYPES

Nearly anything that can be photographed can be made into a hologram image. However there are different characteristics which will produce different results.

As on today, there are four holographic effects which can be captured in a foil image-

1. Holographic patterned foils - These are available in variety of standard styles. Although not actually three-dimensional, the shimmering, prismatic patterns cut into these foils create an image of depth and a rainbow of shifting color as the angle of view changes.

2.Three-Dimensional holograms - These are created from inanimate objects. Here the 3-D illusion is quite realistic as the object seems to rotate in space as the angle of view changes. This types of hologram also creates the illusion of "looking around" a foreground object to reveal background that was not present is a different angle of view.

3. Multiple plane holograms - 'ultiple plane holograms layer two to ;ur images into separate planes creating the illusion of a 3-D scene. This process is highly versatile in that individual layers can incorporate two-dimensional images or three-dimensional shapes. This allows even flat, 2-D artwork or photo subject to take on 3-D characteristics as the planes will appear to "float" at different depths within the image.

4. Stereograms - These are sophisticated holograms of live moving models that incredibly render the illusion of a brief 3-D movie on a two-dimensional surface. In most cases, a motion picture camera is employed to capture a sequence of movement.

USES

Hologram are truly unique expressions of creativity and a highly effective form of visual communication that demands attention. Today holograms are being used for:-

1 Security - Hologram are being used as anti-duplication devices for documents, credentials checks, transportation passes, even tickets, credit card, and identity cards.

Hologram are often included in plastic cards as identity markers. For printing, the type most commonly used is the embossed holograms.

An embossed hologram is a lamination of several film layers, including a base of stable polyester material and two or more layers of metalised foil containing the printed and embossed images plus a protective film layer of coating.

Tne KLUGE WEE & WFD *(above) are web fed presses ideal for high-speed application of hologram foils to checks, forms, securit decuments and tickets. The* KLUGE EHE, EHD *and VS 2028 (below) are sheetfed presses maximized for application of hologram foils to books packaana abels, collateral and more.*

Metallisation is the deposition of vapourised method onto a plastic base to get a pure, even film of a metal just a few microns in thickness Embossing of the foils is normally from a metal master derived of from the original hologram or 3-D design. It is important to note that embossing means a subtle variation in the surface structure of the metalised films which are only a few microns in thickness.

Superimposition of the printed and embossed film will give the unique 3-D effect obtainable only by embossed holograms. Holograms are very difficult to photograph which makes them ideal for security purposes.

2 **Book covers** - Many publisher are using holograms to add further impact book jackets. The famous National Geographic hologram cover of a pre-historic skull that became the magazine's most requested issues.

3 **Print ads, direct mail** - The volume in increasing hologram foils, can really a product stand out .

4 **Packaging** - Those who need high impact packaging in -an over-crowed market. Holograms are also a preferred method of brand protection.

5 **Sports trading cards** - These are already replete with foil, holograms and stereo; images of sports heroes in motion represent the ultimate in trading card value.

NUMBERING WITH MICR INK ON ROTARY PRESSES

The development of better numbering machines, inks, papers, presses, rol impression media and wrap-around blankets have followed one another.

I. INK AGITATION

Originally recommended and now known to be mandatory that the fountain be equipped with an ink agitator. The fountain should be moderately loaded so agitation is effective, but not overloaded. New ink for each shift or running period seems to help. (A few free -flowing inks do not seem to need an agitator).

II. HEAT FOUNTAINS

To insure proper breakdown of the ink, many MICR web presses are now equipped with heated fountains. This means easier flow and assists in better distribution. Correct temperature should be determined under actual shop conditions.

III. DISTRIBUTION OF INK

A minimum of two driven form rollers should be used. In some cases three have been necessary. Two vibrators with two (or three) distributor rollers, have usually produced the best result. Form rollers of approximately

3" (or larger) diameter and approx. 40 durometer rubber (plus or minus normal industry standards) have been most satisfactory. Form and distributor rollers must be concentric with their bearings with less than .001" run out.

Most ink manufacturers have suitable inks to do acceptable MICR Automatic serial numbering and/or static encoding. It appears that most inks have to be " tailored' to specific speeds, size of press, grade of papers, drying time, plant atmospheric conditions and other factors present.

IV. FORM ROLLERS AND IMPRESSION CYLINDERS

There are no substitutes for parallelism and concentricity throughout the system. Concentricity of all rollers and impression cylinder is a must before setting for parallelism. Use suitable gauges for setting form rollers and the impression to check out the numbering section.

V. CAM SETTING

All cams must be accurately set before setting rollers or printing . This will insure that the numbering equipment will operate satisfactorily. When cams are set correctly, eliminate possible broken operating arms or cam damage. By so doing *help* eliminate downtime and waste. Proper setting of cams is especially important, when doing check digit numbering because of critical angular stroke required on these machines.

VI. SETTING FORM ROLLERS

Properly used, this gauge will enable to control the amount of ink laid onto the numbering wheels and eliminate over-inking. It should add life to both numbering machines and the form rollers by preventing " digging in" by the characters on the wheels. Any out-of -parallelism that may exist between the form rollers and the numbering cylinders can be readily seen and adjustments made to correct . Rollers should be set one at a time- use convenient adjusting and/ or micrometer screws if at all possible.

VII. IMPRESSION CYLINDERS

Tne wrap- around available today encompass a wide range of materials including Mylar, Nylon, Teflon, Polyurethane, as well as rubber coated tympan. If the impression cylinder is undercut more than .012 " (.02° " seems to be the greatest advisable), the hardest available tympan (90 durometer or higher) seems advisable. Watch all tympans for debossment markings and shift tympan or cylinder if necessary.

VIII. SETTING IMPRESSION

Be assured the impression cylinder and the form rollers are parallel to the numbering shaft. Also, the debossment possibilities will be eliminated. Control impression and avoid damage to the impression cylinder, wrap-around and/or the numbering equipment. Numbering machines should NEVER be used as gauges for setting form rollers or impression.

IX. SPEEDS

The real question in satisfactory MICR Consecutive Numbering and/ or Encoding on the numbering shaft is speed. What speed is best ? Different things happen at different web speeds. Eccentricity or whip, and vibration develop at certain speeds due to a wide variety of causes.

1. A keyed numbering shaft is one possibility because of the unbalanced condition it creates- another opposing. Keyway- or none at all may cure the condition.
2. Numbering machines unequally spaced around the circumference is another possible cause - counter - balances may be required.
3. Lack of adequate bearing (or bearings of inferior quality) is another cause. Install larger, double row or more accurate bearing.
4. Extremely wide span between the frames of the press may require center supports and/or inboard bearings for the numbering shaft.
5. Undersize numbering shafts are another problem- make as large as practical and fit with precision bearings.
6. Lack of synchronized surface speeds of numbering and impression cylinders can cause slurred numbering. Be sure all speeds are identical. Check for gear backlash. Setting up may seem tedious but if done carefuily will pay dividends and help insure minimum downtime if instructions are followed.

Keep records of form roller and impression settings. These settings may vary, depending upon paper used and press speeds.

Pre-ink all numbering machines by changing wheels with a pick and running the press. Do not set with fingers-body oil prevent uniform ink coverage of figures. Use pre-inker equipment for some models that will ink all the automatic wheels on ten revolutions of your press.

Oil carefully. The correct amount is vital- too much and you have inking problems; too little and have excess wear plus functional problems.

Careful inspection of numbering shafts, holding rings and numbering machines should be done at frequent intervals. All out of tolerance condition should be corrected promptly.

Proper care and storage will increase the period of accurate life for all numbering machines, but MICR equipment must be well cared, for.

Many check printers have found that average numbering machine life has increased several fold when each press has its own complement of machines, and the interchange of parts within machines or between presses is kept to an absolute minimum.

The type of numbering equipment chosen, whether flat base male dovetail on adjustable or fixed position rings, or curved base female dovetail machines mounted directly on the shaft or on an adjustable ring, depends on many factors and could be the subject of another publication. However, the fewest elements in an assembly usually produces the best results.

29

Bar Coding

WHAT IT IS

Bar code is a display of parallel rectangular shaped bars and spaces, arranged in a specific predetermined pattern to represent a corresponding figure, letter of symbol, Bar Code may be likened to a current printed version of the old Morse code.

HOW IT WORKS

Bar Code is a language. It can be read by a light source (scanner) and translated by a program (reader) which will turn the symbols (or language) into usable, useful data.

The data can be harnessed to relay a variety of important information to help any organization regardless of size.

The dynamics and sophistication of the programs turn the data into a rainbow of helpful categories, which result in improved operations and tremendous for the organization using the Bar Code.

SOME MAJOR BAR CODES

6 9 8 7 6 5 4 3 2 1 0 5

NIXDORF CODE

9 8 7 6 5 4 3 2 1 0

MONARCH CODE

0 9 8 7 6 5 4 3 2 1

IBM CODE

IATA CODE

IDENTICON CODE

* 1 2 3 4 5 6 7 8 9 0 *

CODE 39

WHY IT'S USED

There are many reasons why a company selects Bar Coding to improve data entry into its computer network.

- Speed - Reliability
- Control - Durability
- Accuracy - Portability
- Non-Control Reading

Of these, Speed, Control and Accuracy seem to be pre-eminent for conversion to Bar Code in the Graphic Arts Industry. In almost every case, applications have markedly improved the collection of important information and vital data while creating areas of significant savings.

Bar Code is a true phenomenon, bringing valuable new procedures, that translate into important advantages for those organizations that use it.

Where it's Used

Listed below are some of the current Bar Code markets. Keep in mind that there are new applications every day of the week. The dynamics of these and other markets change so quickly and so greatly that even though we can produce such a list, there will be many areas that are in transition and others where unique software can alter aspects of many of the new markets. There is new potential in old market, and new markets surface daily.

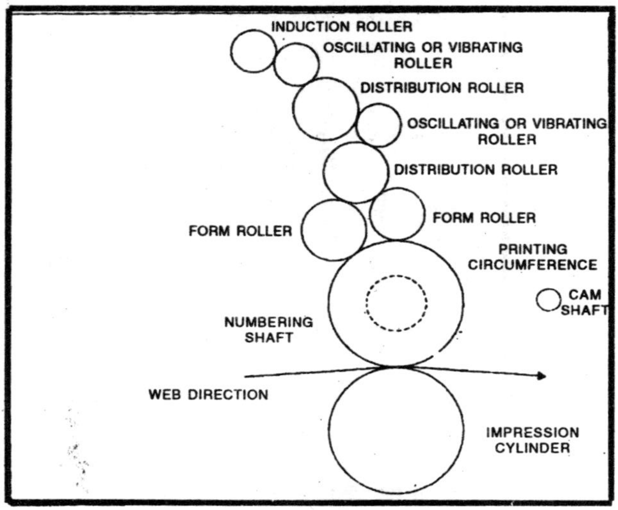

Fig. : BAR CODE READY

TYPE OF BAR CODE AND DESCRIPTION :

Monarch STD Code

Numeric Code. 7-Elements (2 Wide, 5 Narrow) Usually (1) Wide Space (1) Wide Bar, 18-Different Widths.

Monarch (Codebar) AKA: Rational

Numeric Code. 7-Elements (2 Wide, 5 Narrow) Also, (1) Wide space and (1) Bar, 2- different Widths.

Code 39 AKA: Intermec

Alpha/Numeric Code. 9-Elements (3) Always Wide-either bars or spaces. Built in check digit.

Code 49 from : Intermec

Unique New Code. Electronically Reproduced. Designed to carry the most data in the smallest space. Built-in check-digit on both the rows and columns for secure data health industry.

Code 16k From : Laser Lite systems

Similar to code 49, But not as complex. Combines UPC and IBM 128, Check-digit only on each row. Publication Field.

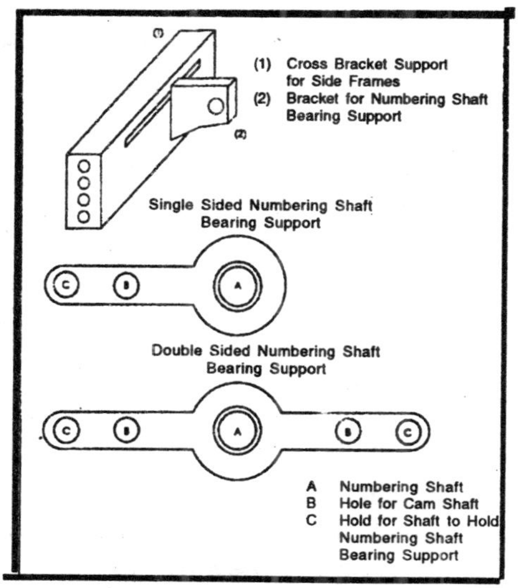

Fig. : BEARING SUPPORT

HOW BAR CODES ARE STRUCTURED

Bar Code symbology (coding, language) falls into (2) different kinds of structures. Currently, all Bar Codes can be called either Discrete or Continuous Codes.

The Continuous Code uses the space between the bars of the printed code carry data. It is normally printed by a direct imaging system and cannot be printed using impact numbering devices (except with a recent Leibinger innovation, which makes this possible)

Examples of Continuous Codes : UPS/EAN, 2 of 5 interleave, Plessy, etc.

As a printer, you will be interested mainly in Discrete Bar Code. These codes do not use the ICG (Intercharacter Gap) as part of their symbology. This allows you greater flexibility in the method selected to print the Bar Code.

Example of Discrete Codes : Intermec Code 39, Identicon (straight) 2 of 5 non-interleave, Monarch —Codabar, IBM, Nixdorf, etc.

CONTINUOUS CODES

The most commonly seen and used symbology is the UPC/EAN (Universal Product Code/European Article Number). This is a fixed code which provides information to an inventory base system, It describes, the product (manufacture, weight, volume, size, etc.). You see the UPC/EAN codes everywhere from grocery to toy stores. The main function is inventory control. The code is fixed for that product and is printed by plate. Since it is a Continuous Code, impact and sequential numbering are not used.

A popular continuous code is the 2 of 5 Interleave (1-2/5). Its main advantage is its ability to conserve space for a given series of sequential numbers. It is a numeric code only (no alpha). Since it is a continuous code by definition, the spaces hold data. It requires approximately 40% less area to print the length of the 1-2/5 than its counterpart, the 2 of 5 non-interleave.

The market area for the 1-2/5 is getting smaller each day, with the book publishing and library industries, the principal users.

Normally, the 1-2/5 is a continuous code, which can only be reproduced using non-impact direct imaging systems. The only deviation is a recent creation of Leibinger's—an electronically—controlled impact numbering system. With this installation, it is possible to number the 1-2/5 Bar Code via impact at speeds in excess of 650 SFM, going one wide on a web press.

DISCRETE BAR CODES

These codes are of special interest to all printers.

INTERMEC CODE 39— is the most popular sequential code used throughout the world today. It is a simple, self contained code that has built-in check digits to make de-coding easier. It is an alpha/numeric code (where most other discrete codes are not). It has 9 elements (6-narrow and 3-wide either bars or spaces). Discrete, by definition, allows for intercharacter gap which lends itself to an impact numbering machine, whether it be for press, collator, letterpress, encoders, etc.

DENSITY vs CPI (Characters Per Inch) as used in describing codes.

DENSITY can have a large variety of characters per inch, depending upon whether or not it is a low, medium or high density Bar Code. For example, a medium density Code 39 could range anywhere from 7.34 CPI to 2.79 CPI.

CPI (Character Per Inch) is a more refined description. It is the combination of all the narrow elements and the wide elements, plus ICG (Intercharacter Gap) of one character set divided into 1.00 inch. This will yield CPI, Example : Code 39, a medium density code, with a density of 5.4 can in reality (once you add up all the narrow, wide and ICG of .053') end up with a CPI of 4.6.

MONARCH CODABAR - is in wide use today, especially in a few specific markets. It is second to Code 39 since the Codabar is only a numeric code. However, it only has 7 elements (usually 2 are wide and 5 are narrow). This, of course, would make it slightly shorter in the length of the printed image (if comparing to Code 39)/ICG, as in the case of Code 39, presents hardly any problems when it comes to the printed image. Since the scanner manufacturing specifications may dictate certain restrictions for the ICG, it

Figure : DISCRETE BAR CODE

does become an issue, which needs to be addressed regardless of the discrete code that is employed.

2 OF 5 NON-INTERLEAVE (Straight)— is used less frequently than the Code or Codabar. It is a numeric code only and has 5 elements (usually 2-wide, 3-narrow). It does have a shorter version of the stop/start code but still has the impression length as great as *40%* more than the 2 of 5 interleave. Only advantage is, it is a shorter impression length code on the printed image, if you were to compare the 2 of 5 non-interleave to either the Code 39 or the Codabar.

HOW DISCRETE BAR CODES ARE MANUFACTURED

For Discrete Codes, there are several possible methods of printing Bar Code sequential numbers through impact devices.

LETTER PRESS :

This is the oldest, most reliable form of numbering. Because of the land heights used in Bar Codes, most machines are over type-high. NOTE : you should select a letter press unit for which you can alter the bed to accept an over type-high letter press numbering machine. (Example : by definition, the moving bed of a cylinder letter press cannot be used for the Bar Code that is over type-high.)

OFF-LINE RIBBON ENCODERS

The ribbon encoder will allow production of various size continuous documents, You only need 1 or 2 Bar Code machines. NOTE : you should select a ribbon encoder that will allow for a larger land height to print the Bar Code image and especially allow for the proper ribbon transfer of the large Bar Code image.

Figure : UPC/EAN CONTINOUS CODES

WEB PRESSES AND COLLATORS :

Even though there are obvious differences in the printing or finishing equipment selected, there are many common attributes that must be in place or modified, to print a good Bar Code image with the impact numbering devices. What follows are general guidelines for the proper preparation of your numbering tower in getting your press or collator "Bar Code Ready."

"BAR CODE READY"

Bard Code numbering on your press or collator is as simple as numbering determine will Gothic tower. It is the initial preparation paratiok speed. This is a diagram of a typical numbering tower to which you can refer in order to make our explanation of " BAR CODE READY" easry to understand.

"BAR CODE READY"

The quality of your equipment is critical. Our experience with our own units is excellent, and we urge you to make certain that all your numbering accessories are made by Leibinger. It will prevent more than half the problems that could occur. What follows are suggestions based on years of in-field experience. When these guidelines were used in conjnection with precision Leibinger equipment, the result was always trouble-free operation and perfect Bar Code imaging.

THE INK TRAIN

Usually we ask that the numbering tower is of MICR quality, ductor roller at least 50 durometer on a Shore scale of (A), minimum of (2) each distribution and oscillation or vibration rollers. This helps break down and smooth out the letter press ink .

You should have no less than (2) form rollers with a durometer of 35 on a Shore scale of (A). They should be rubber covered at least 3/8" to 1/2" thick and with a run out as close to .0015" as possible.

If you leave the specifications up to the roller manufacturer or to the OEM's there is a chance that you will end up with something less than the desired specifications. Softer than 30 durometer would put more ink on the machines, resulting in slower speeds and more frequent cleaning. Harder than 40 durometer would not leave enough ink on the steel wheels, resulting in light print contrast or voids, etc.

NUMBER SHAFT :

The chart below shows the maximum shaft diameter for a given circumference, the total print or land height and desired number of machines around the shaft. It will provide you with general guidelines.

NOTE : These are only general rules but there are other possibilities.

FOR 17" CIRC.

1. 1.500' NUMBERING SHAFT (2) AROUND IS .520" Max.
2. 1,500" NUMBERING SHAFT (4) AROUND IS .400' Max.
3. 1,750" NUMBERING SHAFT (2) AROUND IS .400" Max.
4. 1.750" NUMBERING SHAFT (4) AROUND IS. 380: Max.

ROTARY NUMBERING HEADS, STRAIGHT, WITH LAND HEIGHT OF .600' OR BETTER, WHEN MOUNTING DOWN THROUGH CASE WALL.

1. 21" MAX. 3 AROUND NUMBERING SHAFT DIA. 50 MM
2. 22" MAX. 4 AROUND NUMBERING SHAFT DIA. 60 MM
3. 24" MAX. 4 AROUND NUMBERING SHAFT DIA. 60 MM
4. 25.5" MAX. 5 AROUND NUMBERING SHAFT DIA. 60 MM
5. 28" MAX. 5 AROUND NUMBERING SHAFT DIA. 80 MM

ROTARY NUMBERING HEAD, STRAIGHT, WITH LAND HEIGHT OF AT LEAST .500"

1. 21" MAX. 5 AROUND NUMBERING SHAFT DIA. 50 MM
2. 22" MAX. 6 AROUND NUMBERING SHAFT DIA. 60 MM
3. 24" MAX. 6 AROUND NUMBERING SHAFT DIA. 60 MM
4. 25.5" MAX. 6 AROUND NUMBERING SHAFT DIA. 60 MM
5. 28" MAX. 7 AROUND NUMBERING SHAFT DIA. 80 MM

BEARING SUPPORT :

On applications that are 2 and 3 streams wide or if you should have a wider press from side frame to side frame, should recommend one or more numbering shaft supports. This will help to increase your SFM and reduce numbering machine shaft deflection. Since every press is slightly different in configuration or design, the supports for the numbering shaft may also be slightly different.

Please note : to reach different form layouts it may be most useful to have the support made only 1-3/8" or less in total width. Furthermore, the bracket should be flexible enough to slide into different positions, once the bearing shaft support(s) are in set. You should be able to lock down the shaft support without placing pressure against the numbering shaft, either in an up or down direction.

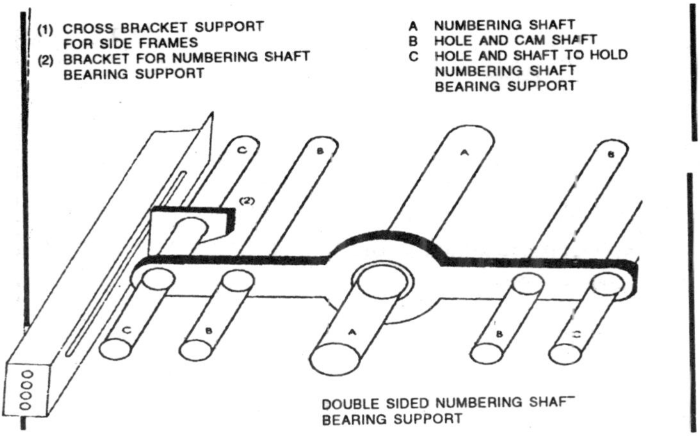

Figure : BEARING SUPPORT

IMPRESSION CYLINDER "BEAR-TO-BEAR" :

The best way to control the paper you are printing is "Bear-to-Bear". This, of course, is a more complex method and most forms printers stay away from it. "Bear-to-Bear' means that each time you change the paper weight or texture you may have to change the blanket material as well as the thickness of the blanket packing material, with which you wrap the impression cylinder. It also means that the undercut on the cylinder has to be great enough to be "Bear-to-Bear and while using different thicknesses of blanket material, you may have trouble printing other products. For example: to accommodate Bar Code products, you may find in going "Bear-to-Bear" and while using different thicknesses of blanket materials, you may have trouble printing other products. For example : to accommodate Bar Code products, you may find in going "Bear-to-Bear, the undercut should be .40" and it varies depending on the paper weights. However, MICR would usually require an undercut of .010" to .020".

Unless you have a designated product, should not recommend the "Bear-to-Bear" technique because of the restriction it places on other product groups.

IMPRESSION CYLINDER

Impression cylinder to accept rubber, sticky-backed blanket materials vs. Impression cylinder accepting mylar blanket using reel rod lock up.

We have found that the best undercuts for bonded papers are usually around .010" to .020" using mylar blanket material to cover the impression cylinder. This is also more desirable of MICR check production.

The best undercut of NCR papers are usually around .035" to .037" using a polyfibron blanket material by W.R. Grace Co. and a thin cover of mylar to prevent the blanket's becoming contaminated from the chemicals in the paper.

From experience, we have found that if you must make a choice of a .0 10" to .020" undercut or an .035" to .037" and you do run check production, you should select the smaller undercut of .010'/.020". This will allow both Bar Code and MICR check production.

BEARING SUPPORT

(1) CROSS BRACKET SUPPORT FOR SIDE FRAMES

(2) BRACKET FOR NUMBERING SHAFT BEARING SUPPORT

SINGLE SIDED NUMBERING SHAFT BEARING SUPPORT

DOUBLE SIDED NUMBERING SHAFT BEARING SUPPORT

A. NUMBERING SHAFT

B. HOLE FOR CAM SHAFT

C. HOLE FOR SHAFT TC HOLD NUMBERING SHAFT BEARING SUPPORT

BEARING SUPPORT

(1) CROSS BRACKET SUPPORT FOR SIDE FRAMES

(2) BRACKET FOR NUMBERING SHAFT BEARING SUPPORT

A. NUMBERING SHAFT

B. HOLE AND CAM SHAFT

C. HOLE AND SHAFT TO HOLD NUMBERING SHAFT BEARING SUPPORT

DOUBLE SIDED NUMBERING SHAFT BEARING SUPPORT

INKS :

We have spent a great deal of time working with ink manufacturers, for we have found that ink composition is very important for a successful Bar Code numbering job. 2 or 3 inks that you should have on hand for the product (s) you are going to manufacture. We have found that certain inks react differently to the paper substrate on which you are printing. Many forms manufacturers do not pay enough attention to this fact and, consequently, produce less than perfect images.

Note : Never use MICR inks with Bar Code heads. The metal particles in the ink will act like sandpaper and cause the wheels to wear prematurely.

"USER FRIENDLY"

Libinger's goal is to manufacture the highest quality impact numbering devices in all areas of numbering and Bar Coding. They have developed a number of innovations, that enhance the performance of the Leibinger Bar Code machines, raising their performance far above anything else in the field. If the term "user friendly" can be used for impact numbering devices, it fits Leibinger Bar Code numbering machines.

Leibinger is the first organization in the field, to create an educational program to make Bar Code printers more aware of their abilities.

As part of its educational process, Leibinger has made professional recommendations, based on field experience, covering numbering tower modifications, cylinder packing materials and inks.

All of this is directed toward insuring swift installations that will allow the end user of Leibinger Bar Code impact numbering devices to be at production speed swiftly without any unforeseen difficulties.

INNOVATIONS :

Always at the forefront of the industry, Leibinger's innovative advances in bar code and numbering combine to make the lot of the printer easier, more productive and more profitable.

The LM 2000 NC Modulus System is a prime example! It was shown to some 150 printers - managers, marketers and technicians - at the Leibinger Bar Code Seminar in Dallas/Ft. Worth the end of February 1990. It was a major component of Leibinger's display at DRUPA 1990. It is the State of the Art in the field, and should prove of great value in the production of checks and related security-conscious documents.

30
Ink-Jet Printing

Ink-jet printing produces an image by directing individual tiny drops of ink, which are given an electrical charge, from an orifice (opening) through a smaller air gap to a printing surface (figure 30.1). This is a nonimpact process and can be used to print on a variety of substrates, including corrugated board, plastics, fabric, and paper. Ink-jet printing was originally used in personalizing direct-mail pieces. This has been traditionally known as direct mail merging; it occurs when a computer inserts database information into personalized printed pieces. This process has become extremely sophisticated and is now referred to as **variable data printing (VDP).** With VDP it is possible to create completely customized documents for individual recipients.

All ink-jet sustem processes rely on computer input to form a dot-matrix image that is made up of individual drops of ink. The major difference between ink-jet systems is whether the ink-spray is *continuous or drop-on-demand.*

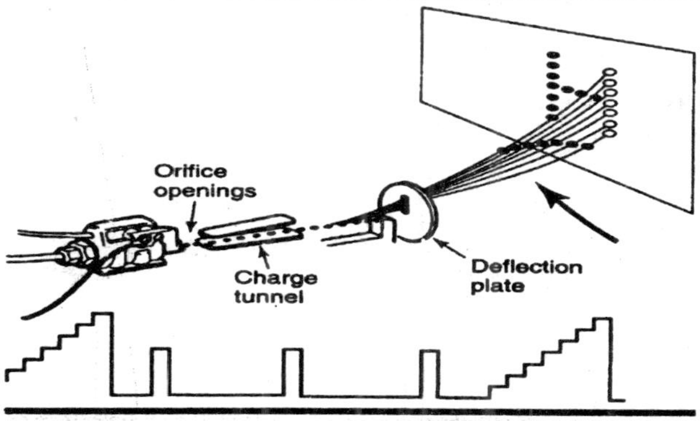

Figure : 30.1 Continuous ink-jet printing

Continuous Ink-jet Technology

In continuous ink-jet technology, each ink orifice sprays ink drops toward the printing surface continuously. When the ink drops leave the orifice, they are subjected to electric charges that vary in strength based on digitized computer information. The electrical charges of various strengths deflect the ink drops from a straight path and move them in the appropriate directions to produce the desired dot-matrix image. Drops not needed in the matrix are given a charge that deflects them into the ink recycling system.

Commonly used in the packaging industry, continuos ink-jet technology, pioneered by Videojet Systems International, is able to print information, such as date and time, serialized numbers, and characters in any directions, in single or numerous lines and in a variety of sizes and can count how many times they print at speeds up to 1,200 feet per minute. This type of ink-jet printing works by continuously emitting a small stream of in through a single ink orifice. When liquid is forced through a small orifice at high speed,it tends to break into a spray containing drops of various sizes. To control drop size and direction, the ink is given a frequency by a vibrating crystal. This frequency regulates the size, shape, and spacing of the emitted drops.

A variation on continuous ink-jet technology uses an **array,** rather than a single orifice (figure 30.2). Lie regular continuous ink-jet technology, this is a noncontact form of printing in which the ink is given a frequency by a viabrating crystal. In an array, 100 to 200 nozzles can generate over 15 million droplets of ink per second, which allows for high-resolution images and flexibility in image creation (figure 30.3). Over 50,000 characters per second can be produced with an orifice array; only about 13,000 characters per second can be produced with a single orifice system.

Drop-On-Demand Systems

In a **drop-on-demand** system the printer produces a drop from the ink orifice only when the drop is needed to form the matrix image. The ink drops are given a frequency in a manner similar to that used for continuous drops. There are two main approaches to drop-on-demand ink-jet technology: thermal (bubble) and piezoelectric technology (figure 30.4).

Thermal Ink-jet Technology

Thermal ink-jet or bubble-jet technology uses heat to force small drops of ink out at high speed onto the substrate. This is accomplished by applying an electrical current to small resistors built into the nozzle. The ink is heated within the nozzle, causing it to vaporize locally and then form a bubble, which provides the force to eject the ink from the nozzle.

Piezoelectric Ink-jet Technology

Piezoelectric ink jet uses electrical impulses based on digitized information to cause individual drops to be squeezed out of the ink orifice and propelled toward the printing surface. This occurs when an electrical pulse is applied to a piezo crystal, causing it to contract slightly and force the ink onto the substrate. Drop-on-demand units have the advantage of being less complicated than continuous-spray systems, because they do not require a dot deflection system to place needed dots or remove unwanted drops. They are also more economical, because ink is provided only where it is needed to form the image (figure 30.2)

Figure 30.2 A continuous ink-jet imaging system utilizing array technology.
Courtesy of Marconi Data systems, Inc.

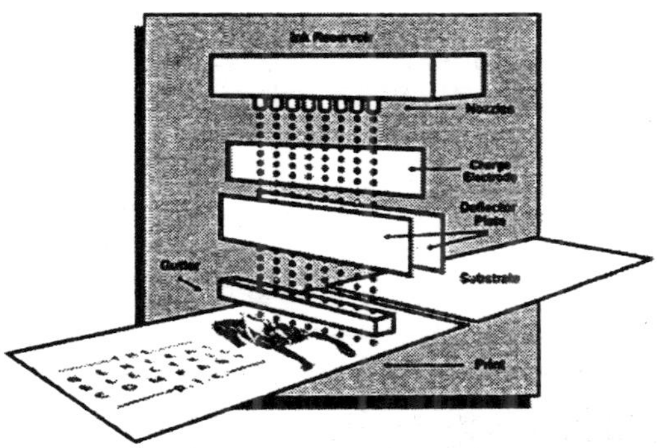

Figure 30.3 Continuous ink-jet array.
This technology can produce high-resolution images at line speeds of up to 1,200 feet per minute.
Courtesy of Marconi Data Systems, Inc.

Ink-jet Applications

Continuous ink-jet technology has proved invaluable in packaging by providing bar codes, manufacturer information, postal codes, and the host of tracking symbols needed to maintain complex databases. It is also used in labeling and for high-volume direct-mail imprinting. Piezoelectric has become the choice for high-end proofers in the printing work-flow, as in the Afga Sherpa proofer series, which provides large-format, six-color contract quality proofs (figure 30.6). With a wider color gamut than a traditional press, acceptable contract color is achievable with the correct color profiles and calibration. And, as fewer customers demand to see halftone dots in their proofs, this technology will most likely continue to replace high-end, expensive halftone dot proofers.

Drop-on demand has grown from a slow process able to print only low-resolution images to a high-quality, high-resolution image technology capable of providing acceptable proofs to customers. This has also been helped by

The difference between thermal ink-jet printers and piezoelectric ink-jet printers is in how the ink is expelled. In a thermal ink-jet printer, resistors built into the nozzle heat up when an electrical charge is applied, which heats up ink in the reservoir, causing a bubble to form. The bubble displaces the ink and forces it onto the paper. In a piezoelectric ink-jet printer, a small electrical current makes an integrated piezo crystal contract, forcing ink from the nozzle.

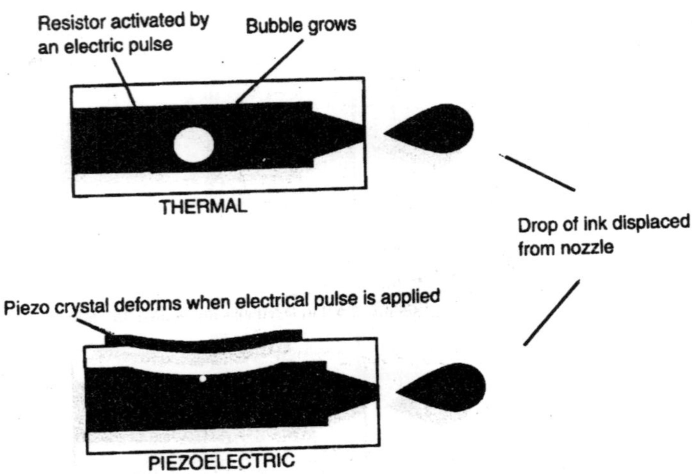

Figure 30.4 Comparison of thermal and piezoelectric ink-jet technology.

Figure 30.5 Drop-on-demand ink-jet printers come in a variety of formats.
Courtesy of Epson America, Inc.

Figure 30.6 A large format ink-jet proofer.
Courtesy of Agfa Corporation.

the development of color profiles that are part of current operating systems (or can be added on) and better color management. Ink-jet printers have become relatively inexpensive and are increasingly being used by customers prior to submitting their work to the printer. In addition, low-end ink-jet printers are invariably bundled with PC purchases now, further ensuring the continued use of this technology.

Ink-jet printers are used extensively in check printing operations. Large corporations produce thousands of pay-roll checks every week. Each has a different name and dollar amount. Ink-jet printing is ideally suited to this application. Check blanks with the company name and bank identification information are preprinted as continuous, perforated rolls or sheets by offset printing. The variable information (name, address, and amount) is added by running the preprinted checks through an ink-jet printer.

One final advantage of ink-jet printing is that the process is nonimpact, which means the image carrier is not forced against the printing substrate. This feature makes it possible to print with ink jet on almost any surface, regardless of surface texture, shape, or pressure resistance. Ink-jet printing can place an image on a plastic container or bubble package (or even an egg yolk) as easily and as quickly as it can on paper, a textured surface such as sandpaper, or a curved surface such as a pill or capsule.

31

Digital Printing

Digital printing has revolutionized the printing industry. It has changed the production cycle and drastically shortened the turnaround time of a job. There are two basic approaches to digital printing: direct imaging or hybrid presses and digital toner-based printing.

Direct Imaging Presses

Direct imaging presses are presses with a digital front end and a conventional ink-based printing unit. Plates are imaged on press from digital data received by networking, essentially incorporating computer-to-plate (C-T-P) unit within the press. There are several advantages to this type of press over a conventional press:

- ★ Because the plate cassette is mounted on the press prior to imaging, there are almost no registration problems.
- ★ The system significantly reduces press make-ready time.
- ★ The plate often uses waterless technology, which is more stable and allows greater ink densities than water-based lithography.
- ★ Material costs are competitive with traditional systems, but the process saves as much as 60 per cent in labor costs.

Plate Exposure

Heidelberg is one of the leaders in direct imaging (DI) presses. The Heidelberg QuickMaster DI is a 13 by 19 inch format press that combines the best of direct imaging technology with offset technology (figure 31.1). It utilizes polyester plates created by Presstek that are imaged through an **ablation** process. The top silicon layer is vaporized by lasers, which expose the ink-receptive base layer. Plates are fed off a supply roll within the cylinder with material sufficient for 35 consecutive print jobs. All four plates are imaged simultaneously and in perfect register in six minutes at a resolution of 1,270 dpi. The plates are also able to be imaged at 2,540 dpi, sufficient for a 200-lpi screen. The waterless ink technology used supports high ink densities and is equipped with 12 ink rollers to provide a deep and even inking of solid areas (figure 31.1).

Figure 31.1 Quickmaster DI.

A four-color digital imaging press. Courtesy of Heidelberg USA, Inc.

The new Speedmaster 74 DI press, a 20 by 29 inch format (four up) DI press, is equipped with automatic plate loading, with imaging the Presste PEARLgold™ plates at 2,400 dpi in under three and a half minutes (figure

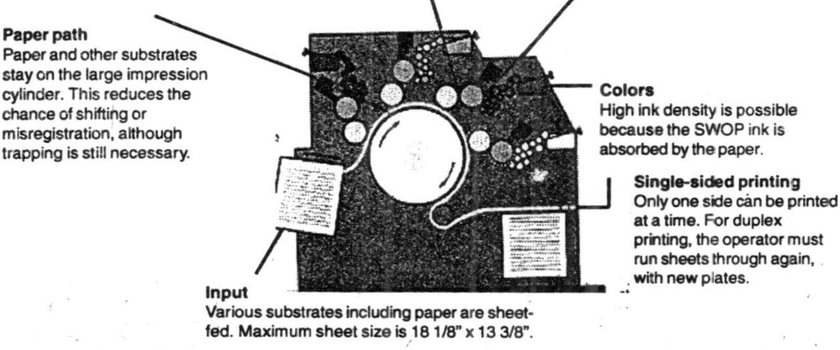

Figure 31.2 Internal diagram of the Quickmaster DI.
Courtesy of Agfa Corporation.

31.3). It uses Creo Squarespot® thermal imaging technology, which allows simultaneous imaging of plates in up to six printing units.

The main limitations of a direct imaging press is that duplex printing can be done only by the operator running the sheets through again with new plates and that once a plate is etched it cannot be altered. Although this technology does not allow for true variable data printing (VDP), certain types of versioning can be done by printing color shells that can then run through a digital printer for personalization. Direct imaging presses are gaining in acceptance due to the fact that the back end is truly an offset press and provides quality output measurable against completely conventional presses, yet with a far faster turn-around time and less manual labor. There are predictions that by the year 2005 purchase of direct imaging presses may outnumber those of conventional presses.

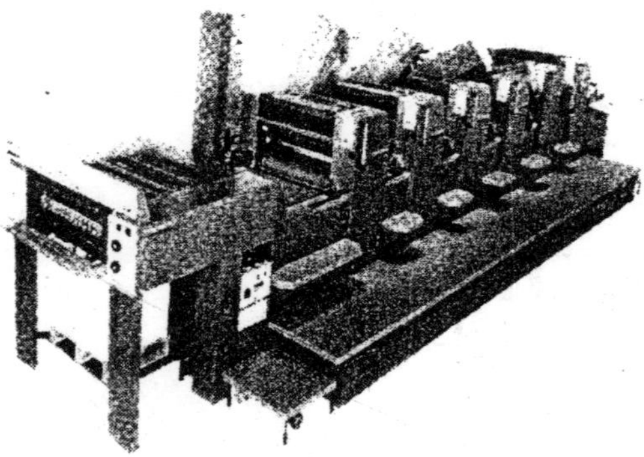

Figure 31.3 A high-end digital imaging press capable of 15,000 sheets per hour. Courtesy of Heidelberg USA Inc.

Digital Toner- Based Printing

Digital toner-based printing has truly caused a revolution in printing. Originally introduced in 1993, it has allowed "full-color pages to be printed directly or remotely from digital page data with the ability to vary the information at every cylinder revolution." (Landa-Indigo) Four-color jobs can now be turned around in hours, rather than days or weeks, and print runs of one are now possible.

Press Configuration

Different vendors have different solutions, but the two basic digital print imaging technologies used are **laser electrophotographic** and **LED electrophotographic**. Digital printers from companies such as Xerox and

Canon use a digitally driven laser beam that reflects off a rotating prism to charge parts of the imaging drum. Toner powder is then attracted to the charged parts and subsequently transferred to the paper. Indigo, a manufacturer of digital printing presses, uses a liquid in called ElectroInk rather than a dry toner. Charged ink is injected at the plate, adhering to any area with an opposite charge. This plate transfers it to the substrate. When the ElectroInk contacts the substrate, it hardens instantly and peels from the blanket completely. This offers two major advatages. First, because the ink dries upon contact with the paper, dimensional stability is guaranteed. Most offset inks dry on the paper by absorption or **aerial oxidation**, which introduces moisture into sheet or leaves a wet ink layer. When dimensional stability is assured, more accurate color is possible.

The second primary advatage, which is also true for LED (light-emitting diode) electrophotographic imaging, is that no residual in remains on the blanket after each rotation. In contrast, conventional ink splits between the paper and blanket and leaves a residue. The lack of residual ink also helps eliminate dot gain. Because unused ink does not build up on the blanket, halftone dot size is assured.

Current imaging technology in the Indigo UltraStream™ uses 12 laser beams to image the photo-imaging plate and can operate at 240 feet per minute. Whereas the printing engine in conventional offset prints just one color, the UltraStream print engine prints all colors, with multiple engines multiplying throughout. The Ultrastream is a cut-sheet seven-color press capable of 180 lpi.

Xeion, one of the leading manufacturers of digital printing presses, both under their name and as on original equipment manufacturer for other vendors, has sheet-fed and web-fed press products. Xeion utilizes an LED-array-based dry toner electrophotography .The LED array exposes the charged photoconductor drum to different levels of light, creating conductive areas on the drum. Toner particles adhere to the charged area, forming the image that will be transferred to the drum. Their sheet-fed solution, the CSP 320 D, uses one-pass-duplex technology that prints both sides of the substrate simultaneously, resulting in accurate back-to-front registration. Sheet size ranges from 8.5 by 11 inches up to 12.6 by 18.5 inches on a variety of standard substrates, with speeds of up to 960 A4 (8.5 by 11 inches) full-color duplex impressions an hour. The CSP 320 D can achieve image quality equivalent to 2,400 dpi with a variety of screenings up to 175 lpi (figure 31.5).

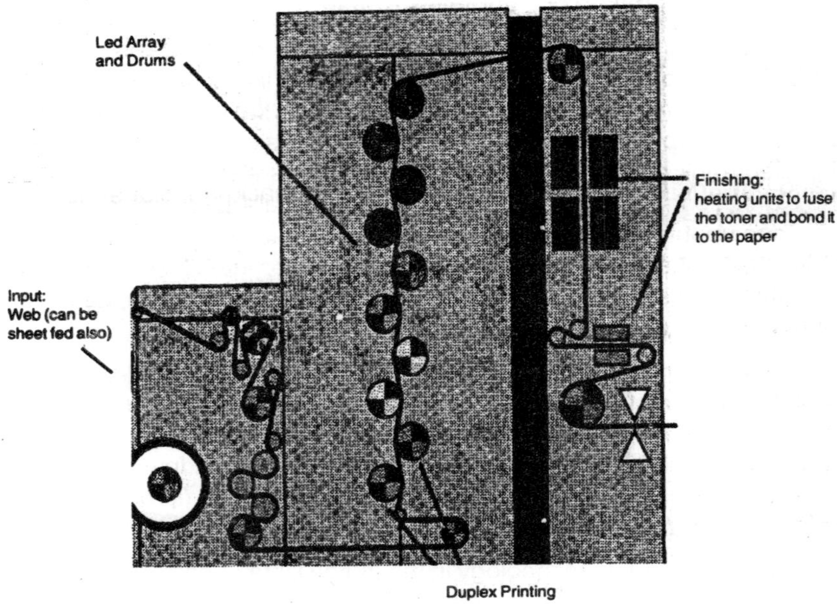

Figure 31.5 Schematic of a Xeikon digital toner-based press.
This system uses dry toner electrophotography.
Courtesy of Xeikon America, Inc.

Implifications of Direct Digital Presses

Before Gutenberg, there was **demand-based publishing.** In demand-based publishing, if people wanted and could afford books they looked for ones to purchase or paid scribes to copy them. Gutenberg created **market-based publishing.** Gutenberg's invention created to situation where the printer decided what to print, prepared many copies, and then sought to convince the market it wanted the product.

For the past 500 years there have been few opportunities for individuals to participate in a demand-based information environment. Someone else decided what to reproduce and distribute.

The introduction of DTP in 1984 was the first step toward individual control and access to focused information. The reality of a true digital press in 1993 took the next step toward these goals, and will continue to transform both the graphic arts industry and our daily lives.

Personalized Magazines and Other Information Pieces

Every home receives hundreds of pounds of magazines and newspapers each year. Much of this material goes unread. There are probably parts of the daily newspaper you do not read. Even special interest magazines, such as *Mac-World, sports Illustrated, or the Worm Runner's Journal,* contain articles of little or no interest to the subscriber. All of these unread, and unwanted, pages come as part of a total package, clutter our homes, and must be discarded.

Digital presses can be used to personalize every publication that comes to your home. Advertisers that sell products in your interest areas can purchase advertising space in your exclusive magazine or newspaper.

In the office community database management can be used to create one-of-a-kind reports, reference manuals, and information books for every employee.

Elimination of Traditional Bookstores

Bookstores sell books. They are everywhere when you walk in the door. Book lovers revel in the environment. However, even the most ardent book lover does not need 200 copies of the same book—one copy suffices.

Large bookstores are large because they need to shelve many copies of every title. If you do not come in and make a purchase then that storage space is wasted. Bookstores of the future may shelve only one sample copy of a work and print a fresh copy on a digital press when you have made your decision to purchase. Personalized information can be included in the copy in the process.

No Out-of-Print Books

Publishers only print and promote books that they can sell. If, after the first year of publication, the buying population of a book dwindles, then the publisher stops printing and promoting that book. The book then becomes out of print and is only available through libraries, used bookstores, or from a friend.

With digital presses a book will never go out of print. Computer archives can store digital information and anyone can order one or many copies of anybook, no matter its current status.

Mass Marketing Replaced by Personalized Marketing

Mass marketing involves sending millions of pieces of mail to millions of individuals with the hope of selling an idea, service, or product to one or two per cent of those individuals. Organizations like the publisher's Clearinghouse have used ink-jet printing to customize direct mail solicitations with considerable success because individuals are more likely to open and read items that contain their names.

Indigo's founder, Benny Landa, offered an image of a customer who wishes to promote two or three dozen out of several thousand company products. However, there is no point to offering lawnmowers to apartment dwellers or diapers to pensioners. The smart company wants to customize individual color fliers so that each consumer sees only products he or she might actually buy. With a direct digital press, every consumer can receive a different catalog. It is now possible to deliver custom catalogs containing items the seller nows matches each buyer's personal interest.

In this example, the customer could program the digital press to deliver custom catalogs containing items it knows match your personal interests. Image databases could also be used to place your photograph next to the motorcycle, car, yatcht, or pogo stick the customer wants you to buy.

Virtual Elimination of Warehousing

Traditional printing relies upon economies of scale. Companies order tens of thousands of copies of books, brochures, technical booklets, reference sheets, promotional items, and training manuals and store them for future use. The cost of warehouse space can be considerable, however.

All print storage space can be replaced by a digital press, computer memory, and a lot of blank paper. When a document is needed it can be simply ordered, printed, and delivered. Demand printing becomes a reality in this way.

Low-cost Access to Quality Color

According to the Xerox Corporation, their newest digital printer, which they refer to as a digital press, has achieved the low operating cost (service plus consumables) of less than 10 cents per page. This will allow retailers to sell copies for close to 50 cents a page, which is thought by some to be the point at which the market demand for color documents will truly take off.

Anyone with a computer and the right a software can afford to prepare and create professional looking documents for personal use with digital printers. Small to medium business can print a variety of short-run promotional pieces at competitive prices and not have to store excess inventory. The ability to personalize a document with every revolution of the drum is already changing the face of advertising today.

Digital printing will have a significant impact on your life and career in printing.

32

Plant Economics of Polythene Bags & Automatic Printing

Rated Plant Capacity	=	150.00 Kgs/Day
LAND & BUILDING		
(Land Area 250 sq.mt.) (On Rental Basis)	=	Rs. 2,500.00

PLANT & MACHINERY

1. Automatic Bag making Plant (including Extruder, Die head, Control board, Air Compressor, Winder) 50 mm
2. Cutting & Sealing Machine
3. Four colour Flexo Printing machine
4. Stereo Hydraulic Press
5. Grinder
6. Weight for printed Matter
7. Lab testing Equipments

FIXED CAPITAL

1. LAND & BUILDING (On Rental Basis)	Rs.	2,500.00
2. PLANT & MACHINERY	Rs.	8,64,000.00
3. OTHER FIXED ASSETS	Rs.	40,000.00
TOTAL	Rs.	9,06,500.00

RAW MATERIALS

1. L.D.P.E. Granules
2. Misc. Polythene additives etc
3. Printing ink and colour

TOTAL WORKING CAPITAL/MONTH

1. RAW MATERIAL	Rs.	1,79,000.00
2. SALARY & WAGES	Rs.	34,615.00
3. UTILITIES & OVERHEADS	Rs.	42,100.00
TOTAL	Rs.	2,55,715.00

TOTAL CAPITAL INVESTMENT

TOTAL FIXED CAPITAL	Rs.	9,06,500.00
TOTAL WORKING CAPITAL FOR 1 MONTHS	Rs.	2,55,715.00
TOTAL	Rs.	11,62,215.00

TURN OVER/ANNUM

By sale of 45000 kg. fresh bags (printed)	Rs.	42,75,000.00

PROFIT SALES RATIO = PROFIT / SALES X 100 = 18%

RATE OF RETURN = OPERATING PROFIT / T.C.I X 100 = 68%

BREAK EVEN POINT (B.E.P) = 50%

33

Plant Economics of Textile Printing (Job-Work)

Rated Plant capacity	=	2000.00 METERS/Day
LAND & BUILDING		
(Land area 1000 sq.mts.)	=	RENTED

PLANT & MACHINERY

1. Long Table
6. Boiler (Mini for Diesel)
8. Screen Printing M/c.
9. Other miscellaneous equipments

FIXED CAPITAL

1. PLANT & MACHINERY	Rs.	2,45,000.00
2. OTHER FIXED ASSETS	Rs.	21,000.00
TOTAL	Rs.	10,16,000.00

RAW MATERIALS

1. Cloth (To Be Supplied By Customer)
2. Kerocene and other chemicals
3. Printing colours

TOTAL WORKING CAPITAL/MONTH

1. RAW MATERIAL	Rs.	1,87,001.00
2. SALARY & WAGES	Rs.	40,370.00
3. UTILITIES & OVERHEADS	Rs.	8,950.00
TOTAL	Rs.	2,36,321.00

TOTAL CAPITAL INVESTMENT

TOTAL FIXED CAPITAL	Rs.	10,16,000.00
TOTAL WORKING CAPITAL FOR 1 MONTH	Rs.	2,36,321.00
TOTAL	Rs.	12,52,321.00

TURN OVER/ANNUM

By Job Work of 6,00,000 Meter printed cloth	Rs. 36,00,000.00

PROFIT SALES RATIO = PROFIT / SALES X 100 = 13%

RATEOF RETURN = OPERATING PROFIT / T.C.I X 100 = 38%

BREAK EVEN POINT (B.E.P) = 52.48 %

34

Plant Economics of Ceramic Tile Printing Medium for Screen Printing

Rated Plant Capacity	=	1.00 Ton/Annum
LAND & BUILDING		
(Land Area 500 Sq.Mts)	=	Rs. 8,75,000.00

PLANT & MACHINERY

1. Mixer (S S) fitted with agitator
2. Filter
3. Lab Testing Equipments
4. Miscellaneous Viz Jigs, fixtures etc.

FIXED CAPITAL

1. LAND & BUILDING	Rs.	8,75,000.00
2. PLANT & MACHINERY	Rs.	1,50,000.00
3. OTHER FIXED ASSETS	Rs.	61,000.00
TOTAL	Rs.	10,86,000.00

RAW MATERIALS

1. Acrylic Resin
2. Turpentine Oil
3. Camphor

4. Plasticizer
5. Phenol - Formaldehyde(PF) Resin
6. Packing Material Drums (MS)

TOTAL WORKING CAPITAL/MONTH

1. RAW MATERIAL	Rs.	21,50,000.00
2. SALARY & WAGES	Rs.	50,625.00
3. UTILITIES & OVERHEADS	Rs.	21,900.00
TOTAL	Rs.	22,22,525.00

TOTAL CAPITAL INVESTMENT

TOTAL FIXED CAPITAL	Rs.	10,86,000.00
TOTAL WORKING CAPITAL FOR 3 MONTHS	Rs.	66,67,575.00
TOTAL	Rs.	77,53,575.00

TURN OVER/ANNUM

Receipts by Sale of 300 MT of Ceramic medium	Rs. 3,00,00,000.00

PROFIT SALES RATIO = PROFIT / SALES X 100 = 8%

RATE OF RETURN = OPERATING PROFIT / T.C.I X 100 = 29%

BREAK EVEN POINT (B.E.P) = 38%

35

Plant Economics of Cloth Printing (Job Work)

Rated Plant Capacity	=	5000.00 Mtr/Day
LAND & BUILDING		
(Land Area 1000 sq.mtr)	=	Rs. 7,50,000.00

PLANT & MACHINERY

1. Printing screens withh various designs
2. Printing tables
3. Dye mixing machine
4. Washing & Treatments tanks (R.C.C. work)
5. Baby boiler
6. Mangles and rollers
7. Drying machine continuous conveyor type
8. Motors and other mis/units like mangles etc.
9. Exposure cabinate
10. Screen stretcher
11. Squeeges
12. Miscellaneous

FIXED CAPITAL

1. LAND & BUILDING	Rs.	7,50,000.00
2. PLANT & MACHINERY	Rs.	6,05,000.00
3. OTHER FIXED ASSETS	Rs.	85,000.00
TOTAL	Rs.	14,40,000.00

TOTAL WORKING CAPITAL/MONTH

1. RAW MATERIAL	Rs.	6,25,000.00
2. SALARY & WAGES	Rs.	64,875.00
3. UTILITIES & OVERHEADS	Rs.	32,750.00
TOTAL	Rs.	7,22,625.00

TOTAL CAPITAL INVESTMENT

TOTAL FIXED CAPITAL	Rs.	14,40,000.00
TOTAL WORKING CAPITAL FOR 3 MONTHS	Rs.	21,67,875.00
TOTAL	Rs.	36,07,875.00

TURN OVER/ANNUM

By screning printing 1500000 meters
of cloth of width 1 meter Rs. 1,20,00,000.00

PROFIT SALES RATIO = PROFIT / SALES X 100 = 20%

RATE OF RETURN = OPERATING PROFIT / T.C.I X 100 = 65%

BREAK EVEN POINT (B.E.P) = 38%

36

Plant Economics of Computer Continuous Stationery

Rated Plant Capacity	=	720000 Nos/Day
LAND & BUILDING (Land Area 1000 sq.mt)	=	Rs. 29,12,500.00

PLANT & MACHINERY

1. *Complete plant comprises–*

   2 Offset printing machine,

   2 letter press,

   1 Coletter,

   1 Slitter

FIXED CAPITAL

1. LAND & BUILDING	Rs.	29,12,500.00
2. PLANT & MACHINERY	Rs.	30,00,000.00
3. OTHER FIXED ASSETS	Rs.	6,75,000.00
TOTAL	Rs.	65,87,500.00

TOTAL WORKING CAPITAL/MONTH

1. RAW MATERIAL	Rs.	17,37,500.00
2. SALARY & WAGES	Rs.	2,30,340.00
3. UTILITIES & OVERHEADS	Rs.	2,36,200.00
TOTAL	Rs.	22,04,040.00

TOTAL CAPITAL INVESTMENT

TOTAL FIXED CAPITAL	Rs.	65,87,500.00
TOTAL WORKING CAPITAL FOR 2 MONTHS	Rs.	44,08,080.00
TOTAL	Rs.	1,09,95,580.00

TURN OVER/ANNUM

By sale of 2,16,000 boxes
(each box containing thousand forms) Rs. 3,78,00,000.00

PROFIT SALES RATIO = PROFIT / SALES X 100 = 24%

RATE OF RETURN = OPERATING PROFIT / T.C.I X 100 = 82%

BREAK EVEN POINT (B.E.P) = 33%

37

Plant Economics of Offset Printing Press

Rated Plant Capacity	=	6800.00 Mag/Day
LAND & BUILDING		
(Land Area 1000 sq.mt)	=	Rs. 21,00,000.00

PLANT & MACHINERY

1. Offset, 6 colour printing machine
2. Letter press printing machine
3. Cutting Machine
4. Stitching Machine
5. Tools & Tackles, Jigs & Fixtures

FIXED CAPITAL

1. LAND & BUILDING	Rs.	21,00,000.00
2. PLANT & MACHINERY	Rs.	58,00,000.00
3. OTHER FIXED ASSETS	Rs.	7,00,000.00
TOTAL	Rs.	86,00,000.00

RAW MATERIALS

1. FOR MAGAZINE OF 100 PAGES/MONTHS
   Printing Paper Size 22"x32"/8, 637 Rims
2. Printing Ink (Black+Coloured) L.S.
3. Art Card for Title pages, 25 Rims
4. Printing Ink for title Pages
5. Matter Writing Cost
6. Photo Composing
7. Processing Black & White Pages (Negative + Positive) L.S.
8. Processing of color pages (Negative + Positives for 8 pages)
9. Processing of title pages
10. For 15000 Copies of Comics Single PATTERN : Printing Ink
11. Printing Paper Size 17"x26", 504 Rims
12. Printing Paper for Title Pages, 32 Rims
13. Artist on Contract
14. Plate making
15. Processing
16. Other Consumable like oil & lubricants

TOTAL WORKING CAPITAL/MONTH

1. RAW MATERIAL	Rs.	6,57,100.00
2. SALARY & WAGES	Rs.	1,07,750.00
3. UTILITIES & OVERHEADS	Rs.	1,35,500.00
TOTAL	Rs.	9,00,350.00

TOTAL CAPITAL INVESTMENT

TOTAL FIXED CAPITAL	Rs.	86,00,000.00
TOTAL WORKING CAPITAL FOR 3 MONTHS	Rs.	27,01,050.00
TOTAL	Rs.	1,13,01,050.00

TURN OVER/ANNUM

1. By sale of 6,00,000 Copies of magazines
2. By sale of 14,40,000 copies of comics
3. Receipt from advertisement

TOTAL Rs. 2,20,20,000.00

PROFIT SALES RATIO = PROFIT / SALES X 100 = 37%

RATE OF RETURN = OPERATING PROFIT / T.C.I X 100 = 73%

BREAK EVEN POINT (B.E.F) = 34%

38

Plant Economics of DTP And Printing Unit

Rated Plant Capacity	=	40000.00 Nos/Day
LAND & BUILDING		
(Land Area 250 sq.mt)	=	Rs. 4,70,000.00

PLANT & MACHINERY

1. H.M.T. Single colour complete with all accessories 1 No.
2. Reprographic camera (Horizontal type) with all accessories 1 No.
3. Plate making eqpt - light table,red lithographer's tape, goldenrod sheets, Scissors, scribing Tools etc.
4. Computer
5. Printer (Laser)
6. Mouse
7. Page Maker & Corel Draw
8. Air conditioner
9. Paper cutting machine

FIXED CAPITAL

1. LAND & BUILDING	Rs.	4,70,000.00
2. PLANT & MACHINERY	Rs.	17,64,500.00
3. OTHER FIXED ASSETS	Rs.	2,50,000.00
TOTAL	Rs.	24,84,500.00

RAW MATERIALS

1. 2000 rim Size of sheet require 23"x 36".
   Each rim contents over 500sheets white paper
2. Ink (For Offset Printing)
3. Phototypesetting paper
4. Graphic art film
5. Rubber Blankets
6. Miscellaenous chemical and materials

TOTAL WORKING CAPITAL/MONTH

1. RAW MATERIAL	Rs.	5,33,000.00
2. SALARY & WAGES	Rs.	32,280.00
3. UTILITIES & OVERHEADS	Rs.	16,400.00
TOTAL	Rs.	5,81,680.00

TOTAL CAPITAL INVESTMENT

TOTAL FIXED CAPITAL	Rs.	24,84,500.00
TOTAL WORKING CAPITAL FOR 3 MONTHS	Rs.	17,45,040.00
TOTAL	Rs.	42,29,540.00

TURN OVER/ANNUM

By sale of 1.2 crores of sheets size 23" x 36" (Including paper charges)	Rs.	1,08,00,000.00

PROFIT SALES RATIO = PROFIT / SALES X 100 = 23%

RATE OF RETURN = OPERATING PROFIT / T.C.I X 100 = 57%

BREAK EVEN POINT (B.E.P) = 40%

39

Plant Economics of Offset Security Printing Press

Rated Plant Capacity	=	6720 Lacs Tickets/Annum
LAND & BUILDING		
(Land Area 600 sq.mtrs.)	=	Rs. 11,85,000.00

PLANT & MACHINERY

1. Four Colour Printing Machine
2. Single colour printing machine
3. Semi-Automatic cutting machine
4. Stitching machine
5. Numbering machine (single phase)
6. Herdelberg Cylinder Letter press
7. Chandler machine (Single phase)
8. Lamination machine

FIXED CAPITAL

1. LAND & BUILDING	Rs.	11,85,000.00
2. PLANT & MACHINERY	Rs.	61,37,000.00
3. OTHER FIXED ASSETS	Rs.	20,50,000.00
TOTAL	Rs.	93,72,000.00

RAW MATERIALS

1. Plants for offset printing
2. Printing Inks
3. Graphic Films
4. Miscellaneous Stores & Consumables

TOTAL WORKING CAPITAL/MONTH

1. RAW MATERIAL	Rs.	2,20,000.00
2. SALARY & WAGES	Rs.	1,24,187.00
3. UTILITIES & OVERHEADS	Rs.	33,500.00
TOTAL	Rs.	3,77,687.00

TOTAL CAPITAL INVESTMENT

TOTAL FIXED CAPITAL	Rs.	93,72,000.00
TOTAL WORKING CAPITAL FOR 2 MONTHS	Rs.	7,55,375.00
TOTAL	Rs.	1,01,27,375.00

TURN OVER/ANNUM

1. By selling of Lottery Tickets
   Printed in 4 colours= 40000 x 2 x
   28 x 300 = 6720 Lakh
2. Earnings from job-work of different printing works

TOTAL Rs. 1,29,74,000.00

PROFIT SALES RATIO = PROFIT / SALES X 100 = 43%

RATE OF RETURN = OPERATING PROFIT / T.C.I X 100 = 55%

BREAK EVEN POINT (B.E.P) = 39%

40

Plant Economics of Roto Gravure Printing (For Flexible Packaging)

Rated Plant Capacity	=	4500.00 Mts/Day
LAND & BUILDING (Land Area 500 sq.mt)	=	Rs. 4,00,000.00

PLANT & MACHINERY

1. 6-color Machine 30" size working width complet motor Blower, drive 16 kwh Heater & Printing cylinder
2. Pouches Cutting and Heating Sealing Machine
3. Small Working tools/and material handling

FIXED CAPITAL

1. LAND & BUILDING	Rs.	4,00,000.00
2. PLANT & MACHINERY	Rs.	6,80,000.00
3. OTHER FIXED ASSETS	Rs.	30,000.00
TOTAL	Rs.	11,10,000.00

RAW MATERIALS

1. Low-Density Polyethelene(L.D.P.E) film and Polyester film
2. Inks etc.
3. Polyurethane based Adhesives
4. Solvents (like thinner, M.E.K. Thyl acetate)

TOTAL WORKING CAPITAL/MONTH

1. RAW MATERIAL	Rs.	2,26,000.00
2. SALARY & WAGES	Rs.	17,160.00
3. UTILITIES & OVERHEADS	Rs.	15,000.00
TOTAL	Rs.	2,58,160.00

TOTAL CAPITAL INVESTMENT

TOTAL FIXED CAPITAL	Rs.	11,10,000.00
TOTAL WORKING CAPITAL FOR 3 MONTHS	Rs.	7,74,480.00
TOTAL	Rs.	18,84,480.00

TURN OVER/ANNUM

Sales of Prtd. Straps for flexible pkg.1350000 mt. legh
(eg. Maggie Noodles,Pan Masala pkt) Rs. 44,55,000.00

PROFIT SALES RATIO = PROFIT / SALES X 100 = 17%

RATE OF RETURN = OPERATING PROFIT / T.C.I X 100 =40%

BREAK EVEN POINT (B.E.P) = 50%

41

Plant Economics of Screen Printing on Paper/Cards

Rated Plant Capacity	= 10000.00 Nos/Day
LAND & BUILDING (Land Area 150 Sq.mts) =	Rs. 2,10,000.00

PLANT & MACHINERY

1. Complete Desk Top Publishing System (including computer, laser printer, mouse, floppies etc.)
2. Automatic Screen Printing Machine
3. Symphony Air Condition Cooler for Computer.

FIXED CAPITAL

1. LAND & BUILDING	Rs.	2,10,000.00
2. PLANT & MACHINERY	Rs.	2,82,000.00
3. OTHER FIXED ASSETS	Rs.	1,02,000.00
TOTAL	Rs.	5,94,000.00

RAW MATERIALS

1. Printing Ink
2. Screen
3. Paper For Visiting Cards, Greeting Cards And Other Type.

TOTAL WORKING CAPITAL/MONTH

1. RAW MATERIAL		Rs.	1,30,000.00
2. SALARY & WAGES		Rs.	14,125.00
3. UTILITIES & OVERHEADS		Rs.	7,620.00
	TOTAL	Rs.	1,51,745.00

TOTAL CAPITAL INVESTMENT

TOTAL FIXED CAPITAL		Rs.	5,94,000.00
TOTAL WORKING CAPITAL FOR 3 MONTHS		Rs.	4,55,235.00
	TOTAL	Rs.	10,49,235.00

TURN OVER/ANNUM

1. Paper Cost (Recovery)
2. Printing Charges(a) Visiting Cards
   Marriage Cd 200/100,Greeting Cd

Rs. 22,50,000.00

PROFIT SALES RATIO = PROFIT / SALES X 100 = 25%

RATE OF RETURN = OPERATING PROFIT / T.C.I X 100 = 68%

BREAK EVEN POINT (B.E.P) = 37%

42

Plant Economics of Heat Transfer Label by Screen Printing Process

Rated Plant Capacity	=	100000.00 Labels/Day
LAND & BUILDING		
(Land Area 150 sq. mts.)	=	Rs. 2,90,000.00

PLANT & MACHINERY

1. Stainless Steel Tank
2. Closed top S.S. tank
3. Screen Printing equipment system
4. Miscellaneous tools, jigs, fixtures

FIXED CAPITAL

1. LAND & BUILDING	Rs.	2,90,000.00
2. PLANT & MACHINERY	Rs.	1,46,000.00
3. OTHER FIXED ASSETS	Rs.	1,15,000.00
TOTAL	Rs.	5,51,000.00

RAW MATERIALS

1. Aniline Oil
2. Potassium chlorate
3. Distilled water
4. Hydrochloric Acid
5. Copper Chloride
6. Latex
7. Sulphur
8. M.B.T.
9. Yellow Pigment
10. Cellophane film
11. Packaging Materials

TOTAL WORKING CAPITAL/MONTH

1. RAW MATERIAL	Rs.	51,213.00
2. SALARY & WAGES	Rs.	16,800.00
3. UTILITIES & OVERHEADS	Rs.	13,000.00
TOTAL	Rs.	81,013.00

TOTAL CAPITAL INVESTMENT

TOTAL FIXED CAPITAL	Rs.	5,51,000.00
TOTAL WORKING CAPITAL FOR 3 MONTHS	Rs.	2,43,040.00
TOTAL	Rs.	7,94,040.00

TURN OVER/ANNUM

By sale of 300 Lakhs stickers	Rs.	30,00,000.00

PROFIT SALES RATIO = PROFIT / SALES X 100 = 61%

RATE OF RETURN = OPERATING PROFIT / T.C.I X 100 = 230%

BREAK EVEN POINT (B.E.P) = 16%

43

Plant Economics of Flexo Graphic & Roto Gravure Printing

Rated Plant Capacity	=	93600.00 Mtr/Day
LAND & BUILDING (Land area 800 sq.mt.)	=	Rs. 7,00,000.00

PLANT & MACHINERY

1. PRINTING M/C :-6 color roto gravure printing machine 30" size working width complete with 2 H.P. blower
2. 5 H.P. driver 16 kwh heater & printing cylinder dia = 150 mm (fully automatic)
3. Slitting and rewinding machine 48" cap. with min= 80 mm slit & max in wind dia=45" with 5 H.P. drive 1 No
4. Roll to Roll making machine 48"
5. Bidsmen lamination plant
6. Gum tape machinery plant
7. Rotary sheet cutter m/c 48"
8. Small working tools/tackels and material handling
9. Composing machine complete with mono key boards mono casters and lines
10. Plate/Cylinder making equipments (for detail see 'Annexure P'

FIXED CAPITAL

1. LAND & BUILDING	Rs.	7,00,000.00
2. PLANT & MACHINERY	Rs.	21,85,001.00
3. OTHER FIXED ASSETS	Rs.	4,64,350.00
TOTAL	Rs.	33,49,351.00

RAW MATERIALS

1. Metal plates
2. Rubber blankets
3. Inks and founts
4. Gum and adhesives etc.
5. Paper : sunlit bond
6. Thermoplastic films

TOTAL WORKING CAPITAL/MONTH

1. RAW MATERIAL	Rs.	15,87,500.00
2. SALARY & WAGES	Rs.	1,06,500.00
3. UTILITIES & OVERHEADS	Rs.	2,00,000.00
TOTAL	Rs.	18,94,000.00

TOTAL CAPITAL INVESTMENT

TOTAL FIXED CAPITAL	Rs.	33,49,351.00
TOTAL WORKING CAPITAL FOR 3 MONTHS	Rs.	56,82,000.00
TOTAL	Rs.	90,31,351.00

TURN OVER/ANNUM

By sale of printed straps for flexible packaging

(eg.Maggie Noodle Pan parag & Nescafe packets etc.)

Rs. 2,81,00,000.00

PROFIT SALES RATIO = PROFIT / SALES X 100 = 10%

RATEOF RETURN = OPERATING PROFIT / T.C.I X 100= 31%

BREAK EVEN POINT (B.E.P) = 59%

44

Plant Economics of Screen Printing Cotton Cloth

Rated Plant Capacity	=	1,50,000 Pieces/Annum
LAND & BUILDING		
(Land Area 500 Sq.mts.)	=	Rs. 3,65,000.00

PLANT & MACHINERY

1. Printing table
2. Screens
3. Gumming machine
4. Washing Machine
5. Guides, Squeezes, etc.
6. Miscellaneous

FIXED CAPITAL

1. LAND & BUILDING	Rs.	3,65,000.00
2. PLANT & MACHINERY	Rs.	1,97,000.00
3. OTHER FIXED ASSETS	Rs.	85,000.00
TOTAL	Rs.	6,47,000.00

RAW MATERIALS

1. Pieces to be printed
2. Pigment, colour, dyes, etc.
3. Miscellaneous

TOTAL WORKING CAPITAL/MONTH

1. RAW MATERIAL	Rs.	5,20,000.00
2. SALARY & WAGES	Rs.	28,125.00
3. UTILITIES & OVERHEADS	Rs.	19,500.00
TOTAL	Rs.	5,67,625.00

TOTAL CAPITAL INVESTMENT

TOTAL FIXED CAPITAL	Rs.	6,47,000.00
TOTAL WORKING CAPITAL FOR 2 MONTHS	Rs.	11,35,250.00
TOTAL	Rs.	17,82,250.00

TURN OVER/ANNUM

By sale of 1,50,000 pieces	Rs.	93,00,000.00

PROFIT SALES RATIO = Profit / Sales x 100 = 24%

RATE OF RETURN = Operating profit /T.C.I x 100 = 123%

BREAK EVEN POINT (B.E.P) = 19%

45

Plant Economics of Hand Block Printing

Rated Plant Capacity	=	100.00 Mts/Day
LAND & BUILDING		
Land Area 25 cents (11,000 Sq.Ft.)]	=	Rental

PLANT & MACHINERY

1. Wooden blocks of different design.
2. Wooden printing rollers.
3. Tieringtable of dimension (2X3 ft) with necessary arrangement.
4. Pigment storage vessels (S.S.)
5. Miscellaneous viz. pitch pins, rubber squeegee etc.

FIXED CAPITAL

1. PLANT & MACHINERY	Rs.	12,500.00
2. OTHER FIXED ASSETS	Rs.	53,000.00
TOTAL	Rs.	65,500.00

RAW MATERIALS

1. Cloth
2. Textile colour pigment paste of different colour.
3. Miscellaneous consumables viz thickening agents etc. (sodium alginate).

TOTAL WORKING CAPITAL/MONTH

1. RAW MATERIAL	Rs.	1,23,000.00
2. SALARY & WAGES	Rs.	37,440.00
3. UTILITIES & OVERHEADS	Rs.	10,800.00
TOTAL	Rs.	1,71,240.00

TOTAL CAPITAL INVESTMENT

TOTAL FIXED CAPITAL	Rs.	65,500.00
TOTAL WORKING CAPITAL FOR 3 MONTHS	Rs.	5,13,720.00
TOTAL	Rs.	5,79,220.00

TURN OVER/ANNUM

By sale of Hand Block Printed cloth 30,000 mts.	Rs.	24,00,000.00

PROFIT SALES RATIO = PROFIT / SALES X 100 = 10%

RATE OF RETURN = OPERATING PROFIT / T.C.I X 100 = 41%

BREAK EVEN POINT (B.E.P) = 59%

46

Plant Economics of Paper Transfer Ink Labels (Like Tattoo)

Rated Plant Capacity	=	1.00 Lac Nos/Day
LAND & BUILDING		
(Land Area 500 Sq.Mts.)	=	Rs. 8,50,000.00

PLANT & MACHINERY

1. Auto flat screen printer,the squeege speed is 900 imprints/hour and the thickness of print is 0.20 m.m
2. Miscellaneous machinery
3. Storage tank (for storing ink, dilution etc.)
4. Slitting machine (automatic)
5. Automatic sticker cutting machine
6. Paper roll unwinding machine (for wax paper and printing paper)
7. Calendering machine
8. Automatic adhesive coating machine

FIXED CAPITAL

1. LAND & BUILDING	Rs.	8,50,000.00
2. PLANT & MACHINERY	Rs.	14,70,000.00
3. OTHER FIXED ASSETS	Rs.	6,90,000.00
TOTAL	Rs.	30,10,000.00

RAW MATERIALS

1. Transfer printing ink
2. Screen
3. Miscellaneous raw materials and consumables
4. Print label making
5. Wax coated paper for label protection
6. Packaging materials

TOTAL WORKING CAPITAL/MONTH

1. RAW MATERIAL	Rs.	85,450.00
2. SALARY & WAGES	Rs.	53,466.00
3. UTILITIES & OVERHEADS	Rs.	72,000.00
TOTAL	Rs.	2,10,916.00

TOTAL CAPITAL INVESTMENT

TOTAL FIXED CAPITAL	Rs.	30,10,000.00
TOTAL WORKING CAPITAL FOR 2 MONTHS	Rs.	4,21,832.00
TOTAL	Rs.	34,31,832.00

TURN OVER/ANNUM

By sale of 300 lacs Print Transfer Labels (Tattoo)	Rs.	45,00,000.00

PROFIT SALES RATIO = PROFIT / SALES X 100 = 25%

RATE OF RETURN = OPERATING PROFIT / T.C.I X 100 = 33%

BREAK EVEN POINT (B.E.P) = 56%

47

Suppliers of Plant & Machineries

Offset Printing Machinery

Budhan Engineerings
6-6-99, Kavadiguda
Secunderabad
Andhra Pradesh-500080
Tel : + 91 (0) 40 27535648
Fax : + 91 (0) 40 27534094
E-mail:
budhan@hd2.dot.net.in

Creed Engineers Pvt. Ltd.
4, South Patel Ngr. Mkt.
New Delhi-110008
Tel : + 91 (0) 11 25767206
Fax : + 91 (0) 11 25767205
E-mail: creed@vsnl.com

Diamond Engineers
6/14, DLF Indl. Area
Moti Ngr.
New Delhi-110015
Tel : + 91 (0) 11 25418353
Fax : + 91 (0) 11 25414571
E-mail: balli@ndf.vsnl.net.in

Friends Machinery Corp.
J-126B, Vishnu Garden
New Delhi-110018
Tel : + 91 (0) 11 25153953
Fax : + 91 (0) 11 25153953
E-mail: friendsmach@
satyamonline.com

Grafitek International
B-262, Chittaranjan Park
New Delhi-110019
Tel : + 91 (0) 11 26283416
Fax : + 91 (0) 11 26480561
E-mail:
taposhc@del2.vsnl.net.in

Laxmi Paper Machine
104, Gali No. 9
Anand Parbat Indl. Area
New Delhi-110005
Tel : + 91 (0) 11 25723587
E-mail: laxmiap2@delhi-
industry.com

C Trivedi & Co.
A-1/16, Phase-II, GIDC
Est., Vatva
Ahmedabad
Gujarat-382445
Tel : + 91 (0) 79 25830552
Fax : + 91 (0) 79 25830553
E-mail:
ctrivedi@ctrivedi.com

Creative Enterprise
C-1, 275-276
GIDC, Naroda
Ahmedabad
Gujarat-382330
Tel : + 91 (0) 79 22813055

Kailash Graphic Machinery
Opp. New Timber Yard
Dabhoi Rd.
Vadodara, Gujarat-390004
Tel : + 91 (0) 265 2561445
Fax : + 91 (0) 265 2561445
E-mail:
kgm@infoquestindia.com

Rajyog Engineering Pvt. Ltd.
893/1, GIDC, Makarpura
Vadodara, Gujarat-390010
Tel : + 91 (0) 265 2642207
Fax : + 91 (0) 265 2641905
E-mail:
rajyog@imprinta17.com

Natraj Industries
Plot No. 65, Sect. 6
Faridabad, Haryana-121006
Tel : + 91 (0) 129 25241403

Fax : + 91 (0) 129 25243403
E-mail:
hitesh@del3.vsnl.net.in

Fair Deal Engineers
66, Ramsarup Cly.
Nr. Sect. 24, Mujessar
Faridabad, Haryana-121005
Tel : + 91 (0) 129 25237143
Fax : + 91 (0) 129 25234314

Graphic Machineries
43 A
DLF Indl. Area, Phase-I
Mathura Rd.
Faridabad, Haryana-121003
Tel : + 91 (0) 129 25279007

Hm Web House Pvt. Ltd.
B-102, Sanjay Cly.,
Sect. 23, NIT
Faridabad, Haryana-121001
Tel : + 91 (0) 129 25440403
Fax : + 91 (0) 129 25238195
E-mail:
webhouse@del3.vsnl.net.in

Kalsi Printmac Pvt. Ltd.
Plot No. 262-D, Sect. 24
Faridabad, Haryana-121005
Tel : + 91 (0) 129 25237174
Fax : + 91 (0) 129 25422471
E-mail: kp-kalsi@usa.net

King Offset Machinery
5D/104, NIT
Faridabad, Haryana-121001
Tel : + 91 (0) 129 25416436
Fax : + 91 (0) 129 25416436

Multitec Aids Pvt. Ltd.
Plot No. 200, Sect. 24
Faridabad, Haryana-121005
Tel : + 91 (0) 129 25234085
Fax : + 91 (0) 129 25233403
E-mail: multitec@vsnl.com

Packwell Industries
339, Sect. 24
Faridabad, Haryana-121005
Tel : + 91 (0) 129 25232603
Fax : + 91 (0) 129 25233255
E-mail:
rollonpackaging@vsnl.net.in

Perfect Graphics
65, Indl. Area (Nr. Hindustan
Vacuum Glass), NIT
Faridabad, Haryana-121001
Tel : + 91 (0) 129 25230475
Fax : + 91 (0) 129 25418698

Prakash Offset Pvt. Ltd.
Raghubir Cly., Sihi Gate
Ballabhgarh
Haryana-121004
Tel : + 91 (0) 1188 25249302
Fax: + 91 (0) 1188 25244536
E-mail:
info@prakashoffset.com

Printcon Engineers
B-53, Rishi Ngr.
Chawla Cly.
Ballabhgarh
Haryana-121004
Tel: + 91 (0)129 25242034
Fax: + 91 (0)129 25310051

Sagar Industries
16/2, Mathura Rd.
Faridabad
Haryana-121002
Tel : + 91 (0)129 25282596

Sahil Graphics
990, Sect. 23, NIT
Faridabad
Haryana-121001
Tel : + 91 (0) 129
25234690
FAX : + 91 (0)129 5247950
E-mail:
sahil_offset@rediffmail.com

SGS Associates
Plot No. 27, Sect. 25
Faridabad, Haryana-121004
Tel : + 91 (0) 129 25231589
Fax : + 91 (0)129 25244467
E-mail:
mudrikaoffset@yahoo.com

Shiva Graphics
Plot No. 31, 16/2
Mathura Rd.
Faridabad, Haryana-121002
Tel : + 91 (0) 129 25268120

Sud & Waren Pvt. Ltd.
Plot No. 72, Indl. Area
Sect.6
Faridabad, Haryana-121006
Tel : + 91 (0) 129 25248464
Fax : + 91 (0) 129 25246141
E-mail: sud@nde.vsnl.net.in

Vee Kay Graphic House
F-Block, Adarsh Ngr.
Nr. Beri Ka Bagh
Malerna Rd.
Ballabhgarh
Haryana-121004
Tel : + 91 (0) 129 25249176
Fax : + 91 (0) 129 25240645

Web Converter (I)
Plot No. 34, 15/6 Mile Stone
Mathura Rd., Sect. 27 A
Faridabad, Haryana-121003
Tel : + 91 (0) 129 25276985
Fax : + 91 (0) 129 25270396
E-mail:
webindia@nda.vsnl.net.in

Web Tech Engg. Pvt. Ltd.
21/1, Sanjay Cly., Nr.
Gouchi Chowk
Sohna Rd.
Faridabad, Haryana-121005
Tel : + 91 (0) 129 25231878

FAX : +91 (0) 129 5442400
E-mail:
info@webtechengg.com

Webco Industries
2258, Air Force Rd.,
Jawahar Cly., NIT
Faridabad, Haryana-121005
Tel : + 91 (0) 129 25237015
Fax : + 91 (0) 129 25239041

Machinery & Spares
Bihar State Co-Op. Bank Bldg.
Main Rd., Bistupur
Jamshedpur
Jharkhand-831001
Tel : + 91 (0) 657 2423480
Fax : + 91 (0) 657 2425870
E-mail:
jsr_machiusa@sancharnet.in

Ambica Machineries
210/1, Akkipet Rd.
Bangalore
Karnataka-560053
Tel : + 91 (0) 80 22872104

Macro Agencies
482
2nd Main Rd., 2nd Stage
Indiranagar
Bangalore
Karnataka-560038
Tel : + 91 (0) 80 25251962
Fax : + 91 (0) 80 25253305
E-mail:
nagaraj@blr.vsnl.net.in.com

Monu Graphics
16/1, 1st Flr.
Opp. Sudha Lodge
Cottonpet Main Rd.
Bangalore
Karnataka-560053
Tel : + 91 (0) 80 22257098

HMT Machine Tools Ltd.
Machine Tools Divn.
HMT Cly. P.O.
Kalamassery
Kerala-683503
Tel : + 91 (0) 248455, 540731
Fax : + 91 (0) 248455, 532166
E-mail:
mtksales@hmtmtm.com

Acme Machinery International
B-3, Royal Indl. Est.
Naigaum Cross Rd.
Wadala
Mumbai
Maharashtra-400031
Tel : + 91 (0) 22 24146345
Fax : + 91 (0) 22 24146928
E-mail:
acmemach@bom4.vsnl.net.in

Acme Machinery (I) Pvt. Ltd.
G-16, Shalimar Indl. Est.
Matunga Labour Camp
Matunga (W) Mumbai
Maharashtra-400019
Tel : + 91 (0) 22 24037398
Fax : + 91 (0) 22 24037270
E-mail:
acme@acmeindia.com

Delux Printing Machinery Co.
2, Shital Sadan, 78/81
Bazar Gate St., Fort
Mumbai
Maharashtra-400001
Tel : + 91 (0) 22 22690540
Fax : + 91 (0) 22 22618463
E-mail:
deluxe_exp@vsnl.com

Deluxe Printing Machinery Co.
75/81, Bazargate St., Fort
Mumbai
Maharashtra-400001
Tel : + 91 (0) 22 22618630
Fax : + 91 (0) 22 22618463
E-mail:
deluxe_exp@vsnl.com

Fieldstar Engg.
Gala No. 2
Worlikar Indl. Est.
Madan Cottage, Tulshet Pada, Lake Rd.
Bhandup (W) Mumbai
Maharashtra-400078
Tel : + 91 (0) 22 25908546

Patel Enterprises Ltd.
B-67, Sanjay Bldg.
No. 5, Mittal Indl. Est.
Marol Naka
Andheri (E) Mumbai
Maharashtra-40005
Tel : + 91 (0) 22 28504601
Fax : + 91 (0) 22 26240545
E-mail: patelent@bol.net.in

Patil Offset Pvt. Ltd.
33, Nilgiri Indl. Est., 183
T J Rd., Sewri (W) Mumbai
Maharashtra-400015
Tel : + 91 (0) 22 24110660
Fax : + 91 (0) 22 24148917
E-mail:
patiloffset@yahoo.com

Sabah Electronics
Hasina Zafruddin Chawl
Ovri Pada
Dahisar (E), Mumbai
Maharashtra-400068
Tel : + 91 (0) 22 28936452

Speed-O-Graph
A H Compound
Walbhat Cross Rd.
S T Yadav Marg

Goregaon (E) Mumbai
Maharashtra-400063
Tel : + 91 (0) 22 26902923
Fax : + 91 (0) 22 28784188
E-mail:
speedo@bom7.vsnl.net.in

MOI Engg. Ltd.
A 7, Phase-I, Indl. Est.
SAS Ngr.
Mohali, Punjab-160051
Tel : + 91 (0) 172 2670628
Fax : + 91 (0) 172 2670630

Poonam Engg. Works
Opp. Mall Mandi,
Plot No. 520, Indl. Area
Amritsar, Punjab-143001
Tel : + 91 (0) 183 2232489

Fuser Graphics Pvt. Ltd.
Old No. 15 (New No. 45)
Dhanakoti Raja St.
Ekkattuthangal
Chennai, Tamil Nadu-600097
Tel : + 91 (0) 44 22323221
Fax : + 91 (0) 44 22325727
E-mail: fge@satyam.net.in

Nelson Group of Companies
57, 58, 59, Great Cotton Rd.
Tuticorin, Tamil Nadu-628002
Tel : + 91 (0) 461 2326726
Fax : + 91 (0) 461 2325194

Batla Engg. Co.
D-316, Sect. 10
Noida,
Uttar Pradesh-201301
Tel : + 91 (0) 120 24547473
Fax : + 91 (0) 120 24518937
E-mail: batla@vsnl.com

Lakshmi Udyog
E-11
Bulandshahr Rd. Indl. Area
Ghaziabad
Uttar Pradesh-201002
Tel : + 91 (0) 120 24702814

Screen Printing Machinery

Metcon Engineers
110, Samrat Bhawan
Ranjit Ngr. Comml. Cmplx.
New Delhi-110008
Tel : + 91 (0) 11 25701038
Fax : + 91 (0) 11 25709527
E-mail:
metcon@del3.vsnl.net.in

Sharpline Engineers
C-71, Laxmi Park
Nangloi
Delhi-110041
Tel : + 91 (0) 11 25945366

ABS Corp.
Nr. Marsden Mills
Rakhial Rd.
Ahmedabad
Gujarat-380021
Tel : + 91 (0) 79 22161302
Fax : + 91 (0) 79 22113187

Abhishek Print Line
E-26, Nehru Ground, NIT
Faridabad, Haryana-121001
Tel : + 91 (0) 129 25416275
FAX : + 91 (0)129 25419983
E-mail:
abhishek@nde.vsnl.net.in

Bonhomie
14/A, Ujagar Indl. Est.
Deonar
Mumbai
Maharashtra-400088
Tel : + 91 (0) 22 25975074
Fax : + 91 (0) 22 25556989
E-mail:
deepak@bonhomie.com

Denzil Machine Tools
12, Swamini Indl. Est.
Opp. Bassein Drugs Ltd.,
Valiv Taluka
Vasai, Thane
Maharashtra-401202
Tel : + 91 (0) 22 28352105
Fax : + 91 (0) 22 28214276
E-mail:
vewsight@bom3.vsnl.net.in

Grafica Flextronica
Plot No. 92, Survey No. 66
Waliv Fata
Vasai (E) (Western Rly.)
Thane
Maharashtra-401208
Tel : + 91 (0) 250 2455435
Fax : + 91 (0) 250 2455434
E-mail:
grafica@bom3.vsnl.net.in

Harish Tex-Mach Pvt. ltd.
1/103 Krishna vatika,
C.S.C., Rd. No. 4
B/h Shakti Ngr.
Dahisar (E)
Mumbai
Maharashtra-400068
Tel : + 91 (0) 22 28949462
Fax : + 91 (0) 22 28949463
E-mail: harish3@vsnl.com

JV Industries
68, Vallalar St.
Venketesa Cly.
Pollachi, Tamil Nadu-642001
Tel : + 91 (0) 4259 227457
Fax : + 91 (0) 4259 227520
E-mail:
jvprint@md3.vsnl.net.in

Nutech Graphic Systems
C-269, Sect. 10
Noida
Uttar Pradesh-201301

Tel : + 91 (0) 118 24549608
Fax : + 91 (0)118 24549608
E-mail: nutechgraphic@mantraonline.com

Bag Printing Machinery

Gujarat Elastomers & Polymer Technologies Ltd.
2/E, 2nd Flr., Trade Centre
Stadium Circle Navrangpura
Ahmedabad
Gujarat-380009
Tel : + 91 (0) 79 26401906
Fax : + 91 (0) 79 26560516
E-mail:
mcpatsonad1@sancharnet.in

Akshardham Industries
Plot No. 2407/1 & 2408/2
Phase-IV, B/h SRS, GIDC
Wadhwan, Gujarat-363035
Tel : + 91 (0) 2752 43525
E-mail:
interpak_ad1@sancharnet.in

Gopal Engg. Works
2/5, B/h HCMC
Kailash Indl. Est., Odhav
Ahmedabad
Gujarat-382415
Tel : + 91 (0) 79 22874969
Fax : + 91 (0) 79 22874969

Technopak Engineering Equipments
D-100, Indl. Est.
Rajajinagar
Bangalore
Karnataka-560044
Tel : + 91 (0) 80 23305182
Fax : + 91 (0) 80 23305182
E-mail: technopak@indiabusinessonline.com

Moscow Machine Tools
C-293, Mukund Ngr.
Ghaziabad
Uttar Pradesh-201001
Tel : + 91 (0) 120 24731630
Fax : + 91 (0)120 24793507
E-mail:
moscowindia@yahoo.co.in

Princessa Overseas Pvt. Ltd.
802, Pragati Tower, 26
Rajendra Place
New Delhi-110008
Tel : + 91 (0) 11 25823626
Fax : + 91 (0) 11 25823627
E-mail:
princessa@mantraonline.com

Flexographic Printing Machinery

Creed Engineers Pvt. Ltd.
4, South Patel Ngr. Mkt.
New Delhi-110008
Tel : + 91 (0) 11 25767206
Fax : + 91 (0) 11 25767205
E-mail: creed@vsnl.com

XL Plastics
303/2, GIDC
Makarpur Indl. Est.
Vadodara
Gujarat-390010
Tel : + 91 (0) 265 2638125
Fax : + 91 (0) 265 2638661
E-mail: info@xlplastic.com

Natraj Industries
Plot No. 65, Sect. 6
Faridabad
Haryana-121006
Tel : + 91 (0) 129 25241403
Fax : + 91 (0)129 25243403
E-mail:
hitesh@del3.vsnl.net.in

Expert Industries Ltd.
274-C, KIADB Indl. Area
Bommsandra
Bangalore
Karnataka-562158
Tel : + 91 (0) 8110 232690
Fax : + 91 (0) 8110 232037
E-mail:
expert@expertindia.com

Packlink Industries
1, Anand Raj Indl. Est.
Nr. Kembros Indl. Est.
Off LBS Marg,
Bhandup (W) Mumbai
Maharashtra-400078
Tel : + 91 (0) 22 25926721
Fax : + 91 (0) 22 25926721

SP International
Ground Flr., 9
Sona Udyog Indl. Est.
Parsi Panchayat Rd.
Andheri (E) Mumbai
Maharashtra-400069
Tel : + 91 (0) 22 28322554
Fax : + 91 (0) 22 26437045

Mahindra Plastic Industries
Opp. Transport Ngr.
Ganesh Ngr., Indl. Area A
St. No. 2, Link Rd.
Ludhiana
Punjab-141003
Tel : + 91 (0) 161 2664337
Fax : + 91 (0) 161 2644624

Gehlot Industries
H-1/122, New Indl. Area
Jodhpur
Rajasthan-342003
Tel : + 91 (0) 291 2740142
Fax : + 91 (0) 291 2740527
E-mail:
surya_palace@satyam.net.in

Carmel Engineering
Plot No. 5, Peniel Ngr.
Viraganur Dam
Madurai
Tamil Nadu-625009
Tel : + 91 (0) 452 2465667
Fax : + 91 (0) 452 2465468
E-mail: carmel123@sify.com

Shaheen Machine Fabric
B-113, Sect. 5
Noida, Uttar Pradesh-01301
Tel : + 91 (0) 120 24552488

Glass Printing Machinery

Princessa Overseas Pvt. Ltd.
802, Pragati Tower, 26
Rajendra Place
New Delhi-110008
Tel : + 91 (0) 11 25823626
Fax : + 91 (0) 11 25823627
E-mail:
princessa@mantraonline.com

Meditek International
C-43
DSIDC Packaging Cmplx.
Kirti Ngr.
New Delhi-110015
Tel : + 91 (0) 11 25901164
Fax : + 91 (0) 11 25901164
E-mail:
rgandhi1@rediffmail.com

Label Printing Machinery

Ankit Machinery
8, Tirupati Est.,
Opp. Bombay Housing
Nr. Rangshala Compound
Bapunagar, Ahmedabad
Gujarat-380024
Tel : + 91 (0) 79 22745481
E-mail:
amflex@wilnetonline.net

Sreerang Technologies Pvt. Ltd.
61, KSRTC Layout,
4th Cross, 1st Main
Phase-II, J P Ngr.
Bangalore
Karnataka-560078
Tel : + 91 (0) 80 26588472
Fax : + 91 (0) 80 26583951

Deluxe Printing Machinery Co.
75/81, Bazargate St., Fort
Mumbai
Maharashtra-400001
Tel : + 91 (0) 22 22618630
Fax : + 91 (0) 22 22618463
E-mail:
deluxe_exp@vsnl.com

Sree Krishna Strapping & Labels Pvt. Ltd.
New No. 48 (Old No. 20/1)
North Crescent Rd., T Ngr.
Chennai
Tamil Nadu-600017
Tel : + 91 (0) 44 28252414
Fax : + 91 (0) 44 28284964
E-mail:
skgmds@satyam.net.in

Jandu Engg Works (Regd.)
B-11, Sect. 58
Noida
Uttar Pradesh-201301
Tel : + 91 (0) 120 24580394
Fax : + 91 (0)120 24588878
E-mail:
jandu@ndf.vsnl.net.in

Surjeet Industries
491, Patparganj Indl. Area
Delhi-110092
Tel : + 91 (0) 11 22143888
Fax : + 91 (0) 11 22427923

Box Case & Carton Printing Machinery

Dianamark
Bible House, 5-2-13/4
Opp. Jabbar Cmplx.
Hyder Basti
Secunderabad
Andhra Pradesh-500003
Tel : + 91 (0) 40 27532548
Fax : + 91 (0) 40 27536451
E-mail:
dianamark@hotmail.com

Princessa Overseas Pvt. Ltd.
802, Pragati Tower, 26
Rajendra Place
New Delhi-110008
Tel : + 91 (0) 11 25823626
Fax : + 91 (0) 11 25823627
E-mail:
princessa@mantraonline.com

Gopal Engg. Works
2/5, B/h HCMC
Kailash Indl. Est.
Odhav
Ahmedabad
Gujarat-382415
Tel : + 91 (0) 79 22874969
Fax : + 91 (0) 79 22874969

Paper Board Machinery Co.
Plot No. 115, Phase-I
DLF Indl. Area
Faridabad
Haryana-121003
Tel : + 91 (0) 129 25275894
Fax : + 91 (0) 129 25290126
E-mail:
bharat@nagpalgroup.com

Acme Machinery (I) Pvt. Ltd.
G-16, Shalimar Indl. Est.
Matunga Labour Camp
Matunga (W) Mumbai
Maharashtra-400019
Tel : + 91 (0) 22 24037398
Fax : + 91 (0) 22 24037270
E-mail:
acme@acmeindia.com

Plastics Printing Machinery

Remica Plastic Machinery Manufacturers
2/AB
Sardar Patel Indl. Est.,
Shahwadi, Narol
Ahmedabad
Gujarat-382405
Tel : + 91 (0) 79 25712741
Fax : + 91 (0) 79 25894234
E-mail:
remica@ad1.vsnl.net.in

Rita Pad Printing Systems Ltd.
Plot No. 44 & 45
Sect. 18
Gurgaon, Haryana-122015
Tel : + 91 (0) 124 26343036
FAX : + 91 (0) 124 26346233
E-mail: info@ritapad.com

Boolani Engineering Corporation
7 & 8, Unique Est.
Off Veer Savarkar Rd.
Mumbai
Maharashtra-400025
Tel : + 91 (0) 22 24302826
Fax : + 91 (0) 22 24229875
E-mail: plastic@vsnl.com

Marsons Print-Graf Machines Pvt. Ltd.
D/9, Udyog Sadan No. 2
MIDC, Andheri (E) Mumbai
Maharashtra-400093
Tel : + 91 (0) 22 28352976
Fax : + 91 (0) 22 28369359
E-mail:
marsons@bom3.vsnl.net.in

Moscow Machine Tools
C-293, Mukund Ngr.
Ghaziabad
Uttar Pradesh-201001
Tel : + 91 (0) 120 24731630
Fax : + 91 (0)120 24793507
E-mail:
moscowindia@yahoo.co.in

Shaheen Machine Fabric
B-113, Sect. 5, Noida
Uttar Pradesh-201301
Tel : + 91 (0) 120 4552488

Princessa Overseas Pvt. Ltd.
802, Pragati Tower, 26
Rajendra Place
New Delhi-110008
Tel : + 91 (0) 11 25823626
Fax : + 91 (0) 11 25823627
E-mail:
princessa@mantraonline.com

Iron Craft Industries
A-43
Gali No. 7, New Rohtak Rd.
Anand Parbat Indl. Area
New Delhi-110005
Tel : + 91 (0) 11 25745588
Fax : + 91 (0) 11 25751367
E-mail:
ironcraftindustries@usa.net

Meditek International
C-43
DSIDC Packaging Cmplx.
Kirti Ngr.
New Delhi-110015
Tel : + 91 (0) 11 25901164
Fax : + 91 (0) 11 25901164
E-mail:
rgandhi1@rediffmail.com

Rotogravure Printing Machinery

Moscow Machine Tools
C-293, Mukund Ngr.
Ghaziabad
Uttar Pradesh-201001
Tel : + 91 (0) 120 24731630
Fax : + 91 (0) 120 24793507
E-mail:
moscowindia@yahoo.co.in

C Trivedi & Co.
A-1/16
Phase-II, GIDC Est.
Vatva, Ahmedabad
Gujarat-382445
Tel : + 91 (0) 79 5830552
Fax : + 91 (0) 79 5830553
E-mail:
ctrivedi@ctrivedi.com

Lanar Enterprises
Plot No. 160
IDA (Expansion), Phase-II
Ramanayyapeta
Kakinada
Andhra Pradesh-533005
Tel : + 91 (0) 884 349783

Anil Engg.
19/73, West Moti Bagh
Nr. Hanuman Mandir
Old Rohtak Rd.
Delhi-110035
Tel : + 91 (0) 11 3527928
Fax : + 91 (0) 11 3544164

Iron Craft Industries
A-43
Gali No. 7, New Rohtak Rd.
Anand Parbat Indl. Area
New Delhi-110005
Tel : + 91 (0) 11 25745588
Fax : + 91 (0) 11 25751367
E-mail:
ironcraftindustries@usa.net

Dev Packaging
R.No.34, 1st Flr.,
Chapra Bldg.
Opp. Plaza Cinema
Dadar (W) Mumbai
Maharashtra-400028
Tel : + 91 (0) 22 24379395
Fax : + 91 (0) 22 24379395

Pragati Polyprint
H-111, Ansal Indl. Est.
Saki Vihar Rd.
Sakinaka, Mumbai
Maharashtra-400072
Tel : + 91 (0) 22 28524427
Fax : + 91 (0) 22 28522426
E-mail:
pragati@boms.vsnl.net.in

Great Cheran Engg.
482-A, Palghat Rd.
Coimbatore
Tamil Nadu-641008
Tel : + 91 (0) 422 2822492
Fax : + 91 (0) 422 2823459

Flex Industries Ltd.
A-107, Sect. IV
Noida
Uttar Pradesh-201301
Tel : + 91 (0) 120 24522558
Fax : + 91 (0) 120 24520877

Das Machinery Corp.
160, Nazrul Islam Avenue
Kolkata
West Bengal-700054
Tel : + 91 (0) 33 23557564
Fax : + 91 (0) 33 23557651

Textile Printing Machinery

Bharat Textile Works
Nagarvel Hanuman Rd.
Rakhial, Ahmedabad
Gujarat-380023
Tel : + 91 (0) 79 22742553
Fax : + 91 (0) 79 22743459

Roopa Engineers
Nr. State Bank of India
B/h Police Comm. Office
Tavdipura, Ahmedabad
Gujarat-380004
Tel : + 91 (0) 79 25625988
Fax : + 91 (0) 79 25622158

Tech Shivam Industries
791, GIDC Est.
Makarpura
Vadodara, Gujarat-390001
Tel : + 91 (0) 265 2642717
Fax : + 91 (0) 265 2640847
E-mail:
techshivam@bnpl.com

Texprint Engineers Ltd.
Texprint Hse.
Chakudia Mahadev Rd.
Rakhial, Ahmedabad,
Gujarat-380023
Tel : + 91 (0) 79 22742808
Fax : + 91 (0) 79 22743459

Jeevee Engineers
A-503, Dheeraj Garden
Nr. Punjab & Maharashtra
Bank, SVP Rd.
Andheri (E) Mumbai
Maharashtra-400093
Tel : + 91 (0) 22 28361693

Himalaya Processors Pvt. Ltd.
37, Shakti Mkt.
Amritsar, Punjab-143006
Tel : + 91 (0) 183 547340

Hindustan Heaters
Vijayalakshmi Indl.
Compound, Nethaji Ngr.
Nanjundapuram Rd.
Coimbatore
Tamil Nadu-641036
Tel : + 91 (0) 422 2320137

Fabcare Equipments
81-A, GH-10, Sunder Apts.
Outer Ring Rd.
Paschim Vihar
New Delhi-110087
Tel : + 91 (0) 11 25180018
Fax : + 91 (0) 11 25475569
E-mail: fabcare@vsnl.com

Meditek International
C-43
DSIDC Packaging Cmplx.
Kirti Ngr.
New Delhi-110015
Tel : + 91 (0) 11 25901164
Fax : + 91 (0) 11 25901164
E-mail:
rgandhi1@rediffmail.com

Hot Foil Printing Services

Plasmould
L/200, GIDC Indl. Est.
Odhav, Ahmedabad
Gujarat-382415
Tel : + 91 (0) 79 22871888
Fax : + 91 (0) 79 22870300
E-mail:
plasmould@web4india.com

Printwell
21, Roopena Agrahara
HSR Main Rd.
Madivala Post
Bangalore
Karnataka-560068
Tel : + 91 (0) 80 5732790

Alankrit Printing & Foil Stamping
105/1, Bhandup Indl. Est.
Co-Op. Soc.
Pannalal Silk Mill Compound
LBS Marg
Mumbai
Maharashtra-400078
Tel : + 91 (0) 22 5651986
E-mail:
alankrit@epages.webindia.com

GMG Pack-Aids Pvt. Ltd.
3-5-152, Narayanguda
Hyderabad
Andhra Pradesh-500029
Tel : + 91 (0) 40 24752030
Fax : + 91 (0) 40 24754256
E-mail: mg01@hotmail.com

48

Suppliers of Raw Materials

Printing Chemicals

Ashok Gum Guar Industries
Plot No. 236
Nr. New Tel. Exch.
GIDC Est., Naroda
Ahmedabad
Gujarat-382330
Tel : + 91 (0) 79 22812312

ATE Enterprises Ltd.
Mistry Chember
Nr. Cama Hotel, Khanpur
Ahmedabad, Gujarat-380001
Tel : + 91 (0) 79 25504032
Fax : + 91 (0) 79 25505694
E-mail:
ateahd@ad1.vsnl.net.in

Hindustan Litho Graphics Co.
45, Maheshwari Indl. Est.
Nr. Maheshwari Mills
Shahibaug Rd.
Ahmedabad
Gujarat-380004
Tel : + 91 (0) 79 25623090
Fax : + 91 (0) 79 25623955
E-mail: hlitho@rediffmail.com

Sharp Tone Graphics Pvt. Ltd.
B/6, Kapidhwaj Avenue,
Nehru Park, Vastrapur
Ahmedabad
Gujarat-380015
Tel : + 91 (0) 79 26750152
Fax : + 91 (0) 79 26760864

Shidimo Interaux Pvt. Ltd.
6th Flr., J K Tower
Nr. Sub-Jail, Ring Rd.
Surat, Gujarat-395002
Tel : + 91 (0) 261 28311422
Fax : + 91(0) 261 28310860
E-mail:
shidimo@icement.net

Auchtel Products Ltd.
142/C, 2nd Flr., Victor
House, N M Joshi Marg
Lower Parel
Mumbai
Maharashtra-400013
Tel : + 91 (0) 22 24933975
Fax : + 91 (0) 22 24939755

Indofil Chemicals Co.
Industrial Products Divn.
Nirlon House
Dr. Annie Besant Rd.
Worli, Mumbai
Maharashtra-400025
Tel : + 91 (0) 22 24937391
Fax : + 91 (0) 22 24935667
E-mail:
indofil@giasbm01.vsnl.net.in

Lastra Niraj Pvt. Ltd.
118, MIDC
Andheri (E), Mumbai
Maharashtra-400093
Tel : + 91 (0) 22 28379069
Fax : + 91 (0) 22 28367654
E-mail:
lastra@bom3.vsnl.net.in

Newstech (I) Pvt. Ltd.
261, Udyog Bhavan
Sonawala Rd.
Goregaon (E), Mumbai
Maharashtra-400025
Tel : + 91 (0) 22 28752494
Fax : + 91 (0) 22 24370004
E-mail:
newstech@vsnl.net.in

NS Chemicals & Consultants Pvt. Ltd.
D-224, Ansa Indl. Est.
Sakinaka
Andheri (E), Mumbai
Maharashtra-400072
Tel : + 91 (0) 22 28523942
Fax : + 91 (0) 22 28523942

Pidilite Industries Ltd.
A Parekh Group Company
Ramkrishna Mandir Rd.
Off Marol Vasanji Rd., P.O. Box 17411
Andheri (E), Mumbai
Maharashtra-400059
Tel : + 91 (0) 22 28387240
Fax : + 91 (0) 22 28384989

Pilot Industrial Chemicals
19, Kapadia Chmbrs.,
51, D Ratanshi Marg
Mumbai
Maharashtra-400009
Tel : + 91 (0) 22 23438581

RSB Chemical Industries
T-136, MIDC
Bhosari, Pune
Maharashtra-411026
Tel : + 91 (0) 20 27120805
Fax : + 91 (0) 20 27122516
E-mail: rsbchem@vsnl.com

Technova Imaging Systems Ltd.
Laxmi Mills Est.
Off Dr. E Moses Rd.
Mahalaxmi, Mumbai
Maharashtra-400011
Tel : + 91 (0) 22 24983390
Fax : + 91 (0) 22 24950322
E-mail:
corp@technovaworld.com

United Ink & Varnish Company Ltd.
37/40, Subhash Rd.
Vile Parle (E), Mumbai
Maharashtra-400057
Tel : + 91 (0) 22 28366020
Fax : + 91 (0) 22 28381337
E-mail:
uninkvar@gjasbm01.vsnl.net.in

Aktive Products
125, Amman Koil St.
Chennai, Tamil Nadu-600079
Tel : + 91 (0) 44 22516627

Asian Reprographics Pvt. Ltd.
115 & 116
SIDCO Indl. Est.
Ambattur, Chennai
Tamil Nadu-600098
Tel : + 91 (0) 44 26251619
Fax : + 91 (0) 44 26251544
E-mail:
asianrepro@vsnl.com

Quality Inorganic Chemicals Pvt. Ltd.
35, Thambuswamy Rd.
Kilpauk, Chennai
Tamil Nadu-600010
Tel : + 91 (0) 44 26427943
Fax : + 91 (0) 44 26427943
E-mail: ranne@vsnl.com

King Chemicals
44 C, Indian Mirror St.
Kolkata
West Bengal-700013
Tel : + 91 (0) 33 22455763
Fax : + 91 (0) 33 22446148
E-mail:
kingchem@cal3.vsnl.net.in

Venus Ethoxyethers Pvt. Ltd.
Plot No. 109-111
Bicholim Indl. Est.
Biocholim
Goa-403504
Tel : + 91 (0) 832 2362337
Fax: + 91 (0) 832 2362257
E-mail:
venus.goa@sma.sril.in

Printing Inks

Resinate Colors Ltd.
578, Phase-II
Indl. Area
Chandigarh (UT)-160002
Tel : + 91 (0) 172 2652631
Fax : + 91 (0) 172 2653631
E-mail:
resinate@satyam.net.in

Asian Ink Co.
618, Suneja Tower-II
District Centre
Janakpuri
New Delhi-110058
Tel : + 91 (0) 11 25593554
Fax : + 91 (0) 11 25593554
E-mail:
chandhok11@vsnl.net

Delite Printing Ink Co.
177-B, Kishan Ganj Mkt.
Delhi-110007
Tel : + 91 (0) 11 23615006

Protex Plastic Pvt. Ltd.
W-18, Phase-II
Okhla Indl. Area
New Delhi-110020
Tel : + 91 (0) 11 26317041
Fax : + 91 (0) 11 26822308
E-mail:
protex@del3.vsnl.net.in

Sakata Inx (India) Ltd.
10th Flr., Devika Tower, 6
Nehru Place
New Delhi-110019
Tel : + 91 (0) 11 26416932
Fax : + 91 (0) 11 26232158
E-mail: sakata@vsnl.com

Atlas Chemical Co.
Nr. Pragati High School
Opp. Anupam Cinema

Khokhra, Bridge End
Ahmedabad
Gujarat-380008
Tel : + 91 (0) 79 22114790
Fax : + 91 (0) 79 22114790
E-mail:
atlasptgink@yahoo.com

Ganesh Inks & Aids
124, Phase-II
Indl. Area, Panchkula
Haryana-134113
Tel : + 91 (0) 172 2591644

Color Coates India
92/A, Main Rd. B
Narayana Pura
Bangalore
Karnataka-560016
Tel : + 91 (0) 80 28512139

Incowax Ltd.
47-D
Bommasandra Indl. Area
Bangalore
Karnataka-562158
Tel : + 91 (0) 8110 233155
Fax : + 91 (0) 8110 232419
E-mail: inks@incowax.com

Media Inks
239/6, 2nd Flr.
Bull Temple Rd.
Bangalore
Karnataka-560018
Tel : + 91 (0) 80 26675125
Fax : + 91 (0) 80 26527332
E-mail: sansetty@vsnl.com

Flexographic Printing Inks

Sicpa India Ltd.
308-312, Mercantile House,
15, K G Marg
New Delhi-110001
Tel : + 91 (0) 11 23355244
Fax : + 91 (0) 11 23314212
E-mail: sicpaho@bol.net.in

Unitech Inks Pvt. Ltd.
39/23, Bhawana Rd.,
Gali No. 15, Samaipur
Delhi-110042
Tel : + 91 (0) 11 27290476
Fax : + 91 (0) 11 27295206

Atlas Chemical Co.
Nr. Pragati High School
Opp. Anupam Cinema
Khokhra, Bridge End
Ahmedabad
Gujarat-380008
Tel : + 91 (0) 79 22114790
Fax : + 91 (0) 79 22114790
E-mail:
atlasptgink@yahoo.com

DMM Printing Inks Pvt. Ltd.
142/A, Sarakki, Phase-I
J P Ngr., Bangalore,
Karnataka-560078
Tel : + 91 (0) 80 26343280
Fax : + 91 (0) 80 26635135

Dolphin Inks Pvt. Ltd.
214, Devendra Indl. Est.,
Lokmanya Ngr.
P.O. Jekegram, Thane
Maharashtra-400606
Tel : + 91 (0) 22 25881110
Fax : + 91 (0) 22 25882128
E-mail:
dipl@bom3.vsnl.net.in

Shil-Inks
Ambekar Lane, 592
Budhwar Peth
Pune, Maharashtra-411002
Tel : + 91 (0) 20 24458932
Fax : + 91 (0) 20 25880851

Madras Printing Inks
146, Nehru ngr.
Old Mahabalipuram Rd.
Thiruvanmiyur, Chennai
Tamil Nadu-600041
Tel : + 91 (0) 44 24482578

Southern Poly Colours
16, G K Indl. Est.
Arcot Rd., Porur
Chennai
Tamil Nadu-600116
Tel : + 91 (0) 44 24767584

Kriskote Industries
28, Mohan Ngr
Anand Indl. Est.
Ghaziabad
Uttar Pradesh-201001
Tel : + 91 (0) 575 2732559

Daman Inks
Gala No.2&3,Survey No.7
& Gala No.104
Srinath Developers
Nr. Dabhel, Daman
Daman (UT)-396210
Tel : + 91 (0) 260 252733

Lithographic Printing Inks

Nutech Inks & Allied Products
39KM Stone
Delhi-Jaipur Rd.
Village Narsingpur
Gurgaon
Haryana-122004
Tel : + 91 (0) 124 26371810
E-mail: nutech@vsnl.com

DMM Printing Inks Pvt. Ltd.
142/A, Sarakki, Phase-I

J P Ngr., Bangalore,
Karnataka-560078
Tel : + 91 (0) 80 26343280
Fax : + 91 (0) 80 26635135

Angel Chemical Industries
CS 1, SIDCO Indl. Est.
Arumbakkam, Chennai
Tamil Nadu-600106
Tel : + 91 (0) 44 4758439
E-mail: pi2979@indiayellow pagesonline-com

Graphi Kolor
No. 1, Veeramamunivar St.
Kandanchavadi
Perungudi Post
Chennai
Tamil Nadu-600096
Tel : + 91 (0) 44 24960976

Madras Printing Inks
146, Nehru ngr.
Old Mahabalipuram Rd.
Thiruvanmiyur
Chennai
Tamil Nadu-600041
Tel : + 91 (0) 44 24482578

Himalaya Enterprises
B-86/A
APIE, Sanat Ngr. Extn.-I
Balanagar, Hyderabad
Andhra Pradesh-500018
Tel : + 91 (0) 40 23771312

Metal Printing Inks

Delhi Packaging Services
B-1/160-A
Vishnu Garden Extn.
New Delhi-110018
Tel : + 91 (0) 11 25155295
Fax : + 91 (0) 11 25557292
E-mail:
packode@delhipackaging.com

Vaibhavi Industries
'Raykamal'
Deghe's Wada
B/h Police Chowki
Dandia Bazar
Vadodara, Gujarat-390001
Tel : + 91 (0) 265 2463317
Fax : + 91 (0) 265 2435345
E-mail:
vaibhaviind@yahoo.com

Rita Pad Printing Systems Ltd.
Plot No. 44 & 45
Sect. 18
Gurgaon, Haryana-122015
Tel : + 91 (0) 124 26343036
Fax : + 91 (0)124 26346233
E-mail: info@ritapad.com

DMM Printing Inks Pvt. Ltd.
142/A
Sarakki, Phase-I
J P Ngr., Bangalore
Karnataka-560078
Tel : + 91 (0) 80 26343280
Fax : + 91 (0) 80 26635135

Eskay Packagings Pvt. Ltd.
51, Raja Indl. Est.
Mulund (W), Mumbai
Maharashtra-400080
Tel : + 91 (0) 22 25646908
Fax : + 91 (0) 22 25670630
E-mail:
kayess@giasbm01.vsnl.net.in

Madras Printing Inks
146, Nehru ngr.
Old Mahabalipuram Rd.
Thiruvanmiyur
Chennai, Tamil Nadu-600041
Tel : + 91 (0) 44 24482578

Quality Inorganic Chemicals Pvt. Ltd.
35, Thambuswamy Rd.
Kilpauk, Chennai
Tamil Nadu-600010
Tel : + 91 (0) 44 26427943
Fax : + 91 (0) 44 26427943
E-mail: ranne@vsnl.com

Numerators India
Mendu Compound, 3/6
Marris Rd.
Aligarh
Uttar Pradesh-202002
Tel : + 91 (0) 571 2509056
Fax : + 91 (0) 571 2502156

Saboo-Berlac Ltd.
Plot No. 81, Phase-II
Indl. Area
Chandigarh (UT)-160002
Tel : + 91 (0) 172 2652559
Fax : + 91 (0) 172 2652302
E-mail: sbl@glide.net.in

Metallic Printing Inks

Rainbow Ink & Varnish Manufacturing Company Ltd.
133-C, Vakola,
Santacruz (E), Mumbai
Maharashtra-400055
Tel : + 91 (0) 22 26183974
Fax : + 91 (0) 22 26116704
E-mail:
rainboink@bom3.vsnl.net.in

Rang Rasayan Ltd.
1421, Maker Chmbrs.
V, 221, Nariman Point
Mumbai
Maharashtra-400021
Tel : + 91 (0) 22 22040232
Fax : + 91 (0) 22 22871476

DMM Printing Inks Pvt. Ltd.
142/A, Sarakki, Phase-I
J P Ngr., Bangalore
Karnataka-560078
Tel : + 91 (0) 80 26343280
Fax : + 91 (0) 80 26635135

Pad Printing Inks

Inks India
14, Glasswala Bldg., 1st Flr.
Turner Rd.
Bandra (W), Mumbai,
Maharashtra-400050
Tel : + 91 (0) 22 26402574
Fax : + 91 (0) 22 26452614

Rainbow Ink & Varnish Manufacturing Company Ltd.
133-C, Vakola
Santacruz (E) Mumbai
Maharashtra-400055
Tel : + 91 (0) 22 26183974
Fax : + 91 (0) 22 26116704
E-mail:
rainboink@bom3.vsnl.net.in

Rita Pad Printing Systems Ltd.
Plot No. 44 & 45, Sect. 18
Gurgaon, Haryana-122015
Tel : + 91 (0) 124 26343036
Fax : + 91 (0) 124 26346233
E-mail: info@ritapad.com

Plastic Film Printing Inks

The Superior Printing Ink Co.
136, SIDCO Indl. Est.
Ambattur, Chennai
Tamil Nadu-600098
Tel : + 91 (0) 44 24910047
Fax : + 91 (0) 44 26252921
E-mail: spico@yahoo.com

Flow Ink & Coatings Pvt. Ltd.
Plot No. 24,25
Changodar Indl. Est. Part II
Bawla Seskhej Rd.
Changodar
Gujarat-382213
Tel : +91 (0) 2717 275671
Fax: +91 (0) 2717 6408143
E-mail:
harekrsna@icenet.net

Screen Printing Inks

Rita Pad Printing Systems Ltd.
Plot No. 44 & 45
Sect. 18
Gurgaon, Haryana-122015
Tel : + 91 (0) 124 26343036
Fax : + 91 (0)124 26346233
E-mail: info@ritapad.com

Colorinks Pvt. Ltd.
136, Laxmi Bldg.
A M Rd.
Kalasipalayam New Extn.
Bangalore
Karnataka-560002
Tel : + 91 (0) 80 22233054
Fax : + 91 (0) 80 22210007
E-mail: sknayak@vsnl.com

Eskay Packagings Pvt. Ltd.
51, Raja Indl. Est.
Mulund (W), Mumbai
Maharashtra-400080
Tel : + 91 (0) 22 25646908
Fax : + 91 (0) 22 25670630
E-mail:
kayess@giasbm01.vsnl.net.in

Inks India
14, Glasswala Bldg., 1st Flr.
Turner Rd.
Bandra (W), Mumbai
Maharashtra-400050
Tel : + 91 (0) 22 26402574
Fax : + 91 (0) 22 26452614

MOK Enterprises
G-30, Kothari Indl. Est.
New Link Rd.
Andheri (W), Mumbai,
Maharashtra-400053
Tel : + 91 (0) 22 26300907
Fax : + 91 (0) 22 26300908

Rainbow Ink & Varnish Manufacturing Company Ltd.
133-C, Vakola
Santacruz (E), Mumbai
Maharashtra-400055
Tel : + 91 (0) 22 26183974
Fax : + 91 (0) 22 26116704
E-mail:
rainboink@bom3.vsnl.net.in

Sericol India Pvt. Ltd.
Gala No.2,
Super Processor Compound
Lal Bag Indl. Est.,
Chivda Galli, Lal Bag
Mumbai
Maharashtra-400025
Tel : + 91 (0) 22 24157025
Fax : + 91 (0) 22 24157026

United Ink & Varnish Company Ltd.
37/40, Subhash Rd.
Vile Parle (E), Mumbai
Maharashtra-400057
Tel : + 91 (0) 22 28366020
Fax : + 91 (0) 22 28381337
E-mail:
uninkvar@gjasbm01.vsnl.net.in

Lavanya Printing Pvt. Ltd.
95, SIDCO Indl. Est.
Thirumazisal, Chennai
Tamil Nadu-602107
Tel : + 91 (0) 44 26290396

Saboo-Berlac Ltd.
Plot No. 81, Phase-II
Indl. Area,
Chandigarh (UT)-160002
Tel : + 91 (0) 172 2652559
Fax : + 91 (0) 172 2652302
E-mail: sbl@glide.net.in

Printing Plates

Creed Engineers Pvt. Ltd.
4, South Patel Ngr. Mkt.
New Delhi-110008
Tel : + 91 (0) 11 25767206
Fax : + 91 (0) 11 25767205
E-mail: creed@vsnl.com

Sharpline Engineering Co.
B8/19, Sect. V, Rohini
Delhi-110085
Tel : + 91 (0) 11 27437286
Fax : + 91 (0) 11 27437286

Spinks India Pvt. Ltd.
Rani Jhansi Rd., 19
Motia Khan
New Delhi-110055
Tel : + 91 (0) 11 23521646
Fax : + 91 (0) 11 23618220
E-mail:
spinks@giasdl01.vsnl.net.in

Ankan Graphics
30, 1st Flr., Adwani Mkt.
Outside Delhi Gate
Shahibag Rd., Ahmedabad
Gujarat-380004
Tel : + 91 (0) 79 25624519

Aster Graphic Systems
9, Shree Geeta Apts.,
2nd Flr.
B/h Union Bank of India
Vasna
Ahmedabad
Gujarat-380007
Tel : + 91 (0) 79 26606522
Fax : + 91 (0) 79 26621984
E-mail:
ags@astergraphics.com

ATE Enterprises Ltd.
Mistry Chember
Nr. Cama Hotel
Khanpur, Ahmedabad
Gujarat-380001
Tel : + 91 (0) 79 25504032
Fax : + 91 (0) 79 25505694
E-mail:
ateahd@ad1.vsnl.net.in

Shivam Moulding Ltd.
49, Patel Soc.
Opp. Ghosha Soc.
Thaltej, Ahmedabad
Gujarat-380054
Tel : + 91 (0) 79 27452379
Fax : + 91 (0) 79 26857180
E-mail: shivam@balzeneton
line.com

Stovec Industries Ltd.
NIDC, Nr. Lambha Village
Village-Post Noral
Ahmedabad
Gujarat-382405
Tel : + 91 (0) 79 25710407
Fax : + 91 (0) 79 25710406
E-mail: stovec@icenet.net

Rita Pad Printing Systems Ltd.
Plot No. 44 & 45, Sect. 18
Gurgaon, Haryana-122015
Tel: + 91 (0) 124 26343036
Fax: + 91 (0) 124 26346233
E-mail: info@ritapad.com

EI DuPont India Ltd.
DLF Plaza Tower, 8th Flr.
Phase-I, DLF Qutab Enclave
Gurgaon, Haryana-122002
Tel : + 91 (0) 124 26540900
Fax : + 91 (0) 124 26540889

Agfa India Pvt. Ltd.
Bayer House
Central Avenue
Hiranandani Gardens
Powai, Mumbai,
Maharashtra-400076
Tel : + 91 (0) 22 25700300
Fax : + 91 (0) 22 25703017
E-mail:
agfa@bom3.vsnl.net.in

Indo Europa Trading Co. Pvt. Ltd.
36, Bhupen Chmbrs., 9
Dalal St., Fort, Mumbai
Maharashtra-400001
Tel : + 91 (0) 22 22674730
Fax : + 91 (0) 22 22672973
E-mail: ietco@vsnl.com

Kodak Polychrome Graphics India Pvt. Ltd.
2nd Flr., Sakhi House
Sion-Trombay Rd.
Chembur, Mumbai
Maharashtra-400071
Tel : + 91 (0) 22 25242098
Fax : + 91 (0) 22 25242899
E-mail:
ojalep@kpgraphics.com

Lastra Niraj Pvt. Ltd.
118, MIDC
Andheri (E), Mumbai
Maharashtra-400093
Tel : + 91 (0) 22 28379069
Fax : + 91 (0) 22 28367654
E-mail:
lastra@bom3.vsnl.net.in

Nippon Enterprises
217, High Tech Indl. Centre
B/h Bank of Baroda,
Caves Rd.
Jogeshwari (E), Mumbai
Maharashtra-400060
Tel : + 91 (0) 22 28345958
Fax : + 91 (0) 22 28375148

Opticolor
206, High Tech Indl. Centre
B/h Bank Of Baroda
Caves Rd.
Jogeshwari (E), Mumbai
Maharashtra-400060
Tel : + 91 (0) 22 28345958
Fax : + 91 (0) 22 28375148
E-mail: cgl@vsnl.net

Technova Imaging Systems Ltd.
Laxmi Mills Est.
Off Dr. E Moses Rd.
Mahalaxmi, Mumbai
Maharashtra-400011
Tel : + 91 (0) 22 24983390
Fax : + 91 (0) 22 24950322
E-mail:
corp@technovaworld.com

Global Printer
A/19
Rasulgarh Indl. Est.
Bhubaneshwar
Orissa-751010
Tel : + 91 (0) 674 585714

Asian Reprographics Pvt. Ltd.
115 & 116
SIDCO Indl. Est.
Ambattur, Chennai
Tamil Nadu-600098
Tel : + 91 (0) 44 26251619
Fax : + 91 (0) 44 26251544
E-mail:
asianrepro@vsnl.com

Maijo Engg. Works
C-183, Sect 10, Noida
Uttar Pradesh-201301
Tel : + 91 (0) 120 24526522

Gnani Industries
1-20-272, 1st Flr.
Rasoolpura
Secunderabad
Andhra Pradesh-500003
Tel : + 91 (0) 40 27901131
Fax : + 91 (0) 40 27818112

EIRI PROCESS TECHNOLOGY BOOKS

Complete Hand Book Of Ayurvedic & Herbal Medicines)

STEEL, S.S., NON FERROUS METALS, ROLLING MILL

01 Modern Technology of Non Ferrous Metals and Metal Extraction

02 Processing Technology of Steels and Stainless Steels

03 Modern Technology of Rolling Mill, Billets, Steel Wire, Galvanized Sheet, Forging & Castings

04 Manufacturing Technology of Non-Ferrous Metal Products

WOOD, VENEER, PLYWOOD, PARTICLE, BOARD, FIBREBOARD, BAMBOO & FOREST PRODUCTS

01 Modern Technology Of Wood, Veneer, Plywood, Particle Board, Fibreboard, Bamboo & Forest Products

BIO GAS PRODUCTION

01 Modern Technology of BIOGAS Production

WASTE BASED PRODUCTS INCLUDING INDUSTRIAL, AGRICULTURE, MEDICAL, MUNICIPALITY, ORGANIC & BIOLOGICAL

01 Technology Of Products From Wastes Industrial, Agriculture, Medical, Municipality, Organic & Biological

02 Products from Waste Technology Handbook

FOOD CHEMICALS, FOOD ADDITIVES, PIGMENTS, AROMA, SWEETENERS AND FOOD EMULSIFIERS

01 Technology of Food Chemicals, Pigments and Food Aroma Compounds

02 Modern Technology of Food Additives, Sweeteners and Food Emulsifiers

SOYBEAN

01 Technology of SOYBEAN Products with Formulae

02 Technology of Soya Milk, Tofu, Hydrlyzate and allied Soyabean Products with Project Profiles

EIRI Multiple Project Reports in CD ROM

01 92 New Lucrative Project Reports In Cd

02. 45 Infotech/It, Hospitality, Hospital, College, School & Club

03. 23 Profitable Construction Projects

04. 56 Agro Based & Food Processing Projects

05. 52 Food, Dairy, Bakery, Confectionery & Snacks Foods

06. 45 New Chemicals & Allied Project Reports

07. 100 Profitable Plastic, Polymers & Allied Projects

08. 44 Textile, Garments, Hosiery & Allied Products In Cd

09. 23 Dairy Farming (Buffalo & Jersey Cow, Dairy Products & Other Milk Processing Ind.

10. 42 Paper, Pulp & Paper Converting Products Reports

EIRI Multiple Project Reports in CD ROM

11. 31 Essential Oils, Perfumes, Flavours & Aromatic Compounds
12. 28 Surgical And Disposable Medical Projects
13. 36 Profitable Printing & Allied Projects Reports
14. 63 Profitable Packaging & Allied Project Reports
15. 99 Profitable Printing & Packaging
16. 42 Electroplating, Anodizing & Allied Projects
17. 31 Profitable Plantation & Cultivation Projects
18. 55 Profitable Products From Agro And Other Industrial Wastes
19. 35 Profitable Gums, Adhesives, Sealants And Resins Projects
20. 54 Paint, Varnish, Solvents, Lacquers, Resins, Enamel, Pigments, Thinners & Powder (54 Project Reports In Cd)
21 43 Iron, Steel, Casting, Fabrication, Wire Drawing & Rolling Mills Projects
22. 26 New Educational Projects (Schools, Colleges, Training/ Management Institute, Hostels Etc.)
23. 12 Mosquitoes Preventive Projects (Coils, Mats, Liquids, Agarbatti, Nets, Spray Etc.)
24. 21 Bakery And Allied Products
25. 67 Rubber And Rubber Goods Industries
26. 37 Aluminium And Aluminium Based Industries
27. 14 Potato And Potato Based Industries
28. 50 Electrical, Electronics And Computer Based Industries
29. 52 Cosmetics (Herbal & Synthetics) Technology Projects (52 Project Reports)
30. 10 Molasses Based Lucrative Projects
31. 24 Fruits/Veg. & Allied Food Dehydration Projects
32. 7 Power Plant Projects (Thermal, Hydro, Solar, Wind Energy And Lignite Coal Based)
33. 35 New Profitable Industries
34. 160 New Export Oriented And Profitable Projects (160 Project Reports In Cd)
35. 5 Ginger & Ginger Processing Industries
36. 13 Fish Farming And Fishery Products
37. 6 Agarbatti And Allied Projects
38. 28 Fruit Juices, Food Dehydration And Allied Products
39. 9 Tea Plantation And Processing Based Industries
40. 14 Roasted/Salted Cashew Nuts, Almonds, Namkeens, Spices Projects
41. 23 Injection Moulded Plastic Products (23 Project Reports)
42. 38 Biofertilizer, Biofuel, Enzyme, Organic Farming & Manure, Protein And Allied Lucrative Projects
43. 10 Maize & Corn Processing Projects
44. 11 Profitable Solar & Solar Based Products
45. 11 Infotech/It Projects
46. 42 Hospitality, Building Material, Power Steels, Alcohol & Other Food Products
47. 24 Soap & Detergents (Scouring Powder, Toilet Cleaner, Laundry Soap, Detergent Cake & Powder, Toilet Soap, Enzyme Based Detergents, Phenyl & Soap Coated Paper)
48. 10 Coconut And Coconut By Products
49. 22 Alcohol, Beer, Imfl, Country Liquor, Wine & Other Related Projects